VOLUME NINETY NINE

ADVANCES IN FOOD AND NUTRITION RESEARCH

Functional Bakery Products: Novel Ingredients and Processing Technology for Personalized Nutrition

VOLUME NINETY NINE

ADVANCES IN FOOD AND NUTRITION RESEARCH

Functional Bakery Products: Novel Ingredients and Processing Technology for Personalized Nutrit

Edited by

WEIBIAO ZHOU
Department of Food Science and Technology, National University of Singapore, Singapore, Singapor

JING GAO
Department of Food Science and Technology, National University of Singapore, Singapore, Singapore

ELSEVIER

ACADEMIC PRESS
An imprint of Elsevier

cademic Press is an imprint of Elsevier
o Hampshire Street, 5th Floor, Cambridge, MA 02139, United States
25 B Street, Suite 1650, San Diego, CA 92101, United States
he Boulevard, Langford Lane, Kidlington, Oxford OX5 1GB, United Kingdom
25 London Wall, London, EC2Y 5AS, United Kingdom

irst edition 2022

ISBN: 978-0-323-85557-0
ISSN: 1043-4526

For information on all Academic Press publications
visit our website at https://www.elsevier.com/books-and-journals

Publisher: Zoe Kruze
Acquisitions Editor: Sam Mahfoudh
Developmental Editor: Jhon Michael Peñano
Production Project Manager: Vijayaraj Purushothaman
Cover Designer: Matthew Limbert

Typeset by STRAIVE, India

Contents

Contributors

Melissa Assad-Bustillos
Food Process Engineering Group, Institute of Food, Nutrition and Health, ETH Zürich, Zürich, Switzerland

Fatma Boukid
Food Safety and Functionality Programme, Food Industry Area, Institute of Agriculture and Food Research and Technology (IRTA), Monells, Catalonia, Spain

Guy Della Valle
INRAE Biopolymères Interactions et Assemblages, Nantes, France

Gilles Feron
Centre des Sciences du Goût et de l'Alimentation, AgroSup Dijon, CNRS, INRAE, Université Bourgogne Franche-Comté, Dijon, France

Jing Gao
Department of Food Science and Technology, National University of Singapore, Singapore, Singapore

Manuel Gómez
Food Technology Area, College of Agricultural Engineering, University of Valladolid, Palencia, Spain

Audrey Hui Si Koh
Department of Food Science and Technology, National University of Singapore, Singapore, Singapore

Suyun Lin
Key Lab for Natural Products and Functional Foods of Jiangxi Province, College of Food Science and Engineering, Jiangxi Agricultural University, Nanchang, China

Martijn Noort
Wageningen University & Research, Wageningen Food & Biobased Research, Wageningen, The Netherlands

Nantawan Therdthai
Department of Product Development, Faculty of Agro-Industry, Kasetsart University, Bangkok, Thailand

Kjeld van Bommel
Equipment for Additive Manufacturing Department, Netherlands Organisation for Applied Scientific Research (TNO), Eindhoven, The Netherlands

Lu Zhang
Wageningen University & Research, Laboratory of Food Process Engineering, Wageningen, The Netherlands

Weibiao Zhou
Department of Food Science and Technology, National University of Singapore, Singapore, Singapore

Preface

This volume of *Advances in Food and Nutrition Research* is compiling seven chapters that describe the latest developments in functional bakery products, with a special focus on novel ingredients and processing technology for personalized nutrition. As consumers become increasingly more health-conscious and have a deeper appreciation of the adage "you are what you eat", demand for functional foods is gaining momentum. As one size does not fit all, developing functional bakery products needs to be tailored to specific dietary requirement or restrictions of the consumers. This book volume is devoted to providing an in-depth review of the recent advances in developing bakery products to address challenges of the reduction of sugar, fat, salt, and gluten, as well as to meet the needs of novel proteins, dietary fibre, and functional ingredients, for targeted consumers (e.g., elderly) and with innovative technology (e.g., 3D printing).

Chapter 1 provides the most recent developments in 3D-printing of bakery products, including printing system, dough formulation and possible personalization for texture, nutrients, and satiety modulation. ***Chapter 2*** presents the development of fibre-enriched bakery products with increased nutritional value and appealing palatability. The trends in dietary fibre supplementation, its impact on dough processing, quality, and physiological functionality of bakery products, as well as approaches to improve the compatibility of dietary fibre in bakery products are discussed. ***Chapter 3*** reviews the application of plant protein in developing new bakery products for a better amino-acid profile and a higher protein intake. Specific functional properties and nutritional values of both conventional and emerging plant protein sources are covered in this chapter. ***Chapter 4*** brings new insights into how to guide the formulation of soft cereal-based foods for elderly population through food oral processing. It provides an overview of the various aspects involved in the oral processing and formulation of soft cereal-based foods, translating them into challenges and opportunities that are of relevance to the design of realistic soft cereal-based foods targeting the elderly while being nutritious and sensory appealing. ***Chapter 5*** focuses on the development of gluten-free bakery products through ingredient and processing innovations. Specifically, the use of gluten-free starch/flour, gluten replacer, fiber, and protein in gluten-free bakery products as well as adaptation of processing methods due to reformulation are discussed. ***Chapter 6*** summarizes the latest development of functional bakery products using phytochemicals, specifically catechins, anthocyanins, fucoidan and quercetin extracted from various plant resources. Detailed examples of their applications as well as the impact on product quality, consumer acceptance and potential health benefits are provided. ***Chapter 7*** covers strategies for the reduction of sugar, salt, and fat in bakery products. It starts with a discussion on the role of each ingredient in baked products. Ingredients and additives that can substitute fat, sugar, and salt to produce desirable bakery characteristics are reviewed in terms of their advantages and disadvantages. Furthermore, alternative processes to reduce the use of fat, sugar and salt are proposed in this chapter.

We are particularly grateful for the excellent contribution by outstanding food scientists and engineers from around the world. This volume is the combined efforts of 13 professionals from 8 countries (i.e., Netherland, China, Spain, Switzerland, France, Singapore, and Thailand) with a variety of background and expertise. Last but not least, we would like to acknowledge the kind encouragement from Professor Fidel Toldra, Editor of the book series *Advances in Food and Nutrition Research* and the high-quality editorial assistance provided by Elsevier for making the publication of this book possible.

Weibiao Zhou
Jing Gao

CHAPTER ONE

Towards the creation of personalized bakery products using 3D food printing

Lu Zhang[a,*], Martijn Noort[b], and Kjeld van Bommel[c]
[a]Wageningen University & Research, Laboratory of Food Process Engineering, Wageningen, The Netherlands
[b]Wageningen University & Research, Wageningen Food & Biobased Research, Wageningen, The Netherlands
[c]Equipment for Additive Manufacturing Department, Netherlands Organisation for Applied Scientific Research (TNO), Eindhoven, The Netherlands
*Corresponding author: e-mail address: lu1.zhang@wur.nl

Contents

Abstract

Bakery products with interesting color, shape and texture have been created using 3D food printing. Current research focuses on the development of new formulations and the optimization of the printing and post-printing treatment processes, in order to obtain high-quality 3D-printed bakery products. Knowledge about food rheology is useful for the development of dough formulations with good 3D-printability. Additives such as hydrocolloids could improve the printability of dough, and novel ingredients are introduced via 3D printing to produce functional bakery products with potential health benefits. One of the main future promises of 3D printing lies in its ability to

Advances in Food and Nutrition Research, Volume 99
ISSN 1043-4526
https://doi.org/10.1016/bs.afnr.2021.11.002

produce bakery products that are personalized in terms of sensorial properties and nutritional composition, in order to meet the preferences and dietary requirements of individual consumers. This chapter addresses the most recent developments in 3D-printed bakery foods and highlights some important research topics to further advance this field.

1. Introduction

3D printing, also known as additive manufacturing, is a technology characterized by a layer-by-layer deposition of materials according to a digital design file. 3D printing of food was first introduced by researchers from Cornell University, and the developed Fab@Home model was used to print paste-like food materials by extrusion (Periard, Schaal, Schaal, Malone, & Lipson, 2007). 3D printing provides a new way of creating food products. Its potential advantages include possibilities to customize food designs that cannot be easily achieved with conventional manufacturing methods (e.g. complex geometry and internal structures); simplifying the supply chain through flexible, local, on-demand food manufacturing (Jayaprakash et al., 2020); broadening the range of ingredients used for food production (e.g. the utilization of food waste/by-stream) (Upprinting Food, 2021); providing food companies with new opportunities to interact with their consumers (Blurhapsody, 2021); and lastly enabling personalized nutrition (Liu & Zhang, 2019) and consumer empowerment (Caulier, Doets, & Noort, 2020).

To date, various 3D food printing technologies have been developed, including extrusion-based printing, binder jetting, inkjet printing and selective laser sintering. Among these technologies, extrusion-based food printing is the most popular method in food printing owing to its simplicity (Sun, Zhou, Yan, Huang, & Lin, 2018) and the fact that extrusion processes are well-known and widely applied in conventional food processes. Food materials that have been used for extrusion-based printing are cookie dough (Pulatsu, Su, Lin, & Lin, 2020), protein gel (Liu, Bhandari, Prakash, Mantihal, & Zhang, 2019; Liu, Zhang, et al., 2019), surimi (Gudjónsdóttir, Napitupulu, & Petty Kristinsson, 2019), tomato paste (Zhu, Stieger, van der Goot, & Schutyser, 2019) and chocolate (Hao, Li, Gong, & Xiong, 2019), to name a few.

Bakery products are primarily based on cereal ingredients. They have an important nutritional role as a main source of macronutrients and essential micronutrients in our daily diet. Besides their nutritious cereal content, sugars and fats are added to many bakery products, which are overconsumed and are considered less healthy. Thus, reducing the amount of those ingredients while keeping the attractiveness of bakery products is an important target for product reformulation for food scientists and the food industry. Also, in recent years, functional bakery products have received increasing attention in scientific studies. Functional foods can be defined as modified food or food ingredients that can provide health benefits to consumers beyond their basic nutrients (Zhang, 2018). The concept of functional foods involves three main aspects, specifically health benefits, nutritional function and technological processes (Bigliardi & Galati, 2013). The technological processes aspect involves (i) the development of functional foods by optimizing conventional processes, e.g. fortifying foods with dietary fiber; (ii) the design of technologies which prevent the deterioration of active ingredients, e.g. (micro)encapsulation; and (iii) the development of new technologies that aim to design and manufacture foods functionalized to the needs of individual consumers, e.g. by applying results from nutrigenomics to create personalized foods through 3D printing.

Novel bakery products and healthy ingredients have been studied in the field of 3D food printing, for example probiotic-containing cookies (Zhang, Lou, & Schutyser, 2018), cereal-based snacks enriched with edible insects (Severini, Azzollini, Albenzio, & Derossi, 2018) or with microalgae (Uribe-Wandurraga et al., 2020), and personalized snacks for soldiers (Caulier et al., 2020). The results from these studies indicate that 3D food printing could be a useful tool for the creation of personalized bakery products with tunable texture, flavor, appearance, satiety perception, and even nutritional content.

In this chapter, we aim to illustrate recent advances and challenges related to the development of 3D-printed personalized bakery products. More specifically, we introduce the principle of extrusion-based 3D printing, the development of extrusion 3D printing systems and the post-processing treatment used to obtain edible products in detail in Section 2. In addition, we will briefly discuss other 3D food printing technologies used for fabricating bakery products. Subsequently, we discuss the recent development of 3D-printed bakery products in Section 3, with a focus on the usage of extrusion-based printing. Additionally, novel ingredients used in 3D-printed bakery products such as probiotics, microalgae and edible insects are addressed in this section. In Section 4, we provide a more detailed discussion on 3D printing of personalized bakery foods. Finally, this chapter is concluded with discussions of remaining challenges and future research directions relevant for the development of personalized bakery products using 3D food printing.

2. 3D printing technology for bakery products

Since the first exploration in 2007, a variety of technologies have been developed for 3D printing food materials, including extrusion-based 3D printing, binder jetting printing, selective sintering printing and inkjet printing (Baiano, 2020). Table 1 describes the printing mechanism of these four printing technologies. Various 3D-printed bakery products have been created using these different technologies, as shown in Table 1. Based on literature, the most widely-studied printing technique for bakery products is extrusion-based 3D printing. Therefore, in this chapter, we will mainly discuss extrusion-based printing of bakery products, and only briefly address the other technologies.

2.1 Extrusion-based 3D printing

2.1.1 Extrusion-based 3D printing of bakery products

Extrusion-based 3D printing is often used to fabricate cereal-based foods. During extrusion-based 3D printing, a semi-solid food material (such as dough) in a syringe is extruded from a nozzle onto the printing platform. The extruded filament of food materials is deposited in a layer-by-layer fashion to form a 3D structure. The extrusion process of the filament can be driven by air pressure, mechanical force, or extrusion rotation in a pneumatic, piston-based or a screw-based system, respectively (Sun et al., 2018). Ideally, food materials used for extrusion-based printing should be pseudoplastic fluids that show shear-thinning behavior so they can be easily extruded. In addition, the materials should have a rapid structural recovery after deposition, in order to form a self-supporting structure. Many food formulations can be extruded at ambient temperature (e.g. dough or vegetable purée printing). For certain food materials such as chocolate or protein gel, the filament is extruded at elevated temperature, and the extruded filament shows a phase transition from a liquid-like to a solid-like structure upon cooling at ambient temperature. The selection of the printing temperature depends on materials' thermal properties such as the melting point of chocolate and the gelation temperature of protein dispersion.

In general, the advantages of extrusion-based printing include the wide choice of food materials and the relative simplicity of the printing system. However, printing of dough formulations and creating complex designs of cereal-based food structures with a high resolution remain challenging. For example, the printed 3D structure of dough may deviate from the digital

Table 1 Principles, schematic diagrams of 3D food printing technologies (Sun, Zhou, Huang, Fuh, & Hong, 2015) and corresponding examples of bakery products that are created: extrusion-based 3D printing (Zhang, Lou, & Schutyser, 2018), binder jetting (Holland, Foster, & Tuck, 2019), selective laser sintering (Noort, Van Bommel, & Renzetti, 2017) and inkjet printing (Foodjet, 2021).

Method	Extrusion-based 3D printing	Binder jetting	Selective sintering printing	Inkjet printing
Schematic diagram	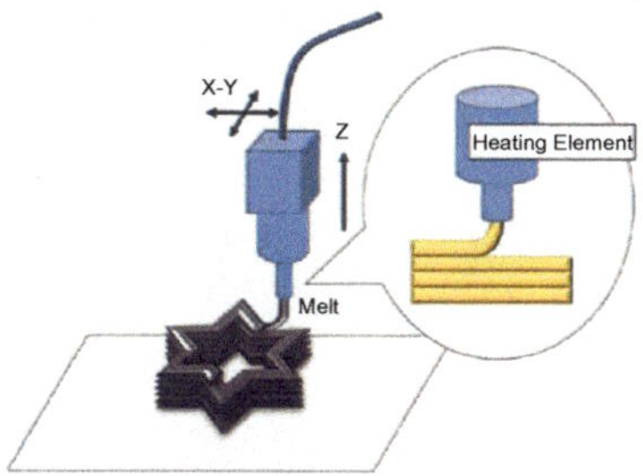	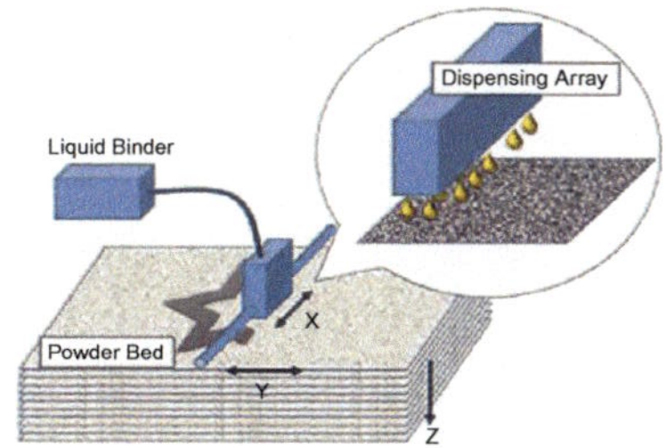	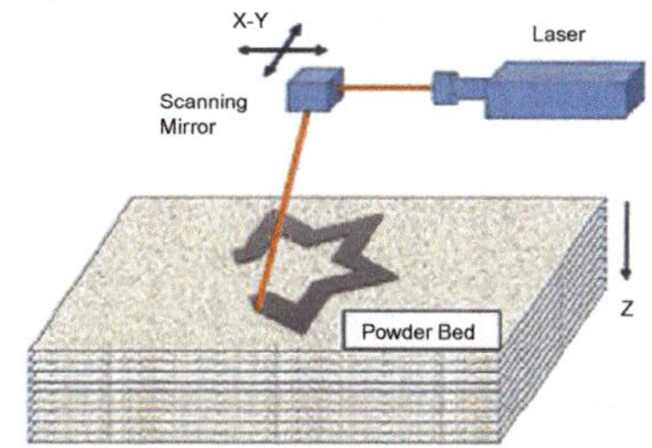	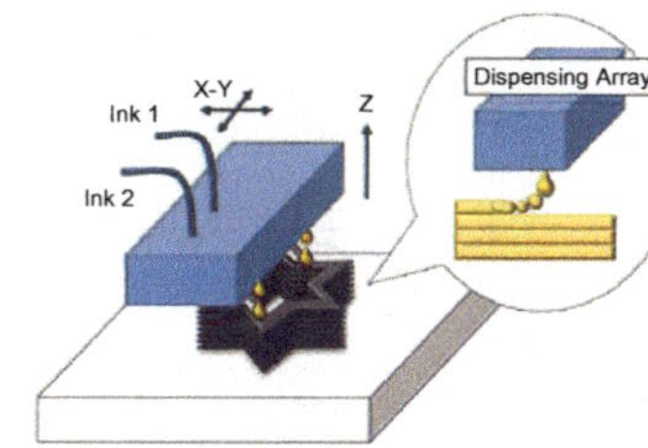
Principle	Semi-solid food material is extruded via a nozzle onto a platform layer by layer; material forms a self-supporting 3D structure. Both room temperature printing and hot-extrusion are used, depending on materials' properties.	Powder particles are distributed evenly on the print bed, and a liquid binder is deposited onto the target area to bind the powders. A new layer of powder is then added to the print bed and the same deposition process is repeated according to the digital design.	Heating source (e.g. laser and hot air) fuses powder particles to form a solid layer while unsintered powder provides support to the printed structure. After printing, sintered structure can be removed from the powder bed.	A stream of droplets/liquid is dispensed from a printhead in a drop-on-demand manner, and solidified to a "2.5 dimensional" image on the surface of the substrates.
Examples	 Cookie containing probiotics	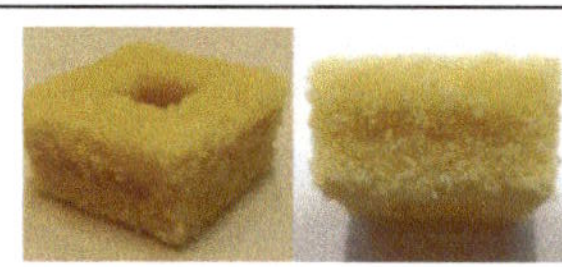 Cake with layered structure	 Cereal foods with anisotropic structure	 Decorated bakery goods (left: cavity filling; right: surface deposition)

TNO, the Netherlands.

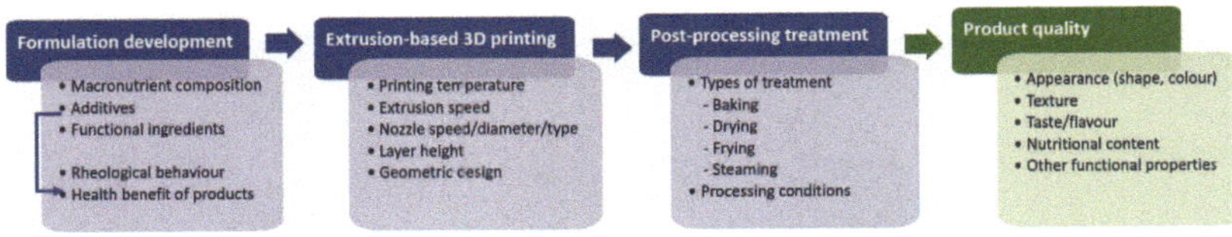

Fig. 1 Three stages of the extrusion-based 3D printing process including formulation development, printing process and post-printing processing; and selective indicators of product quality are listed.

design, which could be caused by imperfections during printing such as lack of continuous structure and loss of layer differentiation (e.g. collapsing of the structure) (Pulatsu et al., 2020). In addition, post-printing heat treatment (e.g. baking) is often required to obtain edible cereal-based foods. Heating, however, can often result in a lowering of viscosity and hence induce flow of the material, which means that the product loses its 3D shape during post-processing (Liu, Zhang, Bhandari, & Wang, 2017). Additionally, expansion of the 3D structure may occur during baking due to the presence of leavening agent in the printing material. Such expansion may result in a signification deformation of the carefully deposited design, possibly leading to a failed printed results after baking. This is especially problematic when aiming for a sophisticated geometrical design with a detailed infill pattern or a specific pore or void percentage. Therefore, when optimizing the production process of 3D-printed bakery foods, one needs to consider the effects of formulation design, printing settings, and the conditions used during post-printing treatment on the properties and quality of the final products (see Fig. 1). Some key factors that influence the performance of each stage during printing are selectively listed. Those factors need to be well controlled in order to achieve a successfully printed products with high quality, which will be elaborated more later in this section.

2.1.2 Extrusion printing systems

As mentioned earlier, three extrusion mechanisms have been developed in extrusion-based 3D printing, i.e. piston-, pneumatic-, or screw-driven (Schwab et al., 2020), among which the piston-driven mechanism is the most commonly used mechanism for food materials (see Fig. 2A).

In brief, in a piston-driven system, a stepper motor is programmed to generate linear motion of the plunger to push the food material out of the nozzle. The extrusion rate (i.e., volumetric flow rate of the material) can be controlled by adjusting the motor speed, which needs to be

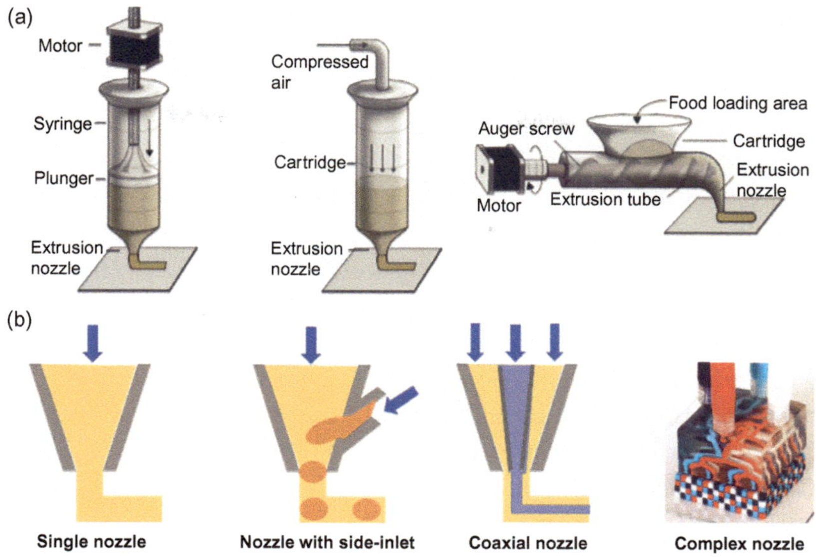

Fig. 2 Extrusion 3D printing system (A) three extrusion mechanisms: piston-, pneumatic-, or screw-driven systems (from left to right) (Sun et al., 2018); (B) schematic drawings of various nozzle configuration designs: single nozzle, nozzle with side-inlet (modified from Schutyser et al., 2018), coaxial-nozzle (modified from Vancauwenberghe, Verboven, Lammertyn, and Nicolaï, 2018), complex nozzle for (non-food) multi-materials (Skylar-Scott, Mueller, Visser, & Lewis, 2019) (from left to right), blue arrows indicate the flow direction of material.

controlled together with the nozzle moving speed to achieve a desired filament formation. This system has been used to print semi-solid or solid food products (Sun et al., 2018).

In pneumatic-driven system, air pressure generated by the pneumatic pump pushes the food material out of the nozzle, and the extrusion rate can be controlled by adjusting the air pressure (see Fig. 2A). This system is relatively more suitable for printing paste or gel-like food materials with a relatively low viscosity due to the limited air pressure (e.g. up to 5.5 bar as reported in Schutyser et al. (2018)). It is also often used in bio-printing applications to print hydrogels.

In screw-driven system, a motor drives the movement of the screw that continuously pushes food materials downwards and through the nozzle. This system is particularly suitable for the extrusion of materials with high viscosity (up to 10^4 Pa s). For example, Mantihal, Prakash, Godoi, and Bhandari (2017) developed a rotary screw extrusion method to 3D print chocolate. In addition, continuous dosing/printing could potentially be realized using a screw-driven system because, in contrast to the other two systems, screw-driven systems can operate using a continuous material supply rather than requiring a filled cartridge/syringe (i.e., batch supply). Nevertheless, the screw-driven system is rarely used for printing food materials so far, which is maybe caused by the difficulties related to cleaning of the system. Moreover, the risk of contamination should be taken into account because the screw is in direct contact with food materials. This is usually not a problem for the other two types of systems because a plastic syringe cap is often placed between the piston/air supply and the food material.

As mentioned earlier, 3D printing technologies create products based on digital design files. For extrusion-based 3D printing, the digital design file defines the printing path planning, which can be generated in three steps: (i) select the configurations of the 3D printer which could be classified as a Cartesian, Delta, Polar, or Scalar printer (Derossi, Caporizzi, Ricci, & Severini, 2019); (ii) create the virtual model of the 3D structure using computer-aided design (CAD) software; (iii) generate commanding geometrical code (G-code) that contains the information for the printer's movement using slicing software (Sli3er open source 3D printing toolbox, 2021). Several important printing parameters can be controlled by the G-code commands, such as the extrusion speed (i.e., movement speed of the syringe plunger or piston), nozzle speed (i.e., movement speed of the nozzle over the printing platform) and layer height (i.e., the distance between the nozzle and the previous printed layer) (see Fig. 1). These parameters significantly influence the printing result of a given dough formulation, which will be discussed in more detail in Section 3. Other parameters such as printing temperature and nozzle configuration also influence the printability of materials and the quality of the final products. Since these two parameters are more related to the design of the printing system, we will discuss their influences here.

Printing temperature is a critical parameter for extrusion-based printing systems (see Fig. 1). Various food materials such as viscous vegetable paste can be printed at room temperature and form a self-supporting 3D structure after deposition, whereas some other materials, such as protein gels, require printing at elevated temperatures (Mantihal et al., 2017). For example, extrusion printing of sodium caseinate has been done at elevated printing temperature which was 5–15 °C higher than the gelation temperature of the sodium caseinate dispersion (Schutyser et al., 2018). A concentrated sodium caseinate dispersion is in a liquid form at a temperature that is above its gelation temperature (i.e., sol-gel transition temperature), making the material extrudable. The gelation temperature of the formulation is related to its protein content, which ranged from 20 to 35 °C as reported in Schutyser et al. (2018). At room temperature, a solid and self-supporting gel is (re)formed due to the cold gelation of the concentrated sodium caseinate (Loveday, Rao, Creamer, & Singh, 2010). One can imagine that the printed structure of the sodium caseinate dispersion will not be self-supporting at room temperature when its gelation temperature is below 25 °C (due to its low protein content). For other viscous food materials such as cookie dough, increasing the temperature reduces its viscosity. Thus, a lower force is needed for the extrusion process, which can be beneficial when the extrusion force of the printing system is limited.

Various nozzle configurations have been developed for extrusion-based printing, such as single nozzle, nozzle with a side-inlet, coaxial nozzle, and complex nozzle designed for the extrusion of multiple formulations (see Fig. 2B). Among these designs, the single nozzle can be applied to print one formulation (which is typically a mixture of multiple ingredients) at a time. It is the most commonly used design for extrusion printing of food materials thanks to its simplicity. This design does not usually allow for controlling the spatial distribution of different formulations within the food matrix. However, in a single-nozzle system, spatial distribution of ingredients can be realized by changing the nozzle configuration (Fig. 2B). For example, Schutyser et al. (2018) designed and tested a dispenser with a side-inlet to introduce a second liquid phase into the deposited main formulation (in this case sodium caseinate gels). By accurately adjusting the dispensing pressure, two phases can be deposited simultaneously. As a result, oil droplets were spatially distributed in the continuous sodium caseinate gel, and a complex multi-layer design was realized.

On the other hand, a so-called coaxial nozzle configuration can be used to realize a different type of ingredient distribution. In coaxial printing, a single filament can be formed, which has core of one formulation and an outer layer or shell of another formulation (Uribe-Wandurraga et al., 2020) (see Fig. 2B). For example, coaxial extrusion is applied to hide the dark green microalgae-enriched dough inside a regular dough to obtain snacks with a more acceptable color (Uribe-Wandurraga et al., 2020).

Moreover, spatial distribution of food materials and complex multi-texture and multi-flavor food products can be realized by using a multi-printhead printer (e.g. dual nozzle extrusion 3D printing). Dual nozzle extrusion 3D printing enables the fabrication of attractive multi-material structures (Liu, Zhang, & Yang, 2018). Also, a company called Nature

Machines developed a printer *Foodini* which has room for up to five food capsules that can automatically exchange as needed to create multi-material structures (NatureMachines, 2021).

Nevertheless, 3D printing of multiple food materials may require a new design of the printing system, because multi-material voxelated structures cannot be easily generated by extruding monolithic cylindrical filaments layer by layer. In the non-food area, researchers have developed a multi-material multi-nozzle 3D (MM3D) printing method (see Fig. 2B). Multiple viscoelastic materials converge at a junction to realize seamless, fast switching between up to eight different materials to continuously deposit heterogeneous voxelated filaments using non-food materials (Skylar-Scott et al., 2019). However, applying such a method in food applications may be challenging due to the complex rheological behavior of food materials.

To summarize, by using different nozzle or printing system designs, 3D-printed food with complex geometric designs, as well as spatial distribution of ingredients can be fabricated. By changing the formulations and the printing settings applied, the macro and/or microstructure of printed food can be controlled. We will have a more detailed discussion on how this can contribute to the personalization of 3D-printed bakery products later in this chapter.

2.1.3 Post-processing treatment

3D-printed bakery products often require a post-printing heat treatment to obtain edible products. One exception is 3D-printed cereal products created by selective laser sintering (SLS, see Section 2.2). Because the food powders are thermally treated with infrared laser beam, in principle no further post-processing is needed (Noort et al., 2017). Some common technologies for the post-processing treatment of 3D-printed bakery foods include baking, drying, frying, steaming and microwave treatment.

Post-processing treatments influence the quality of the final products in terms of color, shape and textural properties. Specifically, thermal fixation of protein and starch components and evaporation of moisture govern the transition of the viscoelastic dough material to the desired texture of baked products, e.g., a soft-sponge-like cake or a crispy-crunchy biscuit. Furthermore, the baked flavor and brown color are resulted from chemical reactions like the Maillard reaction or the caramelization of sugars. Conventional bakery products almost always change their shape drastically during baking, e.g., the expansion of a cake or spreading of a cookie. Similarly, the originally created shape of the 3D-printed structure may collapse (Lille, Kortekangas, Heiniö, & Sozer, 2020) or shrink (Pulatsu et al., 2020) or otherwise change during baking. In some cases, the volume of 3D-printed bakery products could also expand during post-processing treatment. For example, 3D-printed rice dough can swell during steaming due to the high amylopectin content in the waxy rice that was used. This expansion caused instability of the 3D structure (Liu et al., 2020).

In addition, the type of post-processing treatment used may influence the mechanical properties of the final products. For example, Noort et al. (2017) reported a higher hardness of cereal-based structures produced by SLS after post-processing treatment with hot steam as compared to the baked and untreated samples. They attributed the strengthening of the printed structure to the introduction of moisture into the matrix via wet steam which enables reactions such as starch gelatinization and protein denaturation.

With most conventional bakery products spreading and/or expansion occurs during baking. In the case of 3D printed bakery products, such deformation during the post-printing treatment is highly undesired, as it results in the loss of the original 3D printed shape. Three strategies have been investigated to better maintain the 3D shape of the products during post-processing, i.e., (i) reformulating the printing materials, (ii) changing the printing design; and (iii) modifying or optimizing the post-treatment conditions.

(i) Lipton et al. (2010) modified a cookie recipe from Austria to make it printable while retaining the shape during baking by adjusting the ratio of the ingredients, although the authors did not reveal more details of the recipe modification. Differential scanning calorimetry (DSC) and dynamic mechanical thermal analysis (DMTA) can be useful tools to study the influence of composition on structural changes of the cookie matrix during baking (Renzetti & Jurgens, 2016). Noort (2018) used DMTA to reformulate dough based on its thermomechanical properties. Firstly, the ratio of plasticizing ingredients (fat, water and solubilized sugars) over structuring ingredients (starch, protein, fiber) needs to be high enough for the dough to be extrudable at the temperature of printing. However, these plasticizing ingredients contribute to melting of the dough at elevated temperature, and therefore should be minimized. To further constrain the melting process and promote early thermosetting of the dough, gelatinized starch and proteins were added to the formulation. This approach resulted in stable 3D printed dough shapes and avoided collapse during baking, even when the printed walls were relatively thin.

(ii) The printing settings such as infill density and printed layer height have been found to influence the dimensional properties of 3D-printed snacks after baking (Severini, Derossi, & Azzollini, 2016). In addition, the infill density influences the baking time required (Varghese et al., 2020). Longer baking times were required for cookies with a higher infill density because the difference in microstructure influenced the moisture transfer during baking, which should be considered when optimizing the process, to reduce the deformation of the 3D structure during baking.

(iii) An additional cooling step before baking was found to be helpful in retaining the shape of 3D-printed cookies during baking. Yang, Zhang, Fang, and Liu (2019) explored fast-cooling after printing as a way to enhance the shape retention. They found that 3D-printed sample that was subjected to fast-cooling at −65 °C for more than 10 min can maintain its shape during baking, while the non-cooled control sample collapsed during baking. They attributed this phenomenon to the change in rheological behaviors of the dough.

Future research could focus on investigating how printing material composition, printing settings, and post-processing treatment can influence the quality of 3D-printed bakery products. A better quantitative understanding of their complex relationships could contribute to the more rapid development of 3D-printed bakery foods with desired functional and sensorial properties.

2.2 Other 3D printing technologies for bakery products

As mentioned earlier, several technologies have been developed for 3D food printing. We introduced the extrusion-based 3D printing technique in great detail in Section 2.1. In this section, we will briefly describe the other three 3D printing technologies that have been used to fabricate bakery products, i.e. binder jetting, selective sintering printing (SLS) and inkjet printing (see Table 1).

Both binder jetting and selective sintering are powder-based printing technologies. During powder bed printing, powdered food materials are deposited as a thin layer. Powder particles are fused at specified locations according to a digital design, which is achieved either by depositing a liquid binder (in the case of binder jetting) or via sintering by an infrared laser beam or hot air (in the case of selective sintering) (see Table 1). A new layer of powder is deposited after sintering or binding the previous layer of powder. The same binding or sintering process is repeated until a 3D shape is formed within the powder bed. During printing, the unfused powder particles can provide support to the fused parts. After printing, the unfused powder is removed and can be recycled and used for the next printing job.

Powder bed printing shows certain advantages over extrusion-based printing. Specifically, powder bed printing can be performed at a relatively high speed. It also shows greater design freedom in terms of geometry since the unfused powder can provide support to the shape that is created. Powder bed printing has been used to create complex cereal-based structures with designed spatial variations in textures and flavors. For example, a multi-textured cake was created by binder jetting (see Table 1). By varying the printing settings, specifically using higher amounts of the aqueous printing liquid for the middle layers, the top and bottom layers of the printed structure were more porous while the middle layer had a more dense microstructure. After baking, the top and bottom layer of the obtained cake had a softer texture than the middle layer that showed a darker yellow color (Holland et al., 2019). In another example, an anisotropic structure was fabricated by selective laser sintering. Noort et al. (2017) reported a wheat flour-based powder formulation with maltodextrin and palm oil powder as binding components which meets the thermal requirements for sintering using an infrared laser beam. By varying the thickness of the internal product structures during sintering (i.e., 1 mm cell walls in the horizontal direction vs 3 mm in the vertical direction), anisotropic 3D structures containing open cells were created (see Table 1). The anisotropic structure resulted in substantial differences in the mechanical properties of the product depending on the fracturing direction.

One challenge for powder bed printing, however, is the limited choice of food materials with good binding/sintering properties. To date, food materials that have been investigated for powder bed printing are rather limited. Sugar-based and fat-based powder formulations are reported, however, they may be considered less nutritious (Liu et al., 2017). Another challenge for powder bed printing is the printing resolution, especially when a substantial moisture content in the final products is required. In the case of binder jetting, the liquid binder may bleed into neighboring voxels which reduces the printing precision. The fused powder may also shrink or expand during printing, causing undesired changes to the printed structure or even printing failure. Therefore, the development of new powder formulations (or powder-binder combinations) and the optimization of powder bed printing process are interesting research topics, to broaden the range of ingredients and to improve the quality of the final products.

Inkjet printing is another 3D printing technique. It is the most often used technique to realize a graphical decoration, surface filling or cavity deposition of bakery products such as cookies and cakes (see Table 1). Two types of inkjet printing methods exist, namely continuous jet printing and drop-on-demand printing (Liu et al., 2017). Ink is ejected continuously through a piezoelectric crystal vibrating at a constant frequency in a continuous jet printer. Grood and Grood (2013) developed a 3D inkjet food printing technology and commercialized it as FoodJet printing. FoodJet printing can be classified as drop-on-demand deposition. This technology uses an array of pneumatic membrane nozzle-jets which deposit droplets onto a moving object (Godoi & Sangeeta Prakash, 2016), which has been applied to print many different formulations such as pizza sauce, melted chocolate, cheese and butter. For example, by applying multiple "printing-cooling-printing" cycles, a chocolate layer with varied thicknesses can be printed onto a food surface (Zhu, Ribberink, De Wit, Schutyser, & Stieger, 2020). In general, fast printing and good printing quality can be achieved with this technique. However, it might not be suitable for creating very complex 3D structures.

3. Recent development in 3D-printed bakery products

3.1 Developing dough formulations for extrusion-based 3D printing

3.1.1 Printability of dough formulations

In the field of 3D food printing, a good deal of effort has been made to investigate food formulations that are suitable for extrusion-based 3D printing. Hence, the so-called "printability" of food materials is often studied. Printability can be defined as the ability of 3D printing to closely reproduce a target digital design (Telea & Jalba, 2011). In extrusion-based 3D printing, *printability* includes both *extrudability* of the material and *stability* of the printed shape (Liu, Bhandari, et al., 2019; Liu, Zhang, et al., 2019; Schwab et al., 2020). Extrudability assesses the initiation of the flow and the formation of a consistent filament during printing, whereas stability assesses the shape fidelity and shape retention. In this section, we will not include the shape retention during post-processing (see Section 2.1.3) as an indicator of formulations' printability, but only the shape retention over time after printing (before any post-printing treatment happens).

In general, the extrudability is affected by the yield stress of the printing material at the printing temperature, as well as the viscoelasticity and the shear thinning properties of the material. The stability of the printed structure is influenced by the shear recovery, temperature recovery (in the case of printing at elevated temperatures), yield stress at ambient temperature, zero shear viscosity, and the elastic modulus of the material (Liu, Bhandari, et al., 2019; Liu, Zhang, et al., 2019). However, due to the complexity and variations in the composition of food formulations, the quantitative relationship between rheology and printability has not yet been fully established for a wide range of materials. In addition to the aforementioned rheological properties, the printability of food formulations is influenced by printing parameters applied such as layer height, nozzle diameter, printing speed, product geometry, printing pressure/force, and printing temperature (Severini et al., 2018) (see Fig. 1), which will be discussed later in this section.

3.1.2 Influence of macronutrients on 3D-printability

Bakery products generally contain important macronutrients (i.e., proteins, carbohydrates, and lipids) and micronutrients (e.g., vitamins, antioxidants and minerals). The relative volume fractions and type of macronutrients, together with the volume fraction of the solvent (water) and low molecular solutes like sugar in the formulation are the determinants for the rheological properties of dough and its behavior during processing steps (Renzetti & Jurgens, 2016). More specifically, variations in macronutrient composition, e.g., ratio between proteins and carbohydrates, incorporation of lipids and the concentration of thickening and gelling agents, influence the rheological properties of dough formulations (Pérez, Nykvist, Brøgger, Larsen, & Falkeborg, 2019). Therefore, insights into the (quantitative) relationship between the composition of dough and its printing behavior are useful for the development of 3D-printed bakery foods.

Godoi and Sangeeta Prakash (2016) discussed the 3D printability of macronutrients, namely carbohydrates, proteins and lipids. Firstly, the carbohydrate fraction, from small sugars to polymeric carbohydrates, influences the glass transition temperature (T_g) of the food formulations. Controlling the T_g is important for the self-supporting property of the deposited materials. In starch-rich systems with a low moisture content, sugar can act as a plasticizer and co-solute with starch, which can lower the T_g and cause collapsing of the printed structure.

Secondly, another essential component that is often present in bakery products is protein. For example, the bread dough is a heterogeneous matrix in which starch granules are embedded in a continuous gluten network. The presence of the gluten network influences the printability of wheat dough. For example, Zhang, Lou, and Schutyser (2018); Zhang, Taal, Boom, Chen, & Schutyser (2018) reported a stronger gluten network in wheat dough was probably formed by using high-gluten flour. This led to a more elastic dough as indicated by a lower value of the loss factor *tan* δ, as compared to the control dough in which low-gluten flour was used. Due to the increased elasticity of the dough, the extruded filament was stretched by the moving nozzle to form a thread. This significantly distorted the printed 3D structure. It is noteworthy that the printing parameters were not optimized in that study. Optimizing printing parameters such as the printing speed might reduce the occurrence of defects during printing as described above. Alternatively, many studies and applications use soft wheat flour to limit the dough elasticity.

Lastly, the volume fraction, composition and structure of fat or triglyceride in the dough formulations can influence their physiochemical properties, and subsequently influence their printability and the quality of the final baked product (e.g. shape retention during post-printing treatment). For example, Pavičić, Grgić, Ivanov, Novotni, and Herceg (2021) investigated the effect of substituting butter with olive oil on the rheological properties of a whole grain cookie dough and its 3D printability. Their results show that the higher apparent viscosity and consistency coefficient of the olive oil-containing dough resulted in a lower moisture loss during baking and better 3D printing quality, as compared to the butter-containing dough. This clearly exemplifies the importance of ingredient functionality and interactions on dough rheology, printability and texture formation throughout the 3D printing process, and the need to establish physical indicators for successful formulation development.

3.1.3 Influence of additives on 3D-printability

Although cereal-based dough systems often show thixotropic and pseudoplastic behavior which makes dough extrudable, the viscoelasticity of dough influences the shape fidelity of 3D-printed structures after deposition/extrusion, as discussed above. Therefore, in some cases additives are incorporated into the dough formulations to modify their rheological properties and to improve their 3D printability during extrusion-based 3D printing.

One group of ingredients that is often added to the dough formulations are hydrocolloids, a heterogeneous group of long chain polymers, typically, proteins and polysaccharides. Hydrocolloids are characterized by their property of forming viscous dispersions and/or gels when dispersed in water

(Saha & Bhattacharya, 2010). Therefore, they are widely used in food formulations as thickening and gelling agents, which can result in different rheological behaviors and/or texture of the food products. For example, Cohen et al. (2009) demonstrated the approach of combining hydrocolloids (i.e., xanthan gum and gelatin) and flavor agents to independently tune the mouthfeel and flavor of printing formulations. By varying the concentrations of xanthan gum and gelatin, the printing formulations showed varied texture which was indicated qualitatively in a mouthfeel matrix, i.e., from soft to firm, from smooth to granular. Those printing formulations can simulate a broad range of foods, although they contained only a minimal number of ingredients. In another study, a mixture of guar gum and xanthan gum was added to brown rice paste to improve its rheological properties (i.e. viscosity and dynamic viscoelastic behavior) during 3D printing (Huang, Zhang, & Bhandari, 2019). In addition, certain types of proteins have been incorporated in 3D-printed bakery products as additives. For example, adding calcium caseinate to wheat dough may disrupt the continuous gluten network, resulting in a dough with improved extrudability as well as a printed structure with better stability when compared to the regular wheat dough (Zhang, Lou, & Schutyser, 2018; Zhang, Taal, et al., 2018).

3.1.4 Influence of printing settings on 3D-printability

As mentioned earlier, the 3D printability of food formulations is influenced not only by their composition and rheological properties but also by the process parameters applied during printing (see Fig. 1). The latter is somewhat less investigated in 3D printing of bakery goods. Few previous studies optimized the printing parameters to achieve the best printing results of the same materials. For example, Severini et al. (2016) investigated the influences of two printing parameters namely printed layer height and infill percentage on the printability of dough and the quality of cooked snacks. They applied the response surface methodology (RSM) to investigate the effect of these two independent variables on the structural properties of 3D-printed samples (i.e., in terms of diameter and height of a simple cylinder shape). They found the diameter of the uncooked samples increased with the increase of layer height while it decreased with the increase of infill percentage. An optimal deposition was achieved when a layer height of 0.4 mm was used, while the nozzle size was 0.6 mm. They explained that dough was deposited precisely on top of the previous layer at an optimal layer height because of the equilibrium between extrusion speed of dough and nozzle moving speed, which means no under- or over-extrusion occurred.

Similarly, Huang et al. (2019) studied the printability (i.e., dimensional properties, height and diameter) of brown rice paste, as affected by three printing parameters, i.e., nozzle size/diameter, perimeters (number of outer supporting walls), and infill density (solid proportion of the inner part of the printed objects). They fitted the experimental data obtained from printing experiment to a polynomial model, to quantify the relationship between printing parameters and the printability. They found that increasing perimeters or increasing nozzle size resulted in a larger diameter of the cylinder, whereas changing the infill density showed no significant effect. In addition, the height of the printed structure decreased with the increase of nozzle size (indicating collapsing of the structure after printing), while the height increased with the number of perimeters (indicating a more stable structure). It was concluded that using a smaller nozzle size results in good dimensional properties of 3D-printed samples. However, using a smaller nozzle would increase the time needed for the printing job.

To summarize, future research on printability of dough for bakery product should consider the optimization of both food formulations and the printing parameters to achieve the best printing performance possible. Also, the insight into the interaction between food rheology and printing parameters is still limited due to the wide variety of 3D printing systems used in different studies. Currently, the development of new printing formulations and the optimization of printing parameters still largely rely on a trial-and-error approach. In most cases, offline measurements are done to collect data in order to evaluate the printing performance of certain printing materials and/or printing settings applied, such as taking photos of the extruded filament (Fahmy, Becker, & Jekle, 2020) or recording videos of the printing processes (Zhu et al., 2019). In future research, inline assessment of the printing performance can potentially be done by applying sensors (e.g., force or pressure sensor) or other inline measurements based on computer vision (e.g., optical or thermal camera). Data generated during printing experiments can then be processed into predictive models using data-driven approaches. The predictive models can make use of variables derived from mechanistic studies (e.g., rheological behavior of food materials) and the inputs of printing settings, to predict and/or control the quality of the final products (e.g., dimensional properties of the 3D structure). For example, grey-box models were developed to predict the extrudability of food formulations based on parameters that indicate their rheological properties and printing settings applied, which showed promising results (Ma, Schutyser, Boom, & Zhang, 2021).

3.2 3D-printed bakery products containing novel ingredients

3D printing can be used to fabricate novel bakery products with desired nutritional properties. To date, several bakery products with added functional ingredients have been developed using extrusion-based 3D printing. Here, we discuss examples of 3D-printed bakery products containing three types of novel ingredients including probiotics, microalgae and insect protein.

3.2.1 Probiotics

Probiotics are living microorganisms that confer health benefits on the host when consumed in adequate amounts (FAO, WHO, 2002). Bakery products are an emerging category within the functional food segment, which could potentially be used as carriers to deliver functional bioactive ingredients such as probiotic microorganisms to the human gut where they can positively affect the intestinal health of the consumers. However, developing bakery products containing live probiotics is challenging because the baking process may partially or completely inactivate the heat-sensitive bacteria (Zhang, Huang, Ananingsih, Zhou, & Chen, 2014). Previous research has shown that the survival of probiotics in bread dough during baking is affected by baking temperature, baking time and the initial moisture content of the dough (Zhang, Taal, et al., 2018). Specifically, approx. 4–5 log reduction of viable counts of the probiotic bacteria was observed after baking (at 175–235 °C) of bread (i.e., initial mass 5–60 g) for 8 min. Experimental results and kinetic models show that probiotic bacteria are more heat-resistant when the cells are embedded in a matrix with a lower moisture content.

To increase the survival rate of probiotics (*Lactobacillus plantarum* WCFS1) during baking, Zhang, Lou, and Schutyser (2018) used extrusion-based 3D printing to create cereal-based food structures with a large surface-to-volume ratio (see Fig. 3A). More specifically, two 3D printing designs of the products namely "concentric" and "honeycomb" had a surface-to-volume ratio of $9.20\,cm^2/cm^3$ and $7.25\,cm^2/cm^3$, respectively. The survival of probiotics in 3D-printed food structures during baking at three different temperatures (i.e., 145, 175 and 205 °C) was studied. Results show that the baking process of the structure with a higher surface-to-volume ratio (i.e., the honeycomb design) was accelerated as indicated by an increased drying rate. The time needed to reach the required moisture content of the structure was shortened during baking when a "honeycomb" structure was used. As a result, the

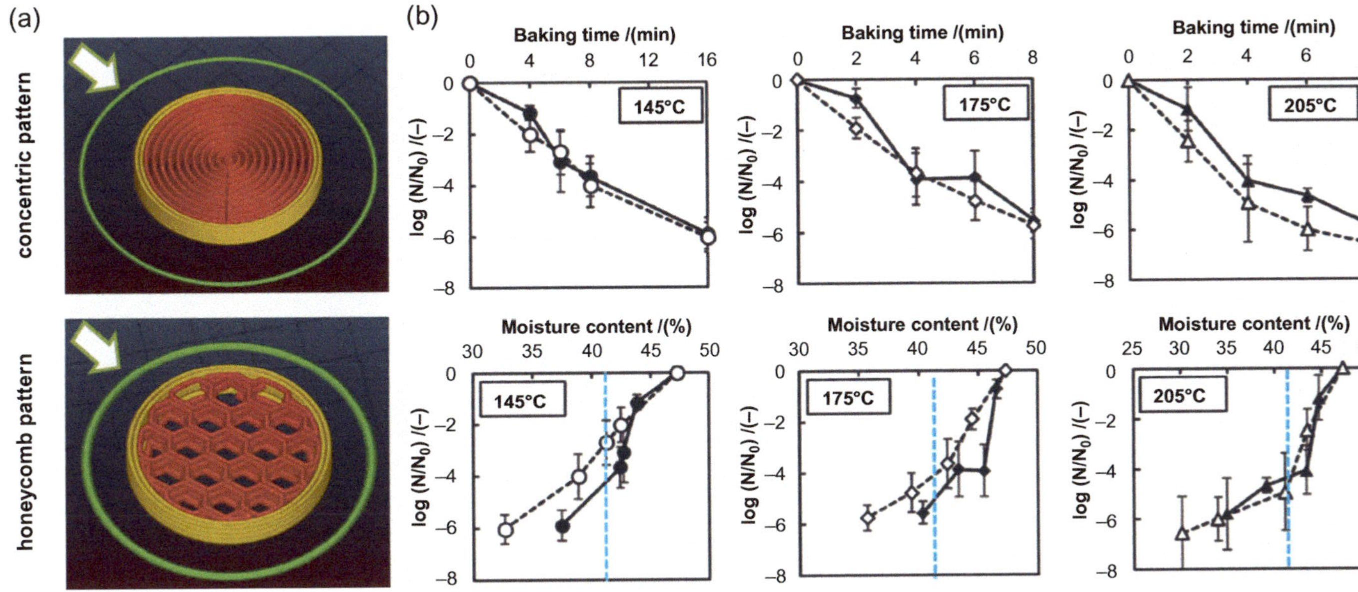

Fig. 3 3D-printed cookies containing probiotics: (A) Printing route for the "concentric" design with 60% infill and the "honeycomb" design with 30% infill. Both designs had a diameter of 40 mm and a height of 3.2 mm. Green circles are the "skirt" printed before the actual printing of the designed structures to obtain consistent flow of the printing materials; (B) Survival rate log(N/N_0) of *L. plantarum* WCFS1 in two 3D-printed structures ($N_0 \approx 10^9$ CFU/g) during baking at 145 °C (○ honeycomb, ● concentric); 175 °C (◊ honeycomb, ◆ concentric); 205 °C (Δ honeycomb, ▲ concentric), which was plotted against the baking time (first row) and the residual moisture content of the object (second row), respectively. Blue lines represent the end point of baking determined by 6% reduction of moisture content. *Figures modified from Zhang, L., Lou, Y., & Schutyser, M.A.I.I. (2018). 3D printing of cereal-based food structures containing probiotics.* Food Structure, 18, *14–22. doi:10.1016/j.foostr.2018.10.002.).*

residual viability of probiotics in the "honeycomb" structure was higher than in the "concentric" structure after baking at 145 °C and 175 °C (see Fig. 3B). The viability of probiotics in the "honeycomb" structure exceeded 10^6 CFU/g after baking at 145 °C for 6 min ($N_0 \approx 10^9$ CFU/g), thus <3 log reduction of viable counts. This research shows 3D printing is a promising technology to produce probiotic bakery products. The viability of probiotics in 3D-printed cereal-based food structures can be further improved by optimizing the structural design, by combining microwave during baking to further shorten the baking time and/or by encapsulating bacteria to enhance their thermo-resistance.

3.2.2 Microalgae biomass

Microalgae biomass can be used as a healthy food ingredient for the production of functional bakery products such as cookies, bread and biscuits with enhanced protein and fiber content (Caporgno & Mathys, 2018; Niccolai et al., 2019). One of the main challenges in the application of microalgae material in bakery products is its dark green color and a slight fish flavor that adversely affect consumers' sensory perception. Uribe-Wandurraga et al. (2020) printed cereal-based snacks fortified with microalgae (*Chlorella vulgaris* and *Arthrospira platensis*). Results show the addition of 4% microalgal biomass significantly increased the shear modulus of the batters, which improved the shape retention of the printed structure. The coaxial printing technique was explored to reduce the negative effect of the dark color on the quality and perception of the fortified snacks (see Fig. 4A): microalgae-enriched batter was deposited as the inner layer and the control batter as the outer layer. These 3D-printed microalgae-enriched snacks may be more acceptable by consumers since the dark green color of microalgae was hidden.

3.2.3 Edible insect

Edible insects are a good source of proteins, fat, vitamins and minerals (Rumpold & Schlüter, 2013), however their appearance limits their consumer acceptance. Severini et al. (2018) incorporated the ground larvae of Yellow mealworms (*Tenebrio molitor*) in wheat dough to print protein-enriched snacks (see Fig. 4B). They mixed wheat flour and mealworm powder into different mass ratios of 100:0, 90:10 and 80:20 on a dry matter basis. The farinograph analysis showed that increasing the content of mealworm powder reduced the water absorption capacity of the dough and shortened its development time. This could be due to the presence of

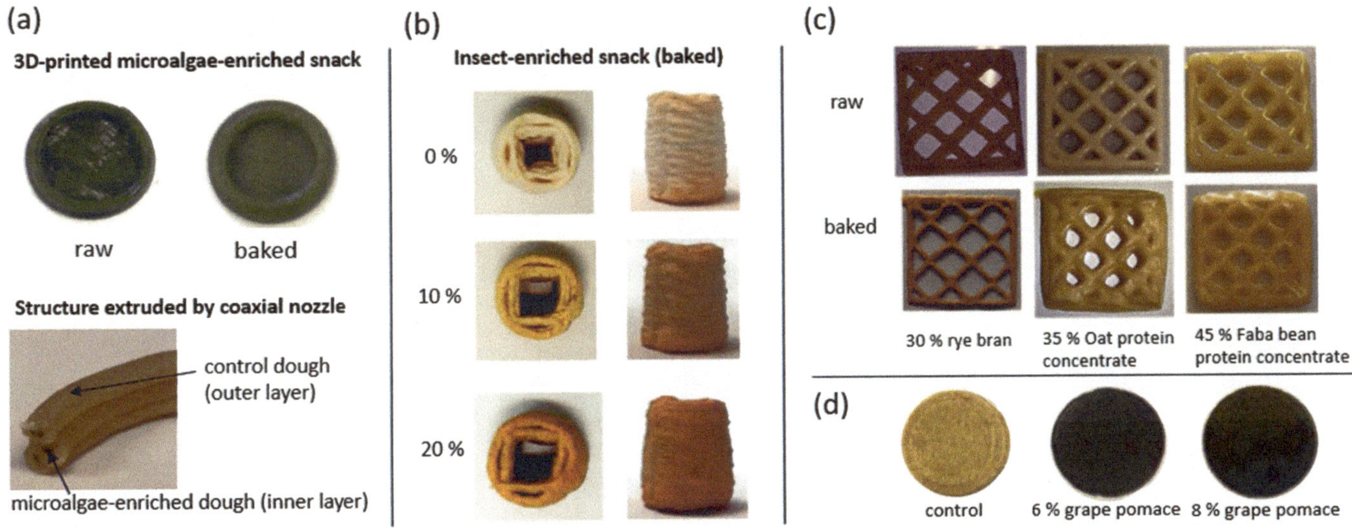

Fig. 4 Examples of 3D-printed bakery products enriched with novel ingredients (A) microalgae; (B) insect protein; (C) dietary fiber and plant protein; (D) food waste or by-products i.e. grape pomace and broken wheat. Please refer to the original publications for detailed information on the dimensions of the printed structures. *Part A: modified from Uribe-Wandurraga, Z. N., Zhang, L., Noort, M.W.J., Schutyser, M.A.I., García-Segovia, P., & Martínez-Monzó, J. (2020). Printability and physicochemical properties of microalgae-enriched 3D-printed snacks.* Food and Bioprocess Technology, 13, *2029–2042. doi:10.1007/s11947-020-02544-4. Part B: Modified from Severini, C., Azzollini, D., Albenzio, M., & Derossi, A. (2018). On printability, quality and nutritional properties of 3D printed cereal based snacks enriched with edible insects.* Food Research International, 106, *666–676. doi:10.1016/j.foodres.2018.01.034. Part C: modified from Lille, M., Nurmela, A., Nordlund, E., Metsä-Kortelainen, S., & Sozer, N. (2017). Applicability of protein and fiber-rich food materials in extrusion-based 3D printing.* Journal of Food Engineering *doi:10.1016/j.jfoodeng.2017.04.034. Part D: modified from Jagadiswaran, B., Alagarasan, V., Palanivelu, P., Theagarajan, R., Moses, J. A., & Anandharamakrishnan, C. (2021). Valorization of food industry waste and by-products using 3D printing: A study on the development of value-added functional cookies.* Future Foods, *100036. doi:10.1016/j.fufo.2021.100036.*

hydrophobic compounds such as fatty acids in ground mealworms. Dough enriched with 20% mealworm was softer than the control, as shown by a significant decrease in the torque value measured during dough mixing (i.e., lower Brabender Units number). As a consequence, the weakening of the dough caused an increase in the mass and a decrease in the height of the printed sample, when the same printing settings were used.

The same study also investigated the post-printing treatment of insect-enriched snacks. After baking at optimal conditions (22 min, 200 °C), the 3D-printed snacks showed a decrease in size due to moisture loss. The final 3D-printed insect-enriched cookies had a protein content of 41.3% and an improved amino acid profile as compared to the control. This study shows that developing novel insect-rich bakery products which have an interesting shape, texture and design using 3D printing could be a good strategy to promote the consumption and acceptance of insect-containing foods (Caporizzi, Derossi, & Severini, 2019).

Besides products containing probiotics, microalgae and insects, other interesting examples of functional 3D-printed bakery products include fiber-rich snacks (Lille et al., 2020) and plant-protein-based snacks (see Fig. 4C) (Lille, Nurmela, Nordlund, Metsä-Kortelainen, & Sozer, 2017). Furthermore, Jagadiswaran et al. (2021) investigated the valorization of by-products generated in food production, specifically grape pomace and broken wheat as main ingredients for 3D printing applications (see Fig. 4D). Adding these ingredients augmented the nutritional value and antioxidative properties of the 3D-printed cookies. To summarize, the aforementioned examples show that 3D food printing is a promising technology to introduce novel food ingredients in bakery products.

4. Personalization in 3D-printed bakery products

Personalized nutrition uses individual-specific information, which can promote dietary behavior change that may result in measurable health benefits (Adams et al., 2020), to sustain the longer-term health of individuals (Celis-Morales et al., 2017). For example, Zeevi et al. (2015) found that personalized dietary interventions could successfully modify elevated postprandial blood glucose levels and its metabolic consequences (i.e., gut microbiota alterations). However, currently existing personalized nutrition platforms generate a personalized dietary advice, a meal planning, or a recommended grocery list. 3D food printing allows for the actual (on-demand) production of personalized foods such as bakery products. Each 3D-printed bakery product could be customized in terms of its nutritional composition, according to the personalized dietary advice. In addition, the appearance (shape and color), flavor, and texture of the bakery products can be customized, according to the preference of individual consumers. For example, interesting bakery products with an unique appearance such as multi-material butterfly cookies were fabricated with 3D printing (Sun et al., 2018), which were found to be attractive to consumers. In this section, we will elaborate on the personalization in 3D-printed bakery products in terms of sensorial properties (with a focus on the texture) and nutritional properties. We will discuss consumers' acceptance of 3D-printed foods and how we can further advance research area in order to realize personalized nutrition.

4.1 Personalized texture

Texture is an important sensorial properties and quality indicator for bakery foods. The texture of bakery products is strongly linked to their micro- and macrostructure. Conventionally, the microstructure of the bakery products could be controlled by changing their formulations (e.g., by varying the amount of the leavening agent), which leads to changes in the texture. However, 3D printing provides new opportunities to control or program the micro/macrostructure of bakery products by optimizing the printing process (e.g., by varying the geometrical designs). For example, Derossi, Caporizzi, Oral, and Severini (2020) found that 3D-printed snacks showed a different microstructure than that of manually-shaped samples after baking. More air was incorporated in the printed structure as shown in the topographic images (see Fig. 5A), which significantly affects the mechanical properties of the cubic cereal-based structures. Therefore, via 3D printing, the micro/macro structure of bakery foods can be rationally designed and controlled, as compared to the "randomly" distributed porous structure obtained by a conventional manufacturing process (Derossi, Caporizzi, Oral, & Severini, 2020; Piovesan et al., 2020; Varghese et al., 2020).

In another example, Derossi, Caporizzi, Paolillo, and Severini (2020) produced cereal snacks with different textures by modulating the number and the position of cubic-shaped pores in a bar-shaped structure via 3D printing (see Fig. 5B). They reported that the hardness of the snacks decreased from 289 N to 84 N as determined by a compression test using a texture analyser, when the relative density of the snacks decreased from 0.569 to 0.401. The Gibson and Ashby model was used to describe the quantitative relationship between the fractural modulus of the printed snacks

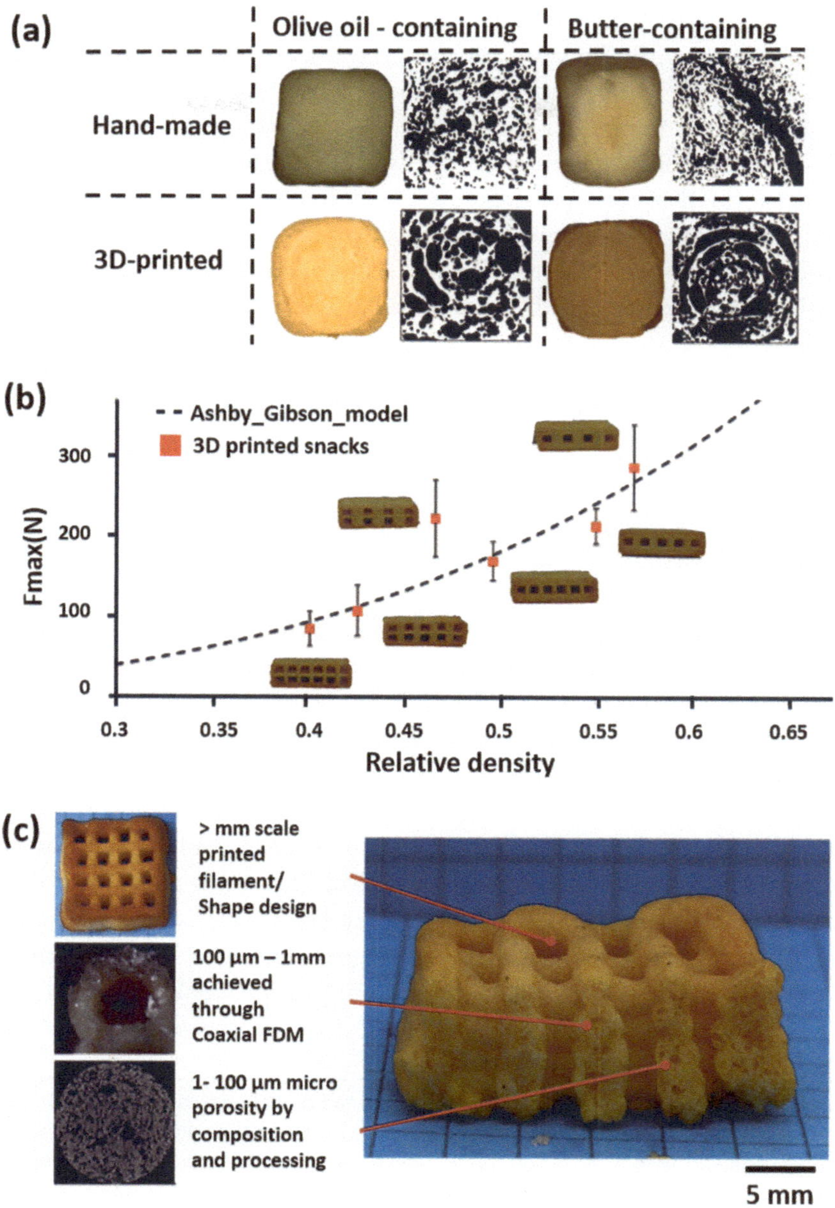

Fig. 5 Design texture of 3D-printed bakery products: (A) digital photos show the top view of hand-made and 3D-printed cereal-based snacks, and larger pores were observed in 3D-printed samples as compared to traditional hand-made samples as shown by the X-ray topographic images (i.e. black pixels indicate void phase of air); (B) relationship between the maximum force to fracture the 3D-printed snacks and their

(Continued)

(i.e., hardness) and their relative density. The fitting result was found satisfactory (see Fig. 5B), however, the heterogeneity of the snack's microstructure is not well represented by this model.

In contrast, in another study, both the macro- and micro-porosity of 3D-printed cookies were considered when constructing a computational model to predict a cookie's texture. X-ray micro-computed tomography (μCT) was used to non-destructively obtain quantitative data on the microstructure of 3D-printed cookies. Finite element modeling was used to link the Young's modulus of 3D-printed cookies to their structural parameters such as wall thickness, (air) cell size and overall porosity. A good match between computational and experimental Young's modulus values was observed (Piovesan et al., 2020). This study indicates that acquiring quantitative data on the micro- and macrostructure of 3D-printed cookies could lead to a more reliable prediction of their texture.

3D printing enables variations in the texture of bakery foods in a single matrix to be designed by structuring at multiple length scales. For example, Noort et al. (2017) created 3D-printed cookies with tunable porosity at three different length scales (see Fig. 5C). First, the overall 3D grid-type design introduced large continuous pores with dimensions of one to several millimeters. This can be combined with the ability to tune the thickness of the dough walls through printing parameters like nozzle diameter and extrusion factor, to determine the first level of porosity. Secondly, by printing with a coaxial nozzle and using air as the core material, hollow dough tubes/filaments could be created that were deposited in a layer-by-layer manner, resulting in the overall 3D shape. The hollow inner channels with a diameter in the 100–1000 μm range allowed for the creation of a second level of porosity. Finally, by tuning the amount of leavening agent in the dough formulation, the third level of porosity (<100 μm), i.e., the micro-porosity of the cookie material after

Fig. 5—Cont'd corresponding relative density, and digital photos show the side view of the samples indicating numbers of macro-pores present; dashed line refer to the fitting results of Gibson & Ashby model which describes the mechanical strength of porous structures and their relative density; (C) structuring cookies at multiple length scales using coaxial extrusion 3D printing (Noort et al., 2017). *Part A: Modified from Derossi, A., Caporizzi, R., Oral, M.O., Severini, C. (2020). Analyzing the effects of 3D printing process per se on the microstructure and mechanical properties of cereal food products.* Innovative Food Science & Emerging Technologies, *66, 102531. doi:10.1016/j.ifset.2020.102531. Part B: Modified from Derossi, A., Caporizzi, R., Paolillo, M., Severini, C. (2020). Programmable texture properties of cereal-based snack mediated by 3D printing technology.* Journal of Food Engineering, *289, 110160. doi:10.1016/j.jfoodeng.2020.110160.*

baking, could be tuned. The overall texture of the baked cookie was the result of the combined structural features at different length scales. This example clearly shows the unique possibilities that 3D printing offers. This multiscale porosity design can control the fracture behavior of cookies in a way which is not feasible by using conventional processing methods. A potential application of this novel technology is designing nutritious bakery foods with desired textural properties for consumers with special dietary requirements. For example, patients who suffer from dysphagia or other diseases related to oral processing of food may benefit from this technology (Cichero, 2015).

4.2 Personalized nutrients

Sugars are often added to some starch-based bakery products such as sweet cookies. The rapidly available glucose leads to periodic elevated plasma glucose and insulin concentrations that are associated to health issues such as diabetes and coronary heart disease (Pinto, Castro, Vicente, Bourbon, & Cerqueira, 2014). Therefore, it is interesting to find a way to reduce the amount of sugar in bakery foods while keeping their attractiveness.

One strategy to reduce the sugar content in foods is to replace sugars by high potency sweeteners. Nevertheless, sugar reduction remains a challenge because those ingredients not only provide a desired taste/flavor (Stieger, 2011) but also contribute to the structure formation and the final texture and color of bakery goods (van der Sman & Renzetti, 2020). Interestingly, recent research has shown that temporal variations of tastant concentration during the consumption of semi-solid or soft food can be introduced by inhomogenously distributing tastant in the food matrix (Stieger & Van de Velde, 2013). For example, granular salt encapsulated with high melting fat was used in bread baking, and the resulted inhomogeneous distribution of salt in bread has been shown to enhance overall saltiness perception substantially (Noort, Bult, & Stieger, 2012), which provides new opportunities of salt reduction in bakery products. Also, increased saltiness perception of bread with reduced sodium content was achieved by inhomogenously distributing salt agglomerates in the matrix (Monteiro, Nakagawa, Pimentel, & Sousa, 2021). Conventional food manufacturing methods are not well able to create products with a specific spatial distribution of certain components in a more controlled manner, while 3D printing allows us to design food products with a specific spatial distribution of ingredients in food matrix (see Figs. 2B and 4A). Therefore, 3D printing could allow us to create innovative healthy foods.

For example, Zhu et al. (2020) used 3D inkjet printing to accurately deposit a chocolate coating varying in thickness (i.e. as a way to change the sugar content) to the surface of rice waffles, to create foods with bite-to-bite contrast in terms of sweetness. They evaluated seven chocolate-coated rice waffles which contained the same amount of chocolate (i.e., sugar) but differed in the thickness of chocolate between three sequential bites. Their results show that sweetness, expected fullness and liking of the waffles were significantly affected by the bite-to-bite variation (i.e., so-called initial boost effect and recency effect). By designing the distribution of chocolate on the rice waffle, inkjet printing technique could potentially be used to create sugar-reduced products in future, as healthier alternatives to conventional ones. A similar concept could in principle be applied to salt-reduced and fat-reduced products without compromising on the taste and texture of bakery foods. In principle, spatial distribution of tastant in bakery products can also be realized by using coaxial extrusion and multi-nozzle extrusion.

4.3 Personalized satiety to modulate eating behavior

Bakery products can be personalized by 3D printing to influence eating behavior and to control the level of satiety after food consumption. Lin et al. (2020) developed a system called FoodFab which allows consumers to modulate their food intake. Via FoodFab, the internal structure of snacks can be varied by using different infill densities during 3D printing. Specifically, three snacks were evaluated, which differed in their infill density (i.e. 39%, 55% and 70%) and size but had the same infill pattern (i.e., honeycomb) and weight. They found that both the chewing time and perceived satiety were significantly affected by the design. The sample with the lowest infill density and hence the largest size required the longest chewing time, and also gave the highest increase in perceived satiety (as indicated by the difference in the level of hunger rated by the participants before and after consuming the sample). It was concluded that the perceived satiety of 3D-printed food can be modulated to create so-called food perception illusions while the amount of ingredients and total calories are kept the same, which could induce health-related behavior change among consumers.

4.4 Consumer acceptance

Novel food technologies such as gene technology, nanotechnology and cultured meat may offer important possibilities for improving food security, safety, and sustainability. However, consumers are often hesitant to accept them (Siegrist & Hartmann, 2020). Consumer acceptance of 3D-printed foods is important for the future development of 3D food printing, which is however rarely researched so far. Only few studies on this topic have been reported in literature (Baiano, 2020) and are often limited to consumer research based on questionnaire. For example, Manstan and McSweeney (2020) investigated consumers' attitudes toward 3D-printed foods in comparison with conventional food products by an online survey. Based on the results, they concluded that consumers are not willing to eat or buy 3D-printed foods, however believe that they are less processed and possess more health benefits than conventional food products. Unfortunately, the reasons why certain consumers do not like 3D-printed food or 3D food technology were not identified in that study.

Caulier et al. (2020) were the first to study consumer attitudes before and after consuming actual 3D printed food products. They studied the acceptance of 3D-printed cereal-based recovery bars in a real life military setting. They found that the initial attitude of these consumers was relatively positive toward 3D-printed food. Repeated consumption of 3D-printed food increased consumers' acceptance. This study suggested that consumer empowerment, degree of personalization, and development of printing technology all play a role in consumer's acceptance toward 3D-printed food. In general, the level of consumer involvement in the innovation of 3D food printing is still rather limited, which could be a focus point of future research (Jayaprakash, Ituarte, & Partanen, 2019).

4.5 Toward personalized nutrition

3D printing technologies are flexible in fabricating new food products, which allows for an easy incorporation of micronutrients such as vitamins in food formulations. For example, a company called Nourished provides its consumers with tailor-made 3D-printed gummy vitamins (Nourished, 2021). It is foreseen that in the future, 3D-printed food can be personalized with respect to composition and texture, in order to meet the requirements of individual consumers based on their nutritional needs and sensorial preferences. Some interesting studies have already been conducted in this area. For example, personalized meals were printed for elderly patients suffering from dysphagia, as demonstrated in the EU project PERFORMANCE (Personalised Food Using Rapid Manufacturing for the Nutrition of Elderly Consumers). A pilot food-printing facility produced personalized meal which contains at least 80% of the vegetable base (e.g., carrot or peas) and the other 20% allows for fortification with proteins, fats and micronutrients according to the dietary requirements of those patients

(PERFORMANCE, 2021). These printed meals provided texturized and natural looking foods which bring enjoyment of eating to the patients whose diets are otherwise often limited to pureed foods.

Consumers' nutritional needs and preferences vary from one person to another. Undoubtfully, realizing personalized nutrition will require the design of foods with varying properties such as (macro/micronutrient) composition and texture. However, changing the composition of food materials will result in changes in their rheological properties hence printing behavior, which might complicate the printing process. Therefore, better understanding of the printing behavior of complex food materials and the development of predictive models (see also Section 3.1) to optimize the printing process will facilitate the future application of personalized nutrition in improving human health. To further advance this field, food scientists should work together with nutritionists and patients themselves, etc., to conduct multidisciplinary research which involves the consumers.

5. Conclusions

Several 3D food printing technologies have been developed and used in fabricating personalized bakery products with interesting appearance, designed texture or even nutritional composition. Studies in extrusion-based 3D printing of bakery products focus on developing printable dough formulations. Additives such as hydrocolloids show beneficial effects in improving the 3D-printability of cereal-based food formulations. Optimizing printing settings could also improve the quality of 3D-printed bakery products. 3D printing technologies provide food scientists and the food industry with new ways i) to incorporate novel ingredients into bakery products to produce functional foods; ii) to control the texture and even satiety of foods by changing the food design; and iii) to alter the sensorial perception (i.e., sweetness, saltiness) of bakery foods by introducing designed spatial distribution of ingredients (e.g., tastant). 3D food printing shows the potential to produce healthier bakery foods without compromising on their attractiveness. More research on 3D printing of bakery products is needed to achieve personalized nutrition and allow efficient manufacturing. The main aspects that need to be considered for future research are highlighted in this chapter: (i) the trial-and-error approach in formulation development and process optimization may be avoided by developing data-driven predictive models; (ii) the printing process should be better controlled to improve its efficiency and the quality of the final products; (iii) consumer behavior and interaction with the 3D printing technology and digital interfaces should be studied to define ways to utilize various food printing technologies' potential to its highest value. The flexible manufacturing by 3D printing based on digital designs and information provides a unique possibility to create and produce personalized nutrition.

Acknowledgments

Lu Zhang would like to acknowledge the financial support from NWO Sectorplan Technik (Mechanical Engineering) on 3D food printing research at the Laboratory of Food Process Engineering, Wageningen University & Research.

References

Adams, S. H., Anthony, J. C., Carvajal, R., Chae, L., Khoo, C. S. H., Latulippe, M. E., et al. (2020). Perspective: Guiding principles for the implementation of personalized nutrition approaches that benefit health and function. *Advances in Nutrition*, *11*, 25–34. https://doi.org/10.1093/advances/nmz086.

Baiano, A. (2020). *3D printed foods: A comprehensive review on technologies, nutritional value, safety, consumer attitude, regulatory framework, and economic and sustainability issues. Vol. 00*. Taylor & Francis. https://doi.org/10.1080/87559129.2020.1762091.

Bigliardi, B., & Galati, F. (2013). Innovation trends in the food industry: The case of functional foods. *Trends in Food Science & Technology*, *31*, 118–129. https://doi.org/10.1016/j.tifs.2013.03.006.

Blurhapsody. (2021). https://blurhapsody.com/. (Accessed 3 May 2021).

Caporgno, M. P., & Mathys, A. (2018). Trends in microalgae incorporation into innovative food products with potential health benefits. *Frontiers in Nutrition*, *5*, 1–10. https://doi.org/10.3389/fnut.2018.00058.

Caporizzi, R., Derossi, A., & Severini, C. (2019). *Cereal-based and insect-enriched printable food*. Elsevier Inc. https://doi.org/10.1016/b978-0-12-814564-7.00004-3.

Caulier, S., Doets, E., & Noort, M. (2020). An exploratory consumer study of 3D printed food perception in a real-life military setting. *Food Quality and Preference*, *86*, 104001. https://doi.org/10.1016/j.foodqual.2020.104001.

Celis-Morales, C., Livingstone, K. M., Marsaux, C. F. M., Macready, A. L., Fallaize, R., O'Donovan, C. B., et al. (2017). Effect of personalized nutrition on health-related behaviour change: Evidence from the Food4Me European randomized controlled trial. *International Journal of Epidemiology*, *46*, 578–588. https://doi.org/10.1093/ije/dyw186.

Cichero, J. A. Y. (2015). 6—Texture-modified meals for hospital patients. *Modifying Food Texture*, *2*, 135–162. https://doi.org/10.1016/B978-1-78242-334-8.00006-7. Sensory Analysis, Consumer Requirements and Preferences.

Cohen, D. L., Lipton, J. I., Cutler, M., Coulter, D., Vesco, A., & Lipson, H. (2009). *Hydrocolloid Printing: A novel platform for customized food production* (pp. 807–817).

Derossi, A., Caporizzi, R., Oral, M. O., & Severini, C. (2020). Analyzing the effects of 3D printing process per se on the microstructure and mechanical properties of cereal food products. *Innovative Food Science & Emerging Technologies*, *66*, 102531. https://doi.org/10.1016/j.ifset.2020.102531.

Derossi, A., Caporizzi, R., Paolillo, M., & Severini, C. (2020). Programmable texture properties of cereal-based snack mediated by 3D printing technology. *Journal of Food Engineering*, *289*, 110160. https://doi.org/10.1016/j.jfoodeng.2020.110160.

Derossi, A., Caporizzi, R., Ricci, I., & Severini, C. (2019). *Critical variables in 3D food printing*. Elsevier Inc. https://doi.org/10.1016/b978-0-12-814564-7.00003-1.

Fahmy, A. R., Becker, T., & Jekle, M. (2020). 3D printing and additive manufacturing of cereal-based materials: Quality analysis of starch-based systems using a camera-based morphological approach. *Innovative Food Science & Emerging Technologies*, *63*, 102384. https://doi.org/10.1016/j.ifset.2020.102384.

FAO, WHO. (2002). *Guidelines for the evaluation of probiotics in food*. https://doi.org/10.1186/s40814-015-0013-3.

Foodjet. (2021). *Foodjet inkjet printing 2021*. https://www.foodjet.com/ (accessed August 11, 2021).

Godoi, F. C., & Sangeeta Prakash, B. R. B. (2016). 3D printing technologies applied for food design: Status and prospects. *Journal of Food Engineering*, *179*, 44–54. https://doi.org/10.1016/j.jfoodeng.2016.01.025.

Grood, J. P. W., & Grood, P. J. (2013). *Method and Device for Dispensing a Liquid*. US 8556392B2.

Gudjónsdóttir, M., Napitupulu, R. J., & Petty Kristinsson, H. T. (2019). Low field NMR for quality monitoring of 3D printed surimi from cod by-products: Effects of the pH-shift method compared with conventional washing. *Magnetic Resonance in Chemistry*, *57*, 638–648. https://doi.org/10.1002/mrc.4855.

Hao, L., Li, Y., Gong, P., & Xiong, W. (2019). *Material, process and business development for 3D chocolate printing*. Elsevier Inc. https://doi.org/10.1016/b978-0-12-814564-7.00008-0.

Holland, S., Foster, T., & Tuck, C. (2019). *Creation of food structures through binder jetting*. *Vol. 1867*. Elsevier Inc. https://doi.org/10.1016/b978-0-12-814564-7.00009-2.

Huang, M. S., Zhang, M., & Bhandari, B. (2019). Assessing the 3D printing precision and texture properties of brown rice induced by infill levels and printing variables. *Food and Bioprocess Technology*, *12*, 1185–1196. https://doi.org/10.1007/s11947-019-02287-x.

Jagadiswaran, B., Alagarasan, V., Palanivelu, P., Theagarajan, R., Moses, J. A., & Anandharamakrishnan, C. (2021). Valorization of food industry waste and by-products using 3D printing: A study on the development of value-added functional cookies. *Future Foods*, *100036*. https://doi.org/10.1016/j.fufo.2021.100036.

Jayaprakash, S., Ituarte, I. F., & Partanen, J. (2019). Prosumer-driven 3D food printing. *Fundamentals of 3D Food Printing & Applications*, *2024*, 331–354. https://doi.org/10.1016/b978-0-12-814564-7.00011-0.

Jayaprakash, S., Paasi, J., Pennanen, K., Ituarte, I. F., Lille, M., Partanen, J., et al. (2020). Techno-economic prospects and desirability of 3d food printing: Perspectives of industrial experts, researchers and consumers. *Foods*, *9*, 1–23. https://doi.org/10.3390/foods9121725.

Lille, M., Kortekangas, A., Heiniö, R. L., & Sozer, N. (2020). Structural and textural characteristics of 3D-printed protein- and dietary fibre-rich snacks made of milk powder and wholegrain rye flour. *Foods*, *9*. https://doi.org/10.3390/foods9111527.

Lille, M., Nurmela, A., Nordlund, E., Metsä-Kortelainen, S., & Sozer, N. (2017). Applicability of protein and fiber-rich food materials in extrusion-based 3D printing. *Journal of Food Engineering*. https://doi.org/10.1016/j.jfoodeng.2017.04.034.

Lin, Y., Punpongsanon, P., Wen, X., Iwai, D., Sato, K. S., Obrist, M., et al. (2020). In *FoodFab: Creating food perception illusions using food 3D printing Proceedings of the 2020CHI Conference on Human Factors in Computing Systems* (pp. 1–13).

Lipton, J., Arnold, D., Nigl, F., Lopez, N., Cohen, D., Noren, N., et al. (2010). In *Multi-material food printing with complex internal structure suitable for conventional post-processing Solid Free. Fabr. Symp* (pp. 809–815).

Liu, Z., Bhandari, B., Prakash, S., Mantihal, S., & Zhang, M. (2019). Linking rheology and printability of a multicomponent gel system of carrageenan-xanthan-starch in extrusion based additive manufacturing. *Food Hydrocolloids*, *87*, 413–424. https://doi.org/10.1016/j.foodhyd.2018.08.026.

Liu, Y., Tang, T., Duan, S., Qin, Z., Zhao, H., Wang, M., et al. (2020). Applicability of rice doughs as promising food materials in extrusion-based 3D printing. *Food and Bioprocess Technology*, *13*, 548–563. https://doi.org/10.1007/s11947-020-02415-y.

Liu, Z., & Zhang, M. (2019). *3D food printing technologies and factors affecting printing precision.* Elsevier Inc. https://doi.org/10.1016/b978-0-12-814564-7.00002-x.

Liu, Z., Zhang, M., Bhandari, B., & Wang, Y. (2017). 3D printing: Printing precision and application in food sector. *Trends in Food Science & Technology*, *69*, 83–94. https://doi.org/10.1016/j.tifs.2017.08.018.

Liu, Y., Zhang, W., Wang, K., Bao, Y., Mac, R. J., & Zhou, P. (2019). Fabrication of gel-like emulsions with whey protein isolate using microfluidization: rheological properties and 3D printing performance. *Food and Bioprocess Technology*, *12*, 1967–1979. https://doi.org/10.1007/s11947-019-02344-5.

Liu, Z., Zhang, M., & Yang, C. (2018). Dual extrusion 3D printing of mashed potatoes/strawberry juice gel. *LWT - Food Science and Technology*, *96*, 589–596.

Loveday, S. M., Rao, M. A., Creamer, L. K., & Singh, H. (2010). Rheological behavior of high-concentration sodium caseinate dispersions. *Journal of Food Science*, *75*, 30–35. https://doi.org/10.1111/j.1750-3841.2009.01493.x.

Ma, Y., Schutyser, M. A. I., Boom, R. M., & Zhang, L. (2021). Predicting the extrudability of complex food materials during 3D printing based on image analysis and gray-box data-driven modelling. *Innovative Food Science & Emerging Technologies*, *73*, 102764. https://doi.org/10.1016/j.ifset.2021.102764.

Manstan, T., & McSweeney, M. B. (2020). Consumers' attitudes towards and acceptance of 3D printed foods in comparison with conventional food products. *International Journal of Food Science and Technology*, *55*, 323–331. https://doi.org/10.1111/ijfs.14292.

Mantihal, S., Prakash, S., Godoi, F. C., & Bhandari, B. (2017). Optimization of chocolate 3D printing by correlating thermal and flow properties with 3D structure modeling. *Innovative Food Science & Emerging Technologies*. https://doi.org/10.1016/j.ifset.2017.09.012.

Monteiro, A. R. G., Nakagawa, A., Pimentel, T. C., & Sousa, I. (2021). Increasing saltiness perception and keeping quality properties of low salt bread using inhomogeneous salt distribution achieved with salt agglomerated by waxy starch. *LWT*, *146*, 111451. https://doi.org/10.1016/j.lwt.2021.111451.

NatureMachine. (2021). *Nature machine Foodini printer 2021*. https://www.naturalmachines.com/foodini (accessed August 10, 2021).

Niccolai, A., Venturi, M., Galli, V., Pini, N., Rodolfi, L., Biondi, N., et al. (2019). Development of new microalgae-based sourdough "crostini": functional effects of Arthrospira platensis (Spirulina) addition. *Scientific Reports*, *9*, 1–12. https://doi.org/10.1038/s41598-019-55840-1.

Noort, M. (2018). *3D printed cereal foods for personalized nutrition, ICC International Conference Grains for wellbeing, 5 - 8 November 2018, Taipei, Chinese Taipei.*

Noort, M. W. J., Bult, J. H. F., & Stieger, M. (2012). Saltiness enhancement by taste contrast in bread prepared with encapsulated salt. *Journal of Cereal Science*, *55*, 218–225. https://doi.org/10.1016/j.jcs.2011.11.012.

Noort, M., Van Bommel, K., & Renzetti, S. (2017). 3D-printed cereal foods. *Cereal Foods World*, *62*, 272–277. https://doi.org/10.1094/CFW-62-6-0272.

Nourished. (2021). *Personalized Vitamins 2021*. https://get-nourished.com/ (accessed August 10, 2021).

Pavičić, T. V., Grgić, T., Ivanov, M., Novotni, D., & Herceg, Z. (2021). Influence of flour and fat type on dough rheology and technological characteristics of 3d-printed cookies. *Foods*, *10*. https://doi.org/10.3390/foods10010193.

Pérez, B., Nykvist, H., Brøgger, A. F., Larsen, M. B., & Falkeborg, M. F. (2019). Impact of macronutrients printability and 3D-printer parameters on 3D-food printing: A review. *Food Chemistry*, *287*, 249–257. https://doi.org/10.1016/j.foodchem.2019.02.090.

PERFORMANCE. (2021). *EU project PERFORMANCE 2021.* https://cordis.europa.eu/project/id/312092 (accessed August 10, 2021).

Periard D, Schaal N, Schaal M, Malone E, Lipson H. Printing Food. Proc. 18th Solid Free. Fabr. Symp. Austin TX, 2007, p. 564–74.

Pinto, D., Castro, I., Vicente, A., Bourbon, A. I., & Cerqueira, M.Â. A. (2014). Functional bakery products: An overview and future perspectives. In W. Zhou, Y. H. Hui, I. De Leyn, M. A. Pagani, C. M. Rosell, J. D. Selman, et al. (Eds.), *Bakery Products Science & Technology* (2nd ed., pp. 431–452). https://doi.org/10.1002/9781118792001.ch25.

Piovesan, A., Vancauwenberghe, V., Aregawi, W., Delele, M. A., Bongaers, E., de Schipper, M., et al. (2020). Designing mechanical properties of 3D printed cookies through computer aided engineering. *Foods*, *9*, 1804. https://doi.org/10.3390/foods9121804.

Pulatsu, E. T., Su, J.-W. W., Lin, J., & Lin, M. (2020). Factors affecting 3D printing and post-processing capacity of cookie dough. *Innovative Food Science & Emerging Technologies*, *61*, 102316. https://doi.org/10.1016/j.ifset.2020.102316.

Renzetti, S., & Jurgens, A. (2016). Rheological and thermal behavior of food matrices during processing and storage: Relevance for textural and nutritional quality of food. *Current Opinion in Food Science*, *9*, 117–125. https://doi.org/10.1016/j.cofs.2016.10.003.

Rumpold, B. A., & Schlüter, O. K. (2013). Potential and challenges of insects as an innovative source for food and feed production. *Innovative Food Science & Emerging Technologies*, *17*, 1–11. https://doi.org/10.1016/j.ifset.2012.11.005.

Saha, D., & Bhattacharya, S. (2010). Hydrocolloids as thickening and gelling agents in food: A critical review. *Journal of Food Science and Technology*, *47*, 587–597. https://doi.org/10.1007/s13197-010-0162-6.

Schutyser, M. A. I. I., Houlder, S., De, W. M., Buijsse, C. A. P. P., Alting, A. C., de Wit, M., et al. (2018). Fused deposition modelling of sodium caseinate dispersions. *Journal of Food Engineering*, *220*, 1–7. https://doi.org/10.1016/j.jfoodeng.2017.02.004.

Schwab, A., Levato, R., D'Este, M., Piluso, S., Eglin, D., & Malda, J. (2020). Printability and shape fidelity of bioinks in 3D bioprinting. *Chemical Reviews*, *120*, 11028–11055. https://doi.org/10.1021/acs.chemrev.0c00084.

Severini, C., Azzollini, D., Albenzio, M., & Derossi, A. (2018). On printability, quality and nutritional properties of 3D printed cereal based snacks enriched with edible insects. *Food Research International*, *106*, 666–676. https://doi.org/10.1016/j.foodres.2018.01.034.

Severini, C., Derossi, A., & Azzollini, D. (2016). Variables affecting the printability of foods: Preliminary tests on cereal-based products. *Innovative Food Science & Emerging Technologies*, *38*, 281–291. https://doi.org/10.1016/j.ifset.2016.10.001.

Siegrist, M., & Hartmann, C. (2020). Consumer acceptance of novel food technologies. *Nature Foods*, *1*, 343–350. https://doi.org/10.1038/s43016-020-0094-x.

Skylar-Scott, M. A., Mueller, J., Visser, C. W., & Lewis, J. A. (2019). Voxelated soft matter via multimaterial multinozzle 3D printing. *Nature*, *575*, 330–335. https://doi.org/10.1038/s41586-019-1736-8.

(2021). *Sli3er open source 3D printing toolbox 2021.* https://slic3r.org/. (Accessed 10 August 2021).

Stieger, M. (2011). Texture-taste interactions: Enhancement of taste intensity by structural modifications of the food matrix. *Procedia Food Science*, *1*, 521–527. https://doi.org/10.1016/j.profoo.2011.09.079.

Stieger, M., & Van de Velde, F. (2013). Microstructure, texture and oral processing: New ways to reduce sugar and salt in foods. *Current Opinion in Colloid & Interface Science*, *18*, 334–348. https://doi.org/10.1016/j.cocis.2013.04.007.

Sun, J., Zhou, W., Huang, D., Fuh, J. Y. H., & Hong, G. S. (2015). An overview of 3D printing technologies for food fabrication. *Food and Bioprocess Technology*, *8*, 1605–1615. https://doi.org/10.1007/s11947-015-1528-6.

Sun, J., Zhou, W., Yan, L., Huang, D., & Lin, L. Y. (2018). Extrusion-based food printing for digitalized food design and nutrition control. *Journal of Food Engineering*, *220*, 1–11. https://doi.org/10.1016/j.jfoodeng.2017.02.028.

Telea, A., & Jalba, A. (2011). In *Voxel-Based Assessment of Printability of 3D Shapes Alexandru Math. Morphol. Its Appl. to Image Signal Process. - 10th Int. Symp. ISMM 2011, Verbania-Intra, Italy*. https://doi.org/10.1007/978-3-642-21569-8.

Upprinting Food. (2021). https://www.upprintingfood.com/. (Accessed 3 May 2021).

Uribe-Wandurraga, Z. N., Zhang, L., Noort, M. W. J., Schutyser, M. A. I., García-Segovia, P., & Martínez-Monzó, J. (2020). Printability and physicochemical properties of microalgae-enriched 3D-printed snacks. *Food and Bioprocess Technology*, *13*, 2029–2042. https://doi.org/10.1007/s11947-020-02544-4.

van der Sman, R. G. M., & Renzetti, S. (2020). Understanding functionality of sucrose in cake for reformulation purposes. *Critical Reviews in Food Science and Nutrition*, *0*, 1–17. https://doi.org/10.1080/10408398.2020.1786003.

Vancauwenberghe, V., Verboven, P., Lammertyn, J., & Nicolaï, B. (2018). Development of a coaxial extrusion deposition for 3D printing of customizable pectin-based food simulant. *Journal of Food Engineering*, *225*, 42–52. https://doi.org/10.1016/j.jfoodeng.2018.01.008.

Varghese, C., Wolodko, J., Chen, L., Doschak, M., Srivastav, P. P., & Roopesh, M. S. (2020). Influence of selected product and process parameters on microstructure, rheological, and textural properties of 3D printed cookies. *Foods*, *9*. https://doi.org/10.3390/foods9070907.

Yang, F., Zhang, M., Fang, Z., & Liu, Y. (2019). Impact of processing parameters and post-treatment on the shape accuracy of 3D-printed baking dough. *International Journal of Food Science and Technology*, *54*, 68–74. https://doi.org/10.1111/ijfs.13904.

Zeevi, D., Korem, T., Zmora, N., Israeli, D., Rothschild, D., Weinberger, A., et al. (2015). Personalized nutrition by prediction of glycemic responses. *Cell*, *163*, 1079–1094. https://doi.org/10.1016/j.cell.2015.11.001.

Zhang, L. (2018). Bio-active Ingredients during mini-bread baking. *Wageningen University & Research*.

Zhang, L., Huang, S., Ananingsih, V. K., Zhou, W., & Chen, X. D. (2014). A study on Bifidobacterium lactis Bb12 viability in bread during baking. *Journal of Food Engineering*, *122*, 33–37. https://doi.org/10.1016/j.jfoodeng.2013.08.029.

Zhang, L., Lou, Y., & Schutyser, M. A. I. I. (2018). 3D printing of cereal-based food structures containing probiotics. *Food Structure*, *18*, 14–22. https://doi.org/10.1016/j.foostr.2018.10.002.

Zhang, L., Taal, M. A., Boom, R. M., Chen, X. D., & Schutyser, M. A. I. (2018). Effect of baking conditions and storage on the viability of Lactobacillus plantarum supplemented to bread. *LWT - Food Science and Technology*, *87*, 318–325. https://doi.org/10.1016/j.lwt.2017.09.005.

Zhu, S., Ribberink, M., De Wit, M., Schutyser, M., & Stieger, M. (2020). Modifying sensory perception of chocolate coated rice waffles through bite-to-bite contrast: An application case study using 3D inkjet printing. *Food & Function*, *11*, 10580–10587. https://doi.org/10.1039/d0fo01787f.

Zhu, S., Stieger, M., van der Goot, A. J., & Schutyser, M. (2019). Extrusion-based 3D printing of food pastes: Correlating rheological properties with printing behaviour. *Innovative Food Science & Emerging Technologies*, *58*, 102214. https://doi.org/10.1016/j.ifset.2019.102214.

CHAPTER TWO

Dietary fiber in bakery products: Source, processing, and function

Suyun Lin*
Key Lab for Natural Products and Functional Foods of Jiangxi Province, College of Food Science and Engineering, Jiangxi Agricultural University, Nanchang, China
*Corresponding author: e-mail address: ncusklinsuyun@163.com

Contents

Advances in Food and Nutrition Research, Volume 99
ISSN 1043-4526
https://doi.org/10.1016/bs.afnr.2021.12.001

Abstract

Bakery products are prevalently consumed foods in the world, and they have been regarded as convenient dietary vehicles for delivering nutritive ingredients into people's diet, of which, dietary fiber (DF) is one of the most popular items. The food industry attempts to produce fiber-enriched bakery products with both increasing nutritional value and appealing palatability. As many new sources of DFs become available, and consumers are moving towards healthier diets, studies of using these DFs as functional ingredients in baked goods are becoming vast. Besides, the nutrition value of DF is commonly accepted, and many investigations have also revealed the health benefits of fiber-enriched bakery products. Thus, this chapter presents an overview of (1) trends in supplementation of DF from various sources, (2) impact of DF on dough processing, quality and physiological functionality of bakery products, and (3) technologies used to improve the compatibility of DF in bakery products.

1. Introduction

People increasingly tend to maintain and promote their health through a diet. Dietary fiber (DF) is one of the well-known dietary factors related to its potential impact on human health; hence, it has been extensively applied as a functional ingredient in processed foods to increase foods' functionality (Yegin, Kopec, Kitts, & Zawistowski, 2020). DF includes many natural and modified substances with different physiochemical properties. Remarkably, it could not be digested in the human gastrointestinal digestion systems which may exhibit various potential physiological effects (Poutanen et al., 2018). Plenty of studies have proved its positive effects to prevent and/or cure diverse illnesses, such as cardiovascular disease, diabetes, and obesity (Chuang et al., 2012; Dahm et al., 2010; Du et al., 2010; Eastwood & Morris, 1992; Maki et al., 2010; Satija & Hu, 2012). Correspondingly, the proper intake of DF has been thought to reduce the incidence and risk of common and chronic illnesses.

Based on the nutritional value of DF, the intake of DF has been widely recommended all over the world. The recommendation DF intake in the United Kingdom is 18 g/day and it is expressed as non-starch polysaccharides (NSP), and the amount is enhanced to 30 g/day in France, Germany, Austria, Switzerland, and China (Stephen et al., 2017; Yang, Yang, Zhai, & Cheng, 2016). In the United States of America, the recommended dietary fiber intake is 14 g/1000 kcal, for an average adult, it means 38 g/day for male and 26 g/day for female (Miller-Jones, 2004). A Mediterranean diet, typical in Spain, Italy and Greece, offers a significant content of DF as it is rich in vegetables, cereals, fruits and legumes; the recommended intake in these countries being 20 g/day for men and 15.7 g/day for women (Capita & Alonso-calleja, 2003). Similar to the opinion of the European Food Safety Authority (EFSA) (EFSA Panel on Dietetic Products & Allergies, 2010), the World Health Organization (WHO) recommends an intake of at least 25 g/day of total DF, which can protect against obesity and its consequences (Lin et al., 2015). However, the current DF consumption of people is usually lower than the recommended value (Rodríguez, Jiménez, Fernández-Bolaños, Guillén, & Heredia, 2006; Stephen et al., 2017). Accordingly, food scientists and technologists are trying to develop fiber-enriched or fiber-fortified novel products.

Bakery product is one of the most widely, easily, and regularly consumed food all over the world and they are ideal delivery vehicles of DF and other bioactive compounds. It is produced from recipes mainly composed of wheat or other cereal flours/powder which are mixed with other ingredients and undergo a heat-processing step in an oven which removes most of the moisture (Cauvain & Young, 2008). As mainly prepared from flour or meal derived from some form of grains, they are considered as important energy sources for people. Nowadays, a wide spectrum of grain-based foods can be found in the food market, and one of the most common bakery products is various types of bread including sourdough, par-baked bread, and steamed bread, and other commonly consumed bakery products including cakes, rolls, biscuits, cookies, muffins, pies, pastries, pretzels, etc. In this chapter, bakery products are extended to include similar well-known grain-based foods such as noodles and pasta.

As the rapid development of the cereal industry, wheat grains are usually highly processed to output refined flour which can manufacture products with better sensory quality. However, this process, in turn, induces a great loss of nutrients. Most of the nutrients including essential amino acids, bioactive compounds, and a large amount of DF, are discarded along with wheat bran (Onipe, Jideani, & Beswa, 2015). Therefore, aiming to produce bakery products rich in DF, extra DF or whole grains flour are greatly needed and used. This chapter summarizes the utilization of DFs into bakery products from diverse sources. To date, the most common sources of DFs in bakery products are bran or whole-grain flour from various cereals and pseudo-cereals, but lately, other sources including fruits and vegetables, as well as legumes and pulses are emerging. Additionally, the addition of DF in bakery products has been remarkably reported to impact dough properties and gluten network formation which are key factors controlling end-product, consequently, changing the rheology, texture, and organoleptic quality of the bakery products. The detrimental impacts on the quality of end-products and accordingly their low acceptability by consumers are the main challenges for fiber fortification in bakery products. Fortunately, techniques and solutions to increase the popularity of fiber-enriched bakery products have been successively exploited. Hence, these relevant contents are also gathered and discussed in this chapter.

2. Dietary fibers (DFs)

The term "dietary fiber" (DF) was first presented in the 1950s, involving the plant cell divided materials (Hipsley, 1953); later, DF was considered to consist of the remnants of edible plant cells, polysaccharides, lignin, and associated substances resistant to digestion by the alimentary enzymes of humans (Trowell, 1972; Trowell et al., 1976). Subsequently, the definition of DF has been discussed for many years. However, its precise definition has not been successfully confirmed until now (Maphosa & Jideani, 2015), basically due to the inconsistency in its content of carbohydrates, its physiological effects and its measuring methods (Slavin, 2013; Westenbrink, Brunt, & van der Kamp, 2013). At present, the DF definition of Codex Alimentarius is broadly acknowledged (Codex, 2009). As summarized by Stephen et al. (2017), DF is not hydrolyzed by the enzymes and absorbed in the human intestine. It is made up of carbohydrate polymers with ≥3 monomeric units (MU), and it includes the following categories: (1) non-starch polysaccharides (NSP) from fruits, vegetables, cereals, and tubers whether intrinsic or extracted, chemically, physically and/or enzymatically modified or synthetic (MU ≥ 10); (2) resistant oligosaccharides (RO) which are non-digestible (MU 3–9); and (3) resistant starch (RS) (MU ≥ 10). It should be noted that these polymers must have a beneficial physiological effect demonstrated by generally accepted scientific evidence to consider the polymer as DF. Additionally, several non-carbohydrates "associated substances"—lignin and substances which present in cell walls and closely associated to carbohydrate polymers, are also considered as DF by the accepted analytical methods (Ktenioudaki & Gallagher, 2012).

Based on their solubility in water, DFs can be broadly classified into soluble DF (SDF) and insoluble DF (IDF). They involve differences in their components, structures, technological functionalities, and physiological effects (Elleuch et al., 2011; Nieman, Rains, Dicklin, & Maki, 2014). SDFs consist of non-cellulosic polysaccharides, such as pectin, inulin, β-glucan, gums, alginate, and oligosaccharides, and they are generally present in plants such as fruits, vegetables, legumes, and pulses. Depending on the ability to form viscous gels after dissolving in water, SDFs can be further divided into viscous fiber and non-viscous fiber. Generally, SDFs are characterized by their capacity to increase viscosity and exhibit various physiological effects, such as delaying gastric emptying, retarding glycemic response, and reducing plasma cholesterol (Kristensen & Bugel, 2011; Malunga, Izydorczyk, & Beta, 2017; Satija & Hu, 2012). After consumption, generally, SDFs are more fermentable in the colon when compared to IDFs. IDFs mainly comprise of cell wall components, such as cellulose, lignin, hemicellulose, and RS (Mudgil & Barak, 2019). They are generally regarded to be beneficial for the human digestive system, because they can regulate the bowel transit-time, increase the fecal bulk and excretion of bile acids, and cause softer stool (Mudgil, 2017; Nieman et al., 2014). These physiological effects of SDFs and IDFs depend on their structural and physical properties, which can result in behavior such as hydration, swelling, and enzymatic attack (Oh, Bae, & Lee, 2014).

Based on the fermentability in the larger intestine, DFs can be classified into rapidly fermentable, slowly fermentable, and non-fermentable DFs (Mudgil & Barak, 2019). Rapidly fermentable DFs (e.g., pectin and guar gum) may have less impact on the fecal bulk compared with other two classifications. Slowly fermentable DFs (e.g., some fibers obtained from psyllium and wheat bran) may continue to impact fecal weight after fermentation. Non-fermentable DFs are considered to have the capacity of increasing fecal bulk and fecal weight in colon, which may reduce the risk of constipation-related issues and colon disease. The different degree and pattern of fermentation of DFs may induce different physiological actions and bowel functions, including fecal weight, stool frequency, colon pH, and energy from non-digestive foods. In addition, fermentable DFs can be fermented in the human large intestine and yield metabolites such as short-chain fatty acids (SCFAs), which have been reported to show various physiological functions, including inhibiting cholesterol synthesis and ultimately reducing serum-cholesterol levels in the blood (Mudgil, Barak, Patel, & Shah, 2018). Considering their various sources, generally, fiber from sources containing more parenchyma tissues (such as vegetables) is easier to be fermentable when compared with sources containing more secondary tissues (such as the hull and bran of cereals).

The health benefit of DFs has led to the development of a large and potential market for fiber-enriched products and ingredients to improve the overall nutritional balance of carbohydrate-rich foods, and in recent years, there has been a trend to find new sources of DF that can be used in the food industry (Chau & Huang, 2003). The main sources of DF supplementation in bakery food has been focused on raw plants materials (e.g., cereal grains, vegetables, fruits, legumes, sugar beet, and various bran) in various forms (e.g., flour and plant extracts) after pre-treatments (e.g., extrusion-cooking, canning, grinding, boiling, and frying). The potential of using DFs for food applications is strongly related to their functional properties, such as their fiber source, type, and degree of processing (Guillon & Champ, 2000; Nelson, 2001).

Table 1 describes an appropriate classification of DF which are usually used in foods according to the main components, water solubility, and fermentability. Aforementioned, DFs are discussed from four subgroups–NSP, RO (MU 3–9), and RS (MU $\geq$ 10), and associated substances.

2.1 Non-starch polysaccharides (NSP)

Cellulose, the most abundant natural polysaccharide, is composed of monomers (glucose) linked by β-D-1,4- glycosidic bonds (Kian, Jawaid, Ariffin, & Karim, 2018), and a native cellulose chain usually consists of 10,000–15,000 units. Cellulose is the main component of the plants cell wall (outer layers of cereals, pulses, etc.), and the molecules are usually combined with other polysaccharides and lignins molecules. Additionally, cellulose is considered as IDF because it can not be digested by human enzymes system in gastrointestinal, but around half of them is fermented in the colon (Mudgil & Barak, 2013). Cellulose has a high capacity of water absorption because of its fibroid structure.

Apart from cellulose, hemicellulose is another common NSP. Structurally, hemicelluloses usually contain a backbone chain (consisting of different monosaccharide units), and the backbone is attached with various side chains, leading to a complex structure. Hemicelluloses are heteropolysaccharides that are formed by various types of monomers. They comprise arabinoxylans (AXs), glucuronoxylans and glucuronoarabinoxylans, heteroxylans, and mixed-linkage β-glucan. They mainly exist in the outer layers of plants and show different solubility and fermentability. For instance, AXs, mainly come from the starchy endosperm and aleurone layer of cereal grains, are one of the major units of soluble and insoluble DF fractions which are shown to exert various health benefits (Izydorczyk & Dexter, 2008). AXs are often bound with phenolic acids, thus, having an antioxidant activity (Rao & Muralikrishna, 2006). Additionally, heteroxylans from psyllium (e.g., plantago ovata, indica, psyllium, and asiatica) seed husk β-glucans are usually soluble and can be

Table 1 Classification of DF according to main components, water solubility, and fermentability.

Subgroup	Main components	Water solubility	Fermentability
NSP (MU ≥ 10)	Cellulose	IDF	Yes/No
	Hemicellulose		
	– Arabinoxylans (AXs)	IDF/SDF	Yes
	– Glucuronoxylans and glucuronoarabinoxylans	IDF	Yes
	– Heteroxylans	IDF/SDF	Yes/No
	– β-glucans (mixed-linkage)	SDF	Yes
	Mannans and heteromannans		
	– Mannans	IDF	Yes
	– Galactomannans	SDF	Yes
	– Glucomannans (e.g., guar gum)	SDF	Yes
	– Galactoglucomannans	–	–
	Pectin	SDF	Yes
	Inulin	SDF	Yes
	Other hydrocolloids (including gums and mucilage)		
	– Xanthan gum	SDF	Yes
	– Alginates	SDF	Yes
	– Agar-agar	SDF	Yes
	– Carrageenan	SDF	Yes
RO (MU 3–9)	α-Galactosides (Raffinose, stachyose, verbascose)	SDF	Yes
	Fructo-oligosaccharides (FOS) Galactooligosaccharides (GOS) Xylo-oligosaccharides (XOS)	SDF	Yes
	Resistant dextrins	SDF	Yes
	Polydextrose	SDF	-

Continued

Table 1 Classification of DF according to main components, water solubility, and fermentability.—cont'd

Subgroup	Main components	Water solubility	Fermentability
RS (MU ≥10)	RS1-physically inaccessible starch	IDF	Yes
	RS2-granular starches	IDF	Yes
	RS3-gelatinised and retrograded starches	IDF	Yes
	RS4-chemically modified	IDF	Yes
Associated substances (non-carbohydrates)	Lignin	IDF	No
	Waxes	IDF	No
	Chitins/chitosan	IDF	No

Notes: MU, monomeric units; AX, arabinoxylans; RS, resistant starch; IDF, insoluble dietary fibre; SDF, soluble dietary fibre; Yes, fermentable; No, non-fermentable.

fermented in the colon. Commercially, pure hemicelluloses are not available because the separation of pure hemicellulose and the associated compounds is extremely difficult and expensive.

Besides, mannans and heteromannans (e.g., galactomannans, glucomannans, and galactoglucomannans) are NSPs that present naturally in plant tissue. Generally, they are often present in plants such as ivory nut, date, green coffee bean seeds, the endosperm of grain legumes, and fungal cell walls. Their solubility and fermentability varied. Specifically, glucomannans, such as guar gum, are mannans substituted with side chains of galactose residues. Guar gum, one of the most promising SDF, is a polygalactomannan mainly extracted from legume plants seeds. It is widely utilized as a thickener and stabilizer in food products such as sauces, syrups, ice cream, instant foods, beverages, confectionaries, and baked goods (Dogan, Kayacier, & Ic, 2007; Gupta, Saurabh, Variyar, & Sharma, 2015).

Pectin is another NSP that naturally exists in cell walls and intracellular layers of plant cells. Pectin is a heteropolysaccharide comprised of polygalacturonic substituted with arabinans, galactan, or arabinogalactan. Pectin is water-soluble, and it cannot be digested by the human enzymes in the gastrointestinal system but can be fermented by microbes in the colon. It can be mainly found in fruit (apple, citrus, etc.) peel, beetroot, rice endosperm, grain legumes, and rapeseed cotyledon. In addition, pectin is regarded to show hypocholesterolemic behavior since its gel-forming ability in water phase. Therefore, pectin is widely used as an important hydrocolloid in food production (Kim, Miller, Lee, & Kim, 2016).

As a type of NSP, inulin is a water-soluble polymer and consists of fructose (units) assembled by β-(2-1)-D-frutosyl fructose linkage. It can not be digested in the small intestine due to the presence of β-configuration of anomeric carbon, but can be fermented in the colon (Apolinário et al., 2014). Inulin naturally presents in numerous plant species, of which, the chicory roots, dahlia, and resurrection plants are the most abundant sources, and they are utilized as the commercial production materials of inulin (Flamm, Glinsmann, Kritchevsky, Prosky, & Roberfroid, 2001). The functionality of inulin is greatly dependent on the polymerization and branching degree. Inulin extracted from plants is reported to have a lower degree of polymerization and less branched structure than that from bacterial (Mudgil & Barak, 2019). Also, because of its physiological function such as prebiotic activity, inulin is also utilized as a functional ingredient in the development of novel food products, which are expected to be favorable to human health, especially to the human digestive system (Shoaib et al., 2016)

Other hydrocolloids including mucilage and various gums (e.g., xanthan gum, alginates, agar-agar, and carrageenan) are also parts of NSPs. Mucilage is mainly secreted via certain cells of the plant as a gelatinous polygalacturonic acid polysaccharide (Western, Skinner, & Haughn, 2000). Gums are water-soluble polymers that can form viscous solutions after dispersing in water, and their polymeric compositions are depended on the type of gums. Gums are usually extracted from plant excretion, plant seeds, and seaweeds. Moreover, the composition and concentration of these gums are highly impacted by plant species (Campo, Kawano, Silva, & Carvalho, 2009; Mudgil & Barak, 2013). Besides, several gums (e.g., xanthan gum, and pullulan gum) can be produced by microorganisms during the fermentation process (Garcı´a-Ochoa, Santos, Casas, & Gómez, 2000; Leathers, 2003). In particular, these gums show resistance against the human enzymatic digestion and can be used as functional DFs. The addition of these hydrocolloids not only can increase the DF content in food products but also are utilized as food additives because of their viscous ability in water (Saha & Bhattacharya, 2010).

2.2 Resistant oligosaccharide (RO)

Resistant oligosaccharides (ROs) refer to the carbohydrates which composed of 3-10 monosaccharide units (MU 3-10), linear or branched, associated by glycosidic linkages. Because of the different types and numbers of the monosaccharides, ROs have various types, such as galactosides (e.g., raffinose, stachyose, and verbascose), fructooligosaccharides (FOSs), galactooligosaccharides (GOSs), and xylooligosaccharides (XOSs) (Zhao et al., 2017). Plants and algae are the richest sources of various oligosaccharides. For instance, FOSs naturally occur in artichoke, garlic, onion, asparagus, chicory, and fermented beverage of plant extract (Okada et al., 2010); GOSs can be found in human milk, cow milk (Goulas, Tzortzis, & Gibson, 2007); XOSs present in hardwood, corncob, wheat straw, rice hull and barley straw (Moure, Gullón, Domínguez, & Parajó, 2006). Some typical feruloylated oligosaccharides could be prepared from plant sources, e.g., wheat, maize bran, sugarcane bagasse, and rice (Ou & Sun, 2014). ROs are mostly soluble (but not viscous) and highly fermentable in the colon. Functional oligosaccharides have been applied to food for many purposes, such as DFs, pharmaceuticals, feeds, cosmetics, immunostimulating agents, and prebiotic compounds. For instance, the consumption of some oligosaccharides may benefit the balance of the intestinal microflora, reduce the gastrointestinal infections, as well as decrease the risk of lifestyle-related diseases (e.g., cardiovascular disease, cancer, obesity, and type 2 diabetes) (Qiang, YongLie, & QianBing, 2009). Therefore, oligosaccharides-enriched foods are considered to be good for metabolic syndromes induced specific disorders.

2.3 Resistant starch (RS)

Resistant starch (RS) is a certain category of starch that can not hydrolyzed by the human digestive enzymatic system and can be partially fermented in the human colon (Stephen et al., 2017). RSs are usually classified into four groups, i.e., RS1, RS2, RS3, and RS4 (Sajilata, Singhal, & Kulkarni, 2006). RS1 is the starch that is physically unavailable, such as the starches embedded or entrapped in plant components (i.e., seeds and partly milled grains and legumes) which are inaccessible and resistant toward hydrolysis of enzymes. In this view, RS1 may become enzymatic available once the enclosed coating or structure is disrupted. RS2 refers to intact starch granules which cannot be degraded by amylase unless it is gelatinized. RS2 usually exists in substances such as unripe bananas, uncooked peas, and potatoes. RS3 is retrograded starch and is formed when gelatinization is followed by a cooling period. Therefore, RS3 usually has a crystalline network in its structure which has been found to be hardly attacked by amylase during digestion. RS3 has been found in cooked, cooled potatoes, and bread where starches are retrograded during food processing (Katina, 2003; Wang, Li, Copeland,

Niu, & Wang, 2015). Finally, RS4 refers to chemically (e.g., crosslinking reaction and conversion) modified starches. As the structure is modified, it can resist the enzymatic hydrolysis during the digestive system. RS4 usually needs to be extra added to bakery products (Kiumarsi et al., 2019; Miyazaki, Van Hung, Maeda, & Morita, 2006). The physiological behavior of RS is similar to soluble and fermentable fiber such as guar gum. Their physiological actions include increasing fecal bulk and lowering colonic pH (Slavin, Stewart, Timm, Grabitske, & Hospattankar, 2010), forming metabolites including short-chain fatty acids (SCFA), reducing postprandial glycemic and insulinemic responses which may have been favorable for the management of diabetes (Fuentes-Zaragoza, Riquelme-Navarrete, Sánchez-Zapata, & Pérez-Álvarez, 2010). RS has been considered a new ingredient for creating fiber-enriched bakery products, including bread, cakes, muffins, pasta, and battered foods (Chung, Donner, & Liu, 2011; Sanz, Salvador, & Fiszman, 2008).

2.4 Associated substances

Lignin, usually associated with cellulose fibers in plant cells, is the second-largest available raw material in nature. Structurally, it consists of phenylpropanoid units connected by carbon-carbon or carbon-oxygen linkages (Li et al., 2018). Like cellulose, lignin is also insoluble in aqueous solutions, and it shows unique resistance toward enzymatic and chemical degradation, as well as human digestive secretions and bacterial degradation. Lignin alone is not commercially separated as a high-fiber ingredient, but it is usually linked with other fiber ingredients in plant cells. As it is hard to be isolated from lignocellulosic material, lignin is usually considered as IDF which has been found to be good for gastrointestinal physiology (Verónica, Montserrat, & Maria, 2012). Lignin mainly presents in the plant cells such as are cereal, celery, and carrots.

As another kind of associated substance, chitin and chitosan are the biological polymers consisting of glucosamine and *N*-acetylated glucosamine monomers, which are connected by β-1-4-glycosidic linkages (Hamed, Özogul, & Regenstein, 2016). Chitin and chitosan have distinct levels of acetylation and solubility. The chitosan molecule is deacetylated while the chitin molecule is extensively acetylated, and the latter is highly insoluble in water but the former shows little water insolubility. Crustacean shell wastes from various sources, such as shrimps, krill, crawfish, lobsters, prawns, and crabs, are the important sources of chitin and chitosan (Champ, Langkilde,

Brouns, Kettlitz, & Collet, 2003). Also, some fungi and brown algae can produce chitin and chitosan. The beneficial biological properties of chitosan in the prevention and treatment of several diseases have been demonstrated (Kumar, Muzzarelli, Muzzarelli, Sashiwa, & Domb, 2004). Chitosan in diets has been reported to markedly decrease the serum cholesterol as investigated *in vivo*, which may be related with its viscosity and capacity of cholesterol-binding (Xia, Liu, Zhang, & Chen, 2011).

3. Trends in fiber-enrichment in bakery products

The DFs applied in bakery products are from various sources and in different forms. Fig. 1 depicts the schematic distribution of the main sources of DFs fortified in bakery products. Whole grains and legumes are considered as good sources of DFs. Especially, bran of cereal and pseudo-cereal (e.g., corn bran, oat bran, wheat bran, rice, rye, and barley bran) are found to contain appreciable amounts of DFs. Other common sources of DFs include fruits peels, vegetables, and plant seed and hull, as well as several commercial hydrocolloids (Arslan, Rakha, Xiaobo, & Mahmood, 2019). DF sources commonly contain both SDF and IDF in different ratios. The interest in foods rich in DF has increased in recent decades, and the importance of this food constituent has led to the development of a large market for

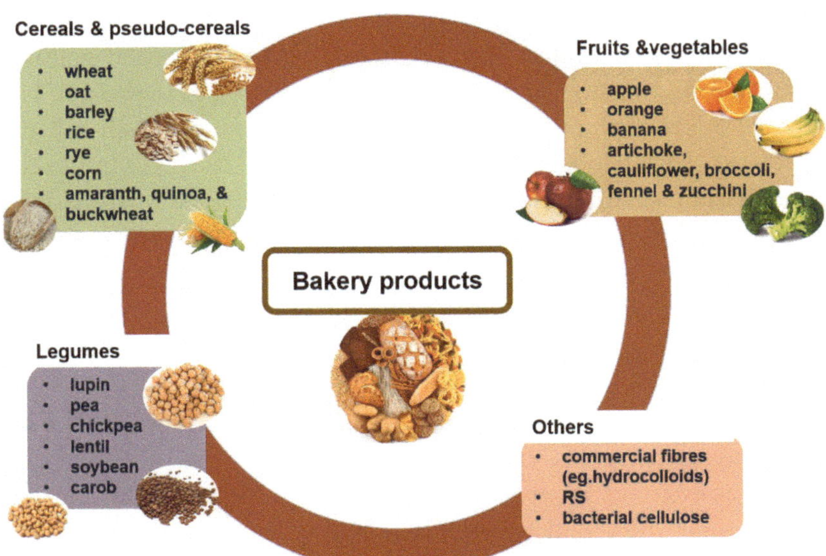

Fig. 1 Schematic distribution of the main sources of DFs fortified in bakery products.

fiber-enriched products and ingredients. A high DF intake has been related to several physiological and metabolic effects (Drzikova, Dongowski, Gebhardt, & Habel, 2005). Nowadays, recovering, recycling, and utilization of industrial by-products are prevalent, which shows an enormous potential in the industries of agriculture and food processing. This is mainly because that the residues or by-products can be reutilized to give an added-value to the main production. By-products produced from fruits and vegetables such as peels, seeds, and bagasse are of particular interest.

3.1 Sources of fiber from cereals and pseudo-cereals

Traditionally, the milling by-products of cereal grains, which have been investigated as possible fiber supplements, are mainly used as livestock feed and their nutritional values are underutilized (Prückler et al., 2014). Cereal and pseudo-cereals are considered as the most common sources of DF that are used in bakery products. The cell walls from the outer layers of grain (bran), regarded as the main DF source of plants, consisting of cellulose and complex xylans and lignin. In addition, the main DFs in their aleurone and endosperm are AXs and β-glucan, and a small portion of cellulose, heteromannans, and esterified phenolic acids. Notably, cereal DF also acts as a natural functional ingredient to deliver phenolic compounds into the gut (Vitaglione, Napolitano, & Fogliano, 2008). For instance, considering as a source of antioxidant dietary fiber and other bioactive compounds, wholegrain cereals and seeds are fortified into bread (Benítez et al., 2018).

As good sources of fiber and phytochemical compounds, the use of whole wheat grain and wheat bran is becoming increasingly valuable to formulate various bakery products. The DF content of the whole wheat grain is about 11.6–12.7% (dry weight), and these DFs are accounts for about 13–25% of whole wheat grain weight. Wheat bran is one of the richest sources of DF, and majority of the DF is NSP, which mainly consists of AXs (accounts for 70% of the NSP), celluloses (24%), and β-glucans (6%) (Maes & Delcour, 2002; Nordlund, Katina, Aura, & Poutanen, 2013). The AXs are either water water-soluble or insoluble, of which, water-extractable arabinoxylan (WE-AX) amounted to 6% of the total AX content in wheat bran (Maes & Delcour, 2002). The content of DFs in wheat varies among species and varieties (Gebruers et al., 2008). For instance, the common wheat was reported to have a higher DF content (11.5–18.3%) than emmer (7.2–12.0%) and einkorn (9.2–12.8%). Their contents of total AXs varied both in flour (1.7–2%) and bran (8.9–18%). Besides, the content of β-glucan also varies in flour (0.3–0.7%) and wheat bran (2.2–2.7%). Additionally, the particle size of bran also impact the properties of DFs. For instance, superfine ground wheat bran (*Triticum aestivum L.*) had higher SDF content than the coarse one (Lin et al., 2021).

Oats have various wholesome components, such as DFs, proteins, and minerals. The main DFs in oats are mixed-linkage (1→3), (1→4) β-D-glucan (β-glucan) and AX, which are mainly located in the aleurone region and slightly in the starchy endosperm (Khan et al., 2016). The DFs in oats were around 10.6–23.4% in different varieties, and the content of β-glucan ranged from 4.5% to 5.6% which was mainly presented in oat bran (Özkaya, Özkaya, Duman, & Turksoy, 2017; Shewry et al., 2008; Sudha, Vetrimani, & Leelavathi, 2007). The water-soluble oat β-glucan is an indigestible polysaccharide, and studies have reported its bioactivities including anticancer activity, increasing immunity, and reducing of blood cholesterol, lipids, and blood glucose (Daou & Zhang, 2012). Thus, oat β-glucan is a functional ingredient for various bakery products, such as bread (Hager et al., 2011), cakes (Kalinga & Mishra, 2009), noodles (Choo & Aziz, 2010), and cookies (Lee, Warner, & Inglett, 2005).

In addition, barley grain, known for its nutritional importance and nutraceutical properties, is also considered to be a good source of SDF and IDF, as well as phytochemicals (Lahouar et al., 2014; Madhujith, Izydorczyk, & Shahidi, 2006). Barley was reported to rank top in terms of DF content (24.6%) when compared to wheat (3–4.6%), millet (14.9%), rye (17.7%) and sorghum (21.0%) (Ragaee, Abdelaal, & Noaman, 2006). Considered as a good source of DF, barley bran was reported to have 45% of total DF (10.8% is SDF) (Sudha, Vetrimani, & Leelavathi, 2007). Additionally, the content of total DFs in barley was 11–34%, of which, 3–20% is SDF and is mostly presented as β-glucan (Makeri, Nkama, & Badau, 2013). The content of β-glucan is varied depending on the cultivar of barley, the degree of extract viscosity, and the amount and category of starch (Andersson et al., 2004; Zheng, Rossnagel, Tyler, & Bhatty, 2000). Moreover, the health benefits of consuming β-glucan-enriched foods include lowering levels of blood glucose and serum total low-density lipoprotein (LDL) cholesterol (Behall, Scholfield, & Hallfrisch, 2006; Izydorczyk & Dexter, 2008; Keenan et al., 2007). The aforementioned health advantages of β-glucan are reported to stem from its ability to increase the viscosity of intestinal content, which in turn relies on the molecular weight and solubility of β-glucan (Hashemi, 2015; Smith, Queenan, Thomas, Fulcher, & Slavin, 2008). β-glucan, as an SDF, also shows the capacity to reduce the risk of hypercholesterolemia, hypoglycemia, and chemically induced colon cancer (Newman & Newman, 2008). Additionally, a certain portion of barley DF is RS, which can be fortified as a functional material to various food products. The content of RS in barley can be enhanced by processing, such as cycles of heating and cooling and enzymatic debranching (Chaturvedi, Sharma, Shukla, Singh, & Yadav, 2011). The health benefit of RS is related to a lower glycemic index (GI) and cholesterol absorption.

As for rice, the nutritional quality of rice is impacted by the milling process involving the removal of different layers of rice. The milling of brown rice usually removes the outer bran layer, seed-coat, aleuronic layer, and embryo to make it taste better. Consequently, the rice bran which has an appreciable amount of crude fiber is produced. The rice bran, containing 40% of total DFs, is considered as an excellent DF source (Sudha, Vetrimani, & Leelavathi, 2007). The DFs in rice bran are mainly composed of noncellulosic polysaccharides (around 24%), cellulose (around 10.3%), and lignin (around 10.7%) (Sera et al., 2005). Moreover, the major parts of SDF in rice bran are AXs and some amount of β-glucans, while cellulose, hemicellulose and insoluble AX constitute the IDF (Lai, Lu, He, & Chen, 2007; Rao & Muralikrishna, 2004). Considering the physiological function of DFs of rice bran, some investigations have demonstrated their antidiabetic and hypocholesterolemic effects (Sohail, Rakha, Butt, Iqbal, & Rashid, 2017). Additionally, rice bran has been applied in all sorts of foods, such as prepared foods, nutraceuticals, snacks, bakery, pasta, gluten-free foods, beverages (Faccin, Miotto, Vieira, Barreto, & Amante, 2009) and meat products (Huang, Shiau, Liu, Chu, & Hwang, 2005).

Rye, as a traditional cereal highly used in whole-grain foods in Europe, is considered a healthy ingredient in the bread market. Rye is rich in DFs, including AX, fructan, and β-glucan, as well as various bioactive components, such as cinnamic acids, alkylresorcinols, lignans, sterols, vitamins, and minerals. When fructan was excluded, the total DF content varies between 15.0% and 21.0%, affected by both genotype and environment of rye (Hansen, Rasmussen, Bach Knudsen, & Hansen, 2003; Ragaee et al., 2006). Total AX content in the rye is around 8.0–12.1%, of which WE-AX is about 2.6–4.1% and β-glucan is between 1.3% and 2.2% (Andersson, Fransson, Tietjen, & Åman, 2009). Rye bran has a higher content of total DF, where the contents of AXs were reported to be 4.4%, 9.3% and 18.2% in the outer, inner endosperm and milled bran fractions, respectively (Rakha, Åman, & Andersson, 2010). Rye is favorable to those suffering from diabetes by preventing a sharp increase in blood sugar. Moreover, the fiber content of rye could decrease the symptoms of irritable bowel syndrome (Åman, Andersson, Rakha, & Andersson, 2010).

Corn bran, a good fiber source, is used as a food ingredient, for instance, as a partial flour replacement in muffins, cookies and bread (Ortiz de Erive, Wang, He, & Chen, 2020; Rose, Inglett, & Liu, 2010). Corn bran (a by-product during the milling of corn) is reported to contain 70.6–86.3% IDF (mainly hemicellulose and cellulose) and 0.2–2.6% SDF depended on different analytical methods, being among the best sources of both DFs and bioactive phytochemicals (Inglett & Chen, 2011). The high DF means only a small amount of corn bran was required to achieve a certain fiber concentration in the final product. Moreover, corn bran supplementation was reported to have a lower possible negative impact on quality of the final product (McKee & Latner, 2000).

Referred to pseudo-cereals (e.g., amaranth, quinoa, and buckwheat), their seeds resemble those of the real cereals (e.g., wheat, rice, and barley) in function and composition. Investigations have indicated that the pseudo-cereals are excellent sources of DFs and bioactive compounds (e.g., flavonoids, phenolic acids, and saponins). Especially, the level of DF in buckwheat seeds is relatively higher than amaranth and quinoa, which consist of comparable fiber content as common cereals (Alvarez-Jubete, Arendt, & Gallagher, 2009). Their DF content and fraction are depended on different varieties and tissues. The total DF in quinoa flour has been reported to be 14.5–16.5% (dry basis), specifically, the SDF is 5.4–6.1% and IDF is 9.1–10.4%. The total DF in amaranth flour and buckwheat flour are 13.6–15.5% and 11.9–13.9%, respectively (Collar & Angioloni, 2014; Mir, Riar, & Singh, 2018). As pseudo-cereals are gluten-free, they are good sources of DFs for persons with celiac disease. Thus, pseudo-cereals are often used in gluten-free products such as gluten-free bread (Mir et al., 2018). For instance, the fortification of quinoa and/or buckwheat seeds into bread was reported to induce a valuable effect on its nutritive properties (Collar & Angioloni, 2014). Moreover, substituting wheat flour with half of buckwheat or whole wheat flour showed an insignificant effect on the volume index and quality characteristics of cakes, but induced significantly lower *in vitro* starch hydrolysis in pastes and cakes (Bae, Lee, Ko, & Lee, 2013).

3.2 Sources of fiber from fruits and vegetables

Fruits and vegetables are also a good source of DF to make up for the nutritional shortages of bakery foods (Garcia-Amezquita, Tejada-Ortigoza,

Serna-Saldivar, & Welti-Chanes, 2018). The DFs in fruits are usually utilized in the form of pomace or peel which are recovered from the juicing process. Otherwise, their by-products will be animal feed and considered as agricultural waste. Therefore, considering their richness of fiber and bioactive compounds, both manufactures and researchers attempt to apply them as functional ingredients into food products. For fruits and vegetables, DF is mainly located in the parenchymatous tissues and cell walls. These DFs have a relatively higher amount of SDF, showing beneficial functions such as gel-formation, water/oil holding capacity, and swelling capacity. Besides, they are rich in micronutrients and can play a vital role in the health of celiac patients. Various fruits and vegetables either supplied as commercial fibers or fibers derived from by-products of plant food processing have been applied in the cereal products to improve their nutritional value. In the last decades, more and more fibers from fruits and vegetables are added to bakery products, such as apple pomace, orange pomace, mango peel, lemon, pea, sugar beet, peach, banana, nuts, cocoa, potato, strawberry and date.

Apple peel or pomace is a left over residue of apple manufacturing industry which produce juice, concentrates, wine, cider, or canned slices. The content of DF, phenolic compounds, and minerals are higher in apple peel than other edible parts (Wolfe & Liu, 2003). Considering its potential in improving general health (Wijngaard & Brunton, 2010), it is attractive to add processed apple peel to food products as a functional ingredient. Apple pomace has been reported to be rich in DF (35–60%), which mainly constitutes pectin (5.5–11.7%), cellulose (7.2–43.6%), hemicellulose (4.2–24.4%), lignin (15.3–23.5%) and gum (Bhushan, Kalia, Sharma, Singh, & Ahuja, 2008). Besides, according to Chen, Rubenthaler, Leung, and Baranowski (1988), on a dry basis, the SDF in apple fiber was mainly water-soluble hemicellulose (19.2%), while the IDF was 15.3% of lignin, 39.9% of cellulose, 8.7% of water-insoluble pectin and 4.3% of water-insoluble hemicellulose. Apple fiber has a higher water holding capacity (9.36 g water/g solid) than wheat bran (5.03 g water/g solid) and oat bran (2.10 g water/g solid). Apple fiber was recommended as a good DF source in bakery products (Rupasinghe, Wang, Pitts, & Astatkie, 2009; Sudha, Baskaran, & Leelavathi, 2007).

Similarly, orange pomace or pulp which contains about 75% total alimentary fiber was utilized in cookies to increase its DF content and nutritional value (Larrea, Chang, & Martinez-Bustos, 2005). Containing about 51% total DF (19% SDF and 32% IDF), mango peel powder was also applied at different levels into wheat flour to make DF-enriched biscuits with improved antioxidant properties (Ajila, Leelavathi, & Prasada Rao, 2008). DF from berries possesses a higher nutritional value than cereal fiber because of the higher content of soluble fiber. By-products of berry processing could also be utilized as a source of technologically and nutritionally unique DF to produce foods with enhanced nutritional value (Alba, Campbell, & Kontogiorgos, 2019). In addition, unripe banana have high content of RS (17.5%) and DF (14.5%), and it was reported that even with a high level (60%) fortification of unripe banana, bread had an acceptable sensory (Juarez-Garcia, Agama-Acevedo, Sáyago-Ayerdi, Rodríguez-Ambriz, & Bello-Pérez, 2006). Green plantain powder is another option to fortify food products with DF, RS type II and phenolics (Sarawong, Gutiérrez, Berghofer, & Schoenlechner, 2014). It showed many advantages for commercial application such as year-long availability, low cost, colorless, neutral in taste, and high yield.

Vegetable powders have been widely used to enrich bakery products. For instance, bread enriched with artichoke, cauliflower, broccoli, fennel, and zucchini showed superior sensory quality, higher antioxidant activity, and lower GI (Saccotelli, Spinelli, Conte, & Nobile, 2018). Compared to control bread, the incorporation of a high level of vegetable flours (15%), in particular with artichoke and zucchini flours, was found to decrease the GI of bread by 41% and 38%, respectively (Saccotelli et al., 2018). The high DF content of vegetable flours was regarded as a conceivable reason (Saccotelli et al., 2018). Moreover, the addition of fruits and vegetables not only improved the DF consumption but also had an impact on the water interactions and bioactive ingredients retention. The addition of fruits and vegetables significantly was reported to increase the functionality of bread, including total phenolic content and antioxidant activity (Betoret & Rosell, 2019).

3.3 Sources of fiber from legumes

The legume family contains many edible plants including soybeans, green beans, peas, peanuts, and many more. Legumes are considered a superb source of various nutrients, such as protein and DFs (Kouris-Blazos & Belski, 2016). Generally, the DF content of legumes are highly dependent on the varieties. For instance, the DF content in the legumes varied from 13% to 35%, such as 35.3% in lupin, 22.6% in common bean, 22.2% in peas, 13.1–21.2% in chickpea, and 17% in lentils (Boukid, Zannini, Carini, & Vittadini, 2019). Also, the ratio and composition of SDFs and IDFs in these fibers are discrepant in different legumes species and impacted by processing methods. Generally, their IDFs mainly consist of cellulose and hemicelluloses, and the SDFs primarily consists of pectin (Rehinan, Rashid, & Shah, 2004) (Benítez et al., 2013). As concluded by Mudgil and Barak (2019), legume fibers are usually considered to have more advantages than cereal fibers, such as having more SDFs. This may result in more formation of volatile fatty acids during fermentation in the colon, which may inhibit cholesterol synthesis and reduce the risk of colorectal cancer. One of the fatty acids, acetic acid, has also been reported to increase the feeling of satiation from fibers via various mechanisms (Frost et al., 2014).

Hence, for the concerned of health and balanced-diet, legumes and pulses are important sources of DFs. Besides, they have proteins with a balanced fraction of essential amino acids when compared with cereal grains, and a spectrum of bioactive components (e.g., phenolic acids and flavonoids). Therefore, they are valuable items to be added in foods (Faris, Takruri, & Issa, 2013). The legumes powder (e.g., soybean powder, chickpea powder, and beans powder) has been used to improve the nutritional value of bakery products (Capriles & Arêas, 2014; Foschia, Horstmann, Arendt, & Zannini, 2017; Mudgil & Barak, 2013). Bread enriched with legume flours showed superior functional and nutritional values as reported in previous studies. For example, Miñarro, Albanell, Aguilar, Guamis, and Capellas (2012) evaluated bread produced from soy flour, carob germ flour, chickpea flour, and pea isolate, of which, chickpea bread showed best physicochemical properties, good sensory characteristics, and softest crumb. Moreover, crackers fabricated with chickpea flour showed good physical and nutritional characteristics that are comparable to commercial products (Han, Janz, & Gerlat, 2010). By adding 10% lupin flour, the bread quality was improved, which could be assigned to the cross-linking between lupin and wheat protein and the delayed staling by lupin fiber (Villarino, Jayasena, Coorey, Chakrabarti-Bell, & Johnson, 2016). The addition of legumes produced bread with acceptable volume, texture and taste, and increased antioxidant activity (Turfani, Narducci, Durazzo, Galli, & Carcea, 2017). Also, the organoleptic acceptability of gluten-free bread could be raised by the addition of leguminous flours (such as carob, chickpea, lupin, and soybean) and better nutritional values of bread were achieved by adding chickpea and soybean (Melini, Melini, Luziatelli, & Ruzzi, 2017).

3.4 Other sources of fiber

As parts of commercially available fibers, some hydrocolloids, such as hydroxypropylmethylcellulose (HPMC), carboxymethylcellulose (CMC), gums (e.g., guar gum, locust bean gum, and xanthan gum), oligosaccharides (e.g., polydextrose and maltodextrins), trehalose, carrageenan and inulin, have been widely used in bakery products. There were applied as a potential ant-staling ingredient and were successfully applied in the commercial production of bread (Das, Raychaudhuri, & Chakraborty, 2015; Peng, Li, Ding, & Yang, 2017). The free water molecules would move from fresh breadcrumbs to the surface and evaporate during the baking process. Almost all the protons on the crust are low in mobility due to the extremely low moisture of the crust (Bosmans et al., 2012). The utilization of hydrocolloids at a low concentration could effectively produce a gel network and fill the gaps in the bread crumb, thus reducing moisture evaporation during production and storage. The application of hydrocolloid provides a strategic role in fabricating the gluten-free (GF) dough workable and improving the texture of the final GF product. Demirkesen et al. (2014) reported that a more cohesive crumb was found in bread prepared with 0.5% gums (xanthan, HPMC and CMC) or gum blends (xanthan-guar and xanthan-locust bean gum). Fiber from psyllium, sugar beet, oat, pea and bamboo can improve texture, gelling, thickening and stabilizing properties to GF bread. Buckwheat, amaranth, quinoa, teff and oat are becoming increasingly popular as ingredients in pasta formulations as they improve the nutritional quality of GF pasta. Besides, chitosan is also reported to be added into bakery products, such as breads (Ghoshal & Mehta, 2018; Kerch et al., 2008), and cookies (Lee, Kim, Lee, & Kim, 2005).

Besides, RS is also a significant fiber added into bakery products. A major advantage of RS against traditional sources of fiber (such as whole grains or bran) is its less negative effect on the texture characterize of food. The desirable physicochemical properties of RS include swelling capacity, viscosity, gel formation and water-holding capacity, which make it favorable in many foods (Fuentes-Zaragoza et al., 2010). Supplementation of RSs in bakery and their effect during baking has been widely studied. For instance, Baixauli, Sanz, Salvador, and Fiszman (2008) found that the addition of RS from maize under 10% level (by weight of total formulation) hardly influenced the height or the number and area of gas cells in the crumb of muffin. Laguna, Salvador, Sanz, and Fiszman (2011) investigated the influence of replacing wheat flour with RS rich ingredient from maize at 20%, 40%, and 60% levels in a biscuit formulation. A higher content of total DFs (9.0–15.1%) was observed in RS-enriched biscuits than control biscuit (4.0%). Korus, Witczak, Ziobro, and Juszczak (2009) found that the addition of corn RS at 20% level increased the total DF of bread by up to 89% as compared to control bread, of which the most pronounced change was observed for IDF (increase 137%), while an only slight increase was found for its SDF (18%). Sanz et al. (2008) reported that the replacement of wheat flour by 20% RS3 increased the total DF content from 5.0% to 13.2%.

Bacterial cellulose (BC), a microbial polysaccharide, has significant potential as a food ingredient given its high purity, in situ change of flavor and color, and the ability to form various shapes and textures. BC is non-toxic, and its hydrophilicity, good sorption ability for liquids, non-allergenicity, biodegradability when modified, have led to applications in the medical industry (Shi, Zhang, Phillips, & Yang, 2014), and utilizations as a food additive. When compared with other DFs, BC has several advantages (Shi et al., 2014). The incorporation of BC improved bread quality, including a higher specific volume, porosity, luminosity, and moisture retention, and more tender, less firm crumb, making it more acceptable to the consumer. It may attribute to the fact that BC presented interweaved thicker gluten filaments (Corral, Cerrutti, Vázquez, & Califano, 2017).

4. Effects of DFs on dough processing

One of the most critical steps for baked food is dough processing, and its key factor is dough rheology. The significance of dough rheology in bread-making has been widely stressed (Dobraszczyk & Salmanowicz, 2008; Ktenioudaki, Butler, & Gallagher, 2011; Stojceska, Butler, Gallagher, & Keehan, 2007). The effect of DF fortification on dough rheology is highly dependent on the type, physiochemical properties and the additional levels of various DFs.

Various empirical and fundamental rheological tests are usually applied to measure the dough rheology. To measure dough mixing properties, empirical rheological tests using Farinograph or the Mixograph are commonly used. The effects of adding various cereal bran (wheat, rice, oat and barley bran) and apple pomace on the rheological properties of dough were determined by the Farinograph test (Sudha, Baskaran, & Leelavathi, 2007; Sudha, Vetrimani, & Leelavathi, 2007). In these cases, the water absorption of the blends and the development time increased while the stability time decreased, implying a decreasing dough strength. Specifically, the increase of water absorption was reported to be mainly induced by the greater number of hydroxyl groups in the fiber structure which allow more water interaction through hydrogen bonding (Rosell, Rojas, & De Barber, 2001). Besides, the increase of dough development time indicates that fiber addition slowed the rate of hydration and the development of gluten.

These results implied the dough weakening effect of fiber addition on dough. Similar results were reported in previous investigations of the addition of various fiber sources, such as mango peel (Ajila et al., 2008), hazelnut testa (Anil, 2007), wheat bran (Sanz Penella, Collar, & Haros, 2008), brewer's spent grain (Stojceska & Ainsworth, 2008), water-extractable pentosans (Wang, Hamer, van Vliet, & Oudgenoeg, 2002), and legume fibers (carob fiber and pea fiber) (Wang, Rosell, & Barber, 2002). On the contrary, a decreased water absorption was observed when various chicory inulin samples with different degrees of polymerisation (up to w/w 7.5%) were added into wheat flour (Peressini & Sensidoni, 2009). Moreover, a significant increase in mixing stability was recorded indicating a positive effect of inulin on dough strength (Peressini & Sensidoni, 2009), which is in agreement with the improvement of dough strength when partially soluble β-glucans and AXs (4%) were added (Izydorczyk, Hussain, & MacGregor, 2001). Nevertheless, Rosell, Santos, and Collar (2006) indicated that the addition of inulin from Fibruline (5%) showed little effect on the water absorption by using Farinograph. Therefore, compared with the IDFs, these SDFs seems to have positive effect on dough strength, but the effect is highly dependent upon their structure and addition level.

The measurement of extensional properties of dough is another well-known method for the determination of dough rheology, usually by using tools of Extensograph, Alveograph and Texture analyzer. Good dough extensibility is required for the expansion of gas cells during the proofing and baking process, and it reduces the premature rupture of cell membranes. Based on Extensograph test, generally, the addition of different types of fiber decreased the dough extensibility, while its effect on the resistance to extension varied (Anil, 2007; Sudha, Baskaran, & Leelavathi, 2007; Sudha, Vetrimani, & Leelavathi, 2007). In addition, Ayadi, Abdelmaksoud, Ennouri, and Attia (2009) determined the extensional properties of dough after adding DFs from various sources utilizing the Alveograph, and they found that the supplementation of cladodes flour (*Opuntia ficus indica*) decreased the dough biaxial extensibility (L) and increased the dough resistance to deformation (P). Also, similar increases in values of P after adding three different sources of fibers (carob fiber, chicory inulin, and pea fiber) were observed, of which, inulin induced the biggest changes (Wang, Rosell, & Barber, 2002). In this study, the dough extensibility was reduced for all fibers except inulin. These results could be explained by the distinct interactions between fiber and gluten proteins, and also the distinct constitutes of each type of fiber. Furthermore, the dough proofing properties determined by rheofermentometer confirmed the result that the addition of inulin produced a more stable dough which can withstand longer proofing times. A similar Rheofermentometer test conducted by Sanz Penella et al. (2008) showed that the addition of wheat bran had a negative impact on dough development during proofing, but this negative effect could be reduced by decreasing the particle size of bran. Besides, by using Kieffer dough and gluten extensibility rig of Texture analyzer, Barros, Alviola, and Rooney (2010) found that dough made from whole wheat flour had lower extensibility than the dough made from refined wheat flour, whereas the resistance to extension did not show as much difference between them.

Moreover, other rheological tests, e.g., texture profile analysis, dough stickiness, and dynamic oscillation measurements are also conducted to measure the rheology of dough with the addition of various fibers such as IDFs (wheat bran, oat bran and cellulose microcrystalline) (Gómez, Moraleja, Oliete, Ruiz, & Caballero, 2010; Jin et al., 2021), inulin (Peressini & Sensidoni, 2009), cocoa fiber (Martínez-Cervera, Salvador, Muguerza, Moulay, & Fiszman, 2011) and barley flour (Sullivan, O'Flaherty, Brunton, Arendt, & Gallagher, 2010), to produce bakery products including cakes, bread, noodle and muffins. Specifically, the supplementation of cocoa fiber enhanced the viscosity of the batter and consistency value (K) while decreased the flow index (n), implying a greater resistance to flow and thus a more entangled structure of the dough, consequently, impacting the manageability of the batter and the quality of muffins (Martínez-Cervera et al., 2011). Similarly, Gómez et al. (2010) found that the addition of cellulose, wheat, and oat fiber increased the values of G' and G'', which indicate a stiffer dough of higher consistency and lower tendency to flow.

Several hydrocolloids, as water-soluble polysaccharides, have been widely studied for their potential application in the baking industry, especially in gluten-free bakery products (Anton & Artfield, 2008). Because they are useful for modifying the rheology and texture of aqueous suspensions which are expected to improve the quality of bakery products. Generally, the addition of various hydrocolloids (Gómez, Ronda, Caballero, Blanco, & Rosell, 2007; Guarda, Rosell, Benedito, & Galotto, 2004; Rosell et al., 2001), such as sodium alginate, carrageenan, pectin, hydroxy-propyl-methylcellulose, locust bean gum, guar gum and xanthan gum, has been reported to have important influences on dough processing and baking properties: they require a supplementary addition of water, and their interactions with the other dough components may induce changes in dough rheological and pasting properties, consequently, affect the quality of the final bakery products, finally impact consumers' acceptability. The various effects of hydrocolloids are based on their chemical structures and specific properties.

In summary, the impact of fiber addition on dough processing is due to its effect on the formation of the gluten network (Gómez, Ronda, Blanco, Caballero, & Apesteguía, 2003; Zhou et al., 2021). Especially, the negative effects of bran addition on gluten network have been concluded by many reports (Ktenioudaki & Gallagher, 2012; Noort, Haaster, Hemery, Schols, & Hamer, 2010; Sivam, Sun-Waterhouse, Quek, & Perera, 2010), including: (1) fiber incorporation dilutes of the gluten forming proteins, (2) preferential water binding of fiber restricts the available water for gluten development, (3) interaction of gluten with active components (e.g., ferulic acid) released from fiber, and (4) particles steric hindrance, or physical barrier.

It should be noted that there does not seem to be a "one-size-fits-all" approach to explain the effect of fiber addition on dough properties. The results are controversial in the literature. For instance, Filipović, Popov, and Filipović (2008) reported a disruption in the gluten matrix and consequently a reduction in the quality of bread after the addition of soluble inulin. Whereas, Polaki, Xasapis, Fasseas, Yanniotis, and Mandala (2010) found that the gluten matrix was significantly developed with the addition of inulin and showed similar pore distributions to control. Therefore, the positive or negative effect of DF addition on the dough processing or products' quality needs to be further investigated, considering various compositions and structures of DF.

5. Effect of DFs on quality of bakery products

Besides dough processing, many studies have concentrated on investigating the impact of fiber addition on the quality of bakery products. Not surprisingly, most results have indicated the negative effect of fiber fortification on the quality of the end-products (Katina, 2003). DFs from various sources have been widely added into flour to produce fiber-enriched breads, biscuits, muffins, cakes and extruded bakery products. Their impacts on the quality (e.g., structure, texture and sensorial property) of end-products are highly related with DFs' source, type, component, addition level, and particle size. These impacts are summarized in Table 2.

Bread is an excellent vehicle of various DFs, e.g., cereal bran (Le Bleis, Chaunier, Chiron, Della Valle, & Saulnier, 2015; Majzoobi, Farahnaky, Nematolahi, Mohammadi, & Taghipour, 2013; Rieder, Holtekjølen,

Table 2 Effect of DFs on quality (including structure, texture and sensory) of bakery products.

Products	DF inclusion	Product quality			References
		Structure	Texture	Sensory	
Bread	20% Wheat bran	Coarser and more homogeneous crumb cells;	↑ Bread density; ↑ Crumb firmness; ↑ Crust stiffness;	–	Le Bleis et al. (2015)
Bread	20% Wheat bran	↓ Lightness of crust;	↑ Bread hardness; ↓ Bread cohesiveness;	↑ Hedonic to taste (add small bran particles); ↓ Hedonic to taste (add large bran particles);	Majzoobi et al. (2013)
Bread	40% Oat bran	↓ Bread volume;	– Crumb firmness;	–	Rieder et al. (2012)
Bread	3% Carob fibre	↓ Loaf volume;	↓ Bread hardness;	Scoring >5 for each specific sensory characteristic and overall acceptability;	Wang, Rosell, and Barber (2002)
	3% Inulin	↓ Loaf volume;	↑ Bread hardness;		
	3% Pea fiber	↓ Loaf volume;	↓ Bread hardness;		
Bread	20% Wheat bran	↓ Specific volume;	↑ Bread hardness;	–	Katina et al. (2006); Salmenkallio-Marttila et al. (2001)
Bread	1% Pectin	↑ Specific volume;	↓ Crumb hardness;	–	Ponzio et al. (2013)
Bread	0.8% Guar gum	↑ Bread porosity; ↑ Specific volume;	↓ Crumb hardness;	↑ Total acceptance;	Hejrani et al. (2017)

Continued

Table 2 Effect of DFs on quality (including structure, texture and sensory) of bakery products.—cont'd

Products	DF inclusion	Product quality: Structure	Texture	Sensory	References
Bread	Pumpkin powder (100 g)	↓ Lightness of crust and crumb;	– Crumb hardness; ↓ Crumb cohesiveness, gumminess and chewiness;	↑ Overall liking; ↑ Willing to eat at home; ↑ Ease of chew and swallow;	Amoah et al. (2021)
Cake	30% Apple pomace	↓ Volume;	↑ Hardness and density;	↓ Hedonic to overall quality;	Sudha et al. (2007b)
Muffin	32% Apple skin powder	↓ Lightness; ↓ volume;	↑ Firmness;	– Overall acceptability (≤24%);	Rupasinghe et al. (2009)
Cake	24% Wheat bran, oat bran, or cellulose microcrystalline	↓ Lightness of crust and crumb; ↑ Volume;	↑ Firmness and chewiness; ↓ Elasticity, cohesiveness and resilience;	Scoring >5 for each specific sensory characteristic and overall acceptability;	Gómez et al. (2010)
Cake	10% Potato peel powder	↓ Lightness;	↓ Cake hardness, chewiness, breaking force and springiness;	– Overall acceptability;	Ben Jeddou et al. (2017)
Cake	20% Cashew-apple fiber	↓ Diameter of cookie; ↑ Height;	↓ Spread ratio;	– Overall acceptability	Ebere et al. (2015)
Cookie	25% Extruded orange pulp	↓ Diameter of cookie; ↑ thickness; ↓ Specific volume;	↑ Breaking strength; ↓ Expansion factor;	– General acceptance (5–15%); ↓ General acceptance (25%);	Larrea et al. (2005)

Cookie	75% Wheat bran	↓ Width; ↑ Thickness; ↓ Lightness;	↓ Spread ratio; ↑ Hardness;	–	Kahraman et al. (2019)
Biscuit	20% Mango peel powder	↓ Diameter of cookie; ↓ Thickness; ↓ Lightness;	– Spread ratio; ↑ Breaking strength;	↓ Overall acceptability;	Ajila et al. (2008)
Cookie	30% Mango peel powder or mango kernel powder	–	–	– Overall acceptability;	Bandyopadhyay et al. (2014)
Biscuit	10% Inulin, potato fiber or β-glucan enriched fraction	––	– Firmness; ↓ Shrinkage; ↑ Height;	–	Brennan and Samyue (2004)
Pasta	10% Inulin	↓ Swelling index; – Cooking losses;	↓ Elasticity and firmness; – A dhesiveness;	–	Brennan et al. (2004)
Bread	10% Hazelnut testa	↓ Lightness of crumb;	–	↓ Overall acceptability;	Anil (2007)
Muffin	10% peach DF	↓ Lightness;	↑ Hardness and chewiness; – Springiness and cohesiveness;	– Acceptability (≤4%); ↓ Acceptability (10%);	Grigelmo-Miguel et al. (2001)

Continued

Table 2 Effect of DFs on quality (including structure, texture and sensory) of bakery products.—cont'd

Products	DF inclusion	Product quality			References
		Structure	Texture	Sensory	
Bread	7.5% Inulin	↓ Specific volume;	↑ Hardness;	– Acceptance (≤5%); ↓ Acceptance (7.5%);	Peressini and Sensidoni (2009)
Cake	20% Cladode powder	↓ Lightness; ↓ Volume;	↑ Hardness;	– Overall acceptability (5%); ↓ Overall acceptability (10–20%);	Ayadi et al. (2009)
Bread	10% Guava pulp powder	↓ Lightness;	↑ Loaf volume; ↓ Hardness;	– Overall acceptability (≤5%); ↓ Overall acceptability (7.5–10%);	Arslan et al. (2017)
Bread	30% RS	↓ Cell area/total area;	↑ Hardness and chewiness; – Springiness; ↓ Cohesiveness;	– Overall acceptability (≤10%); ↓ Overall acceptability (20–30%);	Sanz Penella et al. (2008)
Biscuit	60% RS	↓ Length and width gain; ↑ Lightness;	↓ Crumbliness; ↓ Force at break;	– Overall acceptance (≤20%); ↓ Overall acceptance (40–60%);	Laguna et al. (2011)
Bread	20% Chickpea flour	↓ Lightness of crust and crumb; ↓ Specific volume;	–	– Overall acceptability (≤10%); ↓ Overall acceptability (20–30%);	Mohammed et al. (2012)

Sahlstrøm, & Moldestad, 2012), gums (Salehi, 2020), RSs (Korus et al., 2009; Miyazaki et al., 2006), fruit fiber (Rocha Parra, Ribotta, & Ferrero, 2015), and β-glucans (Ekström, Henningsson Bok, Sjöö, & Östman, 2017; Wang et al., 2017). The fiber fortifications generally induce a smaller loaf volume, denser and less aerated structure, and harder or darker crumb. For instance, by adding carob fiber, inulin or pea fiber into bread (3% w/w), Wang, Rosell, and Barber (2002) found that the supplementation of three kinds of fiber all resulted in a decrease of loaf volume, while the impact on crumb firmness was different. Generally, these fiber-enriched breads were sensory acceptable for the consumer panellist. Similar results were also observed in high-fiber wheat bread fortified with 20% wheat bran (Katina, Salmenkallio-Marttila, Partanen, Forssell, & Autio, 2006; Salmenkallio-Marttila, Katina, & Autio, 2001). On the other hand, higher specific volume and softened bread crumb were observed after the addition of high methoxyl pectin (Ponzio, Ferrero, & Puppo, 2013) and guar gum (Hejrani, Sheikholeslami, Mortazavi, & Davoodi, 2017) into bread. Interestingly, vegetable fiber-enriched breads were found to be instrumentally and subjectively less gummy, cohesive and chewy when compared to white and wheatmeal breads, therefore, having the potential to both improve nutrition and "ease of swallowing" in older people (Amoah, Cairncross, & Rush, 2021).

Besides, the effects of the addition of various DFs on cake quality have also been studied. Sudha, Baskaran, and Leelavathi (2007) found that the volume of the cakes decreased while the density and hardness increased with the increasing substitution of apple pomace (from 0% to 30%), and 20% apple pomace was highly acceptable. Similar results were demonstrated by Rupasinghe et al. (2009) who studied the effect of fiber from apple skin powder (32%) on the quality of muffins. On the contrary, a significant increase in cake volume with the addition of wheat and oat fibers (from 0% to 24%) of different sizes was indicated by Gómez et al. (2010). They found that in comparison to the control, the volume of cake with 24% fiber substitution increased, while the volume decreased when the substitution of fiber reached 36%. They also observed that fine fibers (80 μm) led to cakes with higher volumes than coarse ones (250 μm). Moreover, all cakes obtained relatively high sensory scores although lower than control, indicating that they were not disagreeable for consumers. Ben Jeddou et al. (2017) indicated that the replacement of wheat flour with the potato powders (10%) reduced the hardness of cake significantly, as well as the L^* and b^* values of dough color.

Negative effects have also been documented for biscuits/cookies with the addition of various fibers, such as fibers from apple reside (Ebere, Emelike, & Kiin-Kabari, 2015), orange pulp (Larrea et al., 2005), wheat bran (Kahraman, Aktas-Akyildiz, Ozturk, & Koksel, 2019), and mango fibers (Ajila et al., 2008; Bandyopadhyay, Chakraborty, & Bhattacharyya, 2014). The detrimental effects for biscuits/cookies usually involve in the increased hardness, loss of crispiness and reduced spread ratio. Less pernicious effects were reported by Brennan and Samyue (2004), who found that with the addition of inulin, potato fiber, RS and β-glucan, the physical and textural properties of biscuits showed insignificant changes, and the addition level should be lower than 10%. Additionally, Brennan, Kuri, and Tudorica (2004) found that the pasta containing inulin has a significantly lower dry matter content, firmness, and swelling index.

Sensory properties of bakery products have also been reported to be greatly impacted by fiber fortification, the acceptability of product is highly related with the addition level of fiber. The low acceptability scores of higher fiber-enriched bakery products were related to the changes in appearance, texture as well as flavor. Most sensory analyses implied that for most fibers, the acceptable addition level in bakery products should be lower than 15%. For instance, Ajila et al. (2008) demonstrated that acceptable biscuits with mango flavor were obtained by incorporating 10% mango peel powder. Anil (2007) reported that bread samples produced from 5–10% hydrated hazelnut testa were considered acceptable. Grigelmo-Miguel, Carreras-Boladeras, and Martín-Belloso (2001) found that there were no differences between the acceptability of muffins with peach DF up to 4% and control muffins. Larrea et al. (2005) indicated that cookies of good technological quality and with a good level of acceptance were obtained by replacing up to 15% of the wheat flour with orange pulp. Peressini and Sensidoni (2009) suggested that the addition of inulin at about 5% could be an effective way to produce functional fiber-enriched bread without negatively affecting its sensory acceptance. Wang, Rosell, and Barber (2002) revealed that panellists judged the fiber-enriched breads as acceptable after the addition of commercial fibers (carob fiber, inulin or pea fiber) at 3%. Ayadi et al. (2009) described that the overall acceptability showed that a maximum of 5% cladodes flours can be incorporated to prepare cakes with acceptable quality. In addition, the sensorial analysis showed that bread supplemented with 5% of guava pulp powder showed better overall acceptability than control bread (Arslan, Rakha, Khan, & Zou, 2017).

However, there are some exceptions. The modified starches could be used to substitute up to 20% for wheat flour without deterioration of bread quality (Miyazaki et al., 2006). Similarly, Sanz Penella et al. (2008) reported that RS can be incorporated into bread as a source of DF with acceptable changes in bread quality up to 10–20% of flour replacement. Laguna et al. (2011) reported that the sensory acceptability and consumption intention of biscuits were not significantly reduced when the flour replacement of RS rich ingredients at levels of up to 40% inclusion. Mohammed, Ahmed, and Senge (2012) showed that the bread remained acceptable sensory properties and similar to the control bread (100% wheat flour), even when wheat flour was replaced with chickpea flour and at levels of up to 10%. In addition, sensory evaluation demonstrated that replacement of wheat flour by apple skin powder up to 24% (w/w) in muffins still provided similar overall acceptability as control (Rupasinghe et al., 2009).

6. Effect of DFs on physiological functionality of bakery products

DFs represent a wide spectrum of polysaccharides that can escape digestion in the human gut. It has been widely associated with positive health outcomes, thus, fiber-enriched food products are claimed to be a potential basis for nutritional composition worldwide. They may naturally exist in bakery products (e.g. whole grain products) or be added into bakery products to enhance nutritional values. Their properties which are related to nutritional functions mainly include the particle size and bulk volume, the surface area features, the hydration and rheological properties, and the adsorption or entrapment of minerals and organic molecules. Of which, the viscosity and ion exchange capacity are the main factors influencing metabolic effects (glucose and lipid metabolisms). However, the fermentation pattern, bulking effect and particle size are highly related with their impacts on colonic function. Various treatments can change the physicochemical properties of the fiber, improving both the techno-functional and physiological properties of DFs in foods (Guillon & Champ, 2000).

The physiological effects of DF are mostly related to its behavior during the gastrointestinal tract, especially in the small and large intestine. Nevertheless, the mouth and stomach are important for the mechanical, chemical and enzymatic digestion of food and are also regarded as crucial factors in appetite regulation and microbial defense. Thus, aiming to comprehend the effect of increasing amounts of fiber-enrich food in the diet on the longer-term gut health, the physiological functionality would be discussed from the mouth, oesophagus, stomach, small intestine and large intestine.

6.1 Effect on swallowing and the oesophagus

The major physiological functions of the oesophagus and stomach are discussed and recent evidence relating DFs intake to these actions is introduced by Brownlee (2014). Fortification with DFs and other hydrocolloids is considered the main method to modify the rheological properties of foods (Cichero, 2013; Wendin et al., 2010). This method can formulate fluids/gels with different rheological properties (Sworn, Kerdavid, & Fayos, 2008), which is thought to be in favor of swallowing, thus, regarded as a good way to improve swallowing in dysphagic patients (Brownlee, 2014). There is no consensus on a set rheological template to make food easier to swallow, but the various viscous and gel-forming polysaccharides available do allow the design of products tailored to patient needs. Besides, recent studies found that the wheat bran fortification could impact the bread breakdown pathways and destructuration mechanisms during mastication (Le Bleis, Chaunier, Montigaud, & Della Valle, 2016; Lin et al., 2021), this seems to be a new insight to determine the impact of DFs on swallowing. Additionally, the fortification of DFs into foods also seems to impact oesophageal handling of boluses. Nevertheless, this kind of supplementation is unlikely to show major physiological ramifications on the digestion or handling of a single meal, because the time that a food bolus spends within the oesophagus is minimal (Brownlee, 2014). Notably, consumption of whole-grain foods, such as in the form of wholegrain bread/pasta, has been inversely related to the risk of oral and pharyngeal, as well as of esophageal cancer (Soler et al., 2001).

6.2 Effect on gastric emptying and satiety

The rate of gastric emptying is possibly governed by the stomach, and the viscosity and the chemical composition of digesta are reported to be an important stimulus to drive gastric emptying. Gastric content with high viscosity is considered to induce a lower gastric emptying rate than that of low viscosity. There is evidence indicating that specific dietary components (e.g., DF) could act to affect the gastric emptying rate. Considering the effect of DF-enriched foods, Marciani et al. (2013) found that wholemeal bread decreased gastric emptying when compared to an isocaloric meal of white rice pudding. They demonstrated that the wholemeal bread formed a homogenous chyme within the stomach, which might decrease the gastric emptying time through delaying the separation of solid and liquid components of the meal into the proximal and distal stomach, respectively.

Furthermore, they attributed the underlying reason to the NSPs (AXs) in the wholemeal bread, because it could bind a large amount of water and greatly increase the viscosity of the solutions in gastric. These investigations indicated that the physiological actions of DF are driven by its form and physicochemical properties. Although several model gut systems were developed and used to simulate digestion (Houghton et al., 2014; Ramasamy, Venema, Gruppen, & Schols, 2014), in fact, it is very hard to recreate the gastric and intestinal motility effectively. Thus, innovative investigations, especially methods with minimal invasion like Magnetic Resonance Imaging (MRI), are invaluable in the design of fiber-based formulations with targeted physiological actions.

Gastric emptying rates are considered to be crucial determinants of postprandial feelings of satiety, which could be linked to meal rheology (Brownlee, 2014; Hlebowicz, 2009). Another important physiological function of DF in the gut is involved in prolonging satiety or inducing satiation. Touyarou, Sulmont-Rossé, Gagnaire, Issanchou, and Brondel (2012) studied the influence of 15-days of consumption of bread enriched with isolated oat and rye fibers as breakfast on their satiety during the morning, and they found that the fiber-enriched breads decreased the subsequent hunger compared to non-fiber enriched breads. Moreover, satiety ratings were likely to remain relatively constant over the 15 days, although, the intra-individual variation was found (Touyarou et al., 2012). These results show that eating food products involving long satiation should be developed to a long-term eating habit when people aim to manage weight.

6.3 Effect on digestion and absorption

Classically, the physicochemical properties of DFs are crucial for their function of decreasing the rate of intestinal uptake of a range of nutrients. Generally, during the digestive tract, DFs show a buffering effect that links excess acid in the stomach, increases the fecal bulk and stimulates intestinal evacuation; besides, they provide a favorable environment for the growth of the beneficial intestinal flora. Fiber can also bind diverse substances including cholesterol (Jenkins, Kendall, & Ransom, 1998; Jiménez-Escrig & Sánchez-Muniz, 2000). It has been reported that these specific properties of DF play an important role in the prevention and treatment of obesity, atherosclerosis, coronary heart diseases, colorectal cancer and diabetes (Peters et al., 2003; Terry et al., 2001). Consumption of certain fiber, particularly highly viscous SDFs, is usually associated with moderate postprandial glycemic responses, a property of importance in the dietetic treatment of diabetes (Jenkins et al., 1998; Nieman et al., 2014). Two common ways used to measure the effect of DFs on digestion and absorption of bakery products are *in vitro* starch digestion and *in vivo* GI which are discussed in the following section and summarized in Table 3.

6.3.1 Effect on in vitro *digestion*

Fiber-enriched foods are often recommended to diabetics because they are supposed to decrease glycaemic response to the food (Guillon & Champ, 2000). The relationship between the rate of starch digestion and predicted or expected glycaemic index (pGI or eGI) has been established by studies of *in vitro* amylolytic hydrolysis (Englyst, Vinoy, Englyst, & Lang, 2003). Therefore, the effect of DFs on the *in vitro* digestion of macronutrients (especially starch) are widely studied. For bakery products, the physiological effect of DFs is not only reflected in its intrinsic physiological actions but also reflected in its influence on the quality of product such as macro/micro-structure which may further modify the product's *in vitro* digestion. The type and amount of added fiber, food structure, degree of porosity and surface area accessible to the enzyme from the digestive tract seem to contribute to the total degradability of food (Tudorică, Kuri, & Brennan, 2002).

For instance, Brennan, Blake, Ellis, and Schofield (1996) found that the *in vitro* hydrolysis of starch in bread with added guar gum (10%) was attenuated significantly in comparison to control bread (without guar gum). Therefore, they highlighted the concept that guar gum acts as a physical "barrier" to α-amylase–starch interactions and/or subsequent release of hydrolyzed compounds. Besides, the effect of guar gum on digesta viscosity which may reduce the rise in postprandial glycaemia is also pointed out. This result was confirmed by Sasaki (2018), and they found that another commercial gum—xanthan gum resulted in a more pronounced suppressive effect on starch digestion than guar gum. Similarly, the decrease in the extent of starch digestion was also found in pasta added with guar gum. Tudorică et al. (2002) found that an increased glucose release was observed in pasta supplemented with insoluble pea fiber (7.5–15.0%), whereas, a significant decrease was shown in pasta containing soluble guar gum (3.0–10.0%). They demonstrated that the different results may be due to the fact that pea fiber breaks the integrity of the pasta matrix which lead to an open appearance with starch granules "uncovered" and exposed to enzymatic attack. However, the interaction between guar gum and the protein – starch matrix in pasta follows a complex pattern which makes the starch granules be entrapped within a protein – fiber

Table 3 Effect of DFs on starch digestion and absorption of bakery products.

Bakery products (DF inclusion)	Nutrition outcomes	References
***In vitro* test**		
Bread (10% guar gum)	↓ *In vitro* hydrolysis of starch	Brennan et al. (1996)
Bread (1–2% xanthan gum or guar gum)	↓ *In vitro* starch digestion	Sasaki (2018)
Pasta (7.5–15% pea fibre)	↑ Glucose release and available starch	Tudorică et al. (2002)
Pasta (3–10% guar gum)	↓ Glucose release and available starch	Tudorică et al. (2002)
Bread (10% carboxymethylcellulose or locust bean gum)	↓ Digestible starch and eGI, ↑RS	Angioloni and Collar (2011)
Pasta (10% inulin)	↓ Release of sugar and pGI	Brennan et al. (2004)
Muffin (50% inulin)	↓ Predicted glycaemic response	Gao et al. (2016)
bread (20–40% chickpea flour)	↑RS; ↓pGI;	Utrilla-Coello et al. (2007)
Bread (10% banana flour and 0.8% CMC)	↓pGI	Ho et al. (2015)
Bread (30% chempedak seed flour)	↓ *In vitro* starch hydrolysis and eGI	Zabidi and Aziz (2009)
Noodle (20–40% pea flour)	↓ RDS, ↑SDS	Bharath Kumar and Prabhasankar (2015)
Pasta (10–30% sprouted ragi and banana flour)	↓ Available carbohydrates, ↑mineral content	Krishnan and Prabhasankar (2010)
Pasta (30% black gram flour)	↓ *In vitro* starch hydrolysis	Madhumitha and Prabhasankar (2011)
Bread (2.5–7.5% dry pomelo)	↑RS, ↓pGI	Reshmi et al. (2017)
***In vivo* test**		
Bread (20% RS)	↓ Postprandial blood glucose response and total cholesterol.	Brites et al. (2011)

Continued

Table 3 Effect of DFs on starch digestion and absorption of bakery products.—cont'd

Bakery products (DF inclusion)	Nutrition outcomes	References
Bread (3.5% β-glucans)	↓ GI and glucose iPeak	Ekström et al. (2017)
Bread (40% barley flour)	↓ GI and postprandial blood glucose response	Finocchiaro et al. (2012)
Bread (2.5–10% malva nut gum)	↓ pGI and GI	Phimolsiripol et al. (2017)
Pasta (25% chickpea)	↓ Hyperglycemia peak and GI	Goñi and Valentı´n-Gamazo (2003)
Bread & biscuit (3% fibre mix)	↓ GI	Marangoni and Poli (2008)

Notes: GI, glycaemic index; pGI, predicted glycaemic index; eGI, expected glycaemic index; RDS, rapidly digestible starch; SDS, slowly digestible starch; RS, resistant starch.

network and reduce the accessibility of starch to amylases. In addition, Angioloni and Collar (2011) found that the interactions between fortified fibers (cellulose derivatives, galactomannans) and endogenous biopolymers (fructo- and gluco-oligosaccharides) are reported to induce an extremely compact structure and restrict the availability of water during gelatinization. Subsequently, a delay on enzyme attack to starch granules was observed which may result in higher RS, lower digestible starch and a lower *in vitro* eGI. Inulin also has the potential to regulate glycemic response without compromising the pasta quality. Brennan et al. (2004) found that the pasta containing inulin had an insignificant change in adhesiveness and elasticity compared to the control pasta, but they showed a slower release of sugars during *in vitro* starch digestion, and a 15% reduction pGI. Gao, Brennan, Mason, and Brennan (2016) also found that an incorporation of 50% inulin in muffins decreased the predicted glycaemic response without affecting their textural properties.

Besides, DFs were also added into bakery products in the form of flour. For instance, Utrilla-Coello, Osorio-Díaz, and Bello-Pérez (2007) found that the breads added with 20–40% chickpea had a higher RS content and a lower pGI than control bread. Ho, Tan, Abdul Aziz, and Bhat (2015) reported that 10% banana pseudo-stem flour and 0.8% CMC significantly lowered pGI of bread from 80.1 to 77.5. Zabidi and Aziz (2009) found that bread with 30% chempedak (Artocarpus integer) seed flour showed significantly lower *in vitro* starch hydrolysis, resulting in a lower the eGI (59.6) than control bread (63.4). Bharath Kumar and Prabhasankar

(2015) reported that with an increasing addition level of pea flour, the content of rapidly digestible starch (RDS) decreased significantly while the content of slowly digestible starch (SDS) increased during the *in vitro* digestion of noodles. In addition, Madhumitha and Prabhasankar (2011) found that pasta added with 30% black gram flour exhibited lower *in vitro* starch hydrolysis.

Different ratio of SDF and IDF showed distinct impact on the *in vitro* starch digestion and pGI of fiber-enriched bakery products. With the increase of IDF ratio (or reduction of SDF ratio) in 3 g fiber-enriched cakes, SDS contents increased, and the RDS contents decreased. However, the cakes containing 3 g of total DF (the same ratio of SDF and IDF) showed a low pGI and acceptable quality attributes. Especially, the influence of total DF amount and IDF ratio on reducing the pGI value was more effective than the SDF ratio (Oh et al., 2014).

In summary, the impact of DFs on the degree of starch digestion and the pGI evaluated using *in vitro* digestion models have been studied for various bakery products, including pasta, biscuits, and breads (Angioloni & Collar, 2011; Brennan & Samyue, 2004; Englyst et al., 2003; Gelencsér, Gál, Hódsági, & Salgó, 2008; Gularte, de la Hera, Gómez, & Rosell, 2012; Reshmi, Sudha, & Shashirekha, 2017; Vitali, Dragojević, & Šebečić, 2009). The results showed that these effects are dependent upon the types of DFs and starch source (e.g., cereal and legume). However, no clear trend could be defined the ideal type and the amount of DFs *in vitro* starch digestion, thus more investigations are needed in the future.

6.3.2 Effect on GI

DFs were reported to impact the digestion of starch by reducing the rate of starch degradation and preventing excess glucose absorption (Brennan, 2005). Furthermore, various *in vivo* investigations confirmed the effect of DFs on reducing the GI of bakery products. Brites, Trigo, Carrapiço, Alviña, and Bessa (2011) found that the addition of RS into bread could reduce the *in vivo* postprandial glycaemic responses in rats. For *in vivo* human testing, Ekström et al. (2017) indicated that β-glucan has a potential to be used when tailoring the glycaemic profile of bread products, by recruiting 13 volunteers to consume bread with β-glucans (at the level of w/w 3.5%) from whole grain barley flour, coarse barley fiber and oat fiber and determining their postprandial glucose and inulin responses. Similarly, the direct incorporation of barley flour at suitable levels in bread was also found to reduce the GI and glycemic responses in human (Finocchiaro et al., 2012).

Besides, Phimolsiripol et al. (2017) showed that bread containing malva nut gum (7.5%) had a generally lower GI (72.8) compared to control bread (100). The addition of 10% sourdough with a lower molar ratio of lactic to acetic acid was also another option for generating bread with medium to low GI (Novotni et al., 2011). Goñi and Valentı´n-Gamazo (2003) reported that the incorporation of chickpea flour (25%) can reduce GI of pasta by 19% in healthy volunteers. Marangoni and Poli (2008) found that the addition of the fiber mixture composed of inulin (70%), guar gum (20%), glucomannan (5%) and wheat fiber (5%) to flour reduced the GI of bread and biscuit by 21% and 41%, respectively. Scazzina, Siebenhandl-Ehn, and Pellegrini (2013) summarized the potential effect of various DFs to reduce the GI in bread. They concluded that the presence of intact structures not accessible to human amylases, as well as a reduced pH that may delay gastric emptying or create a barrier to starch digestion, seems to be more effective than DF per se in improving glucose metabolism, irrespective of the type of cereal. Moreover, the high contents of indigestible carbohydrates and SDFs in products (e.g., bread) not only show a positive effect on reducing the acute glycaemic response but also show a "second-meal effect" as colonic fermentation might contribute to improved late glucose regulation. As indicated by Nilsson, OÖstman, Granfeldt, and Björck (2008), the metabolites produced during the colonic fermentation of indigestible carbohydrates (i.e., SCFA) may enter the systemic circulation and exert positive effects, including benefits on glucose metabolism. Therefore, this kind of physiological function of fiber-enriched bakery products is very beneficial for those who want to have good body weight management.

6.4 Effect on the colonic microflora

The large intestine is host to a large and diverse resident microflora. DF, as one of the major dry weight constitutions of digesta reaching the caecum, is reported to show a profound effect on the amount and diversity of bacteria in the colon. Because bacteria need certain enzymes to disrupt the saccharide bonds of the diverse range of DFs, therefore, DFs will clearly impact microfloral population dynamics. The presence of any fermentable DFs is likely to lead to an increase in microfloral bifidobacterial and *Lactobacillus* levels which can ferment carbohydrates (Brownlee, 2011). In fact, the physiological actions of DFs in the large intestine are depended on their chemical structure (composition and degree of polymerisation) and their physicochemical properties (solubility and water-binding capacity). In the colon, several non-fermentable IDFs (e.g. cellulose and lignin) are involved with a dilution of intestinal components, resulting in an increase of fecal mass and a decrease of transit time (Glitsø, Brunsgaard, Højsgaard, Sandström, & Bach Knudsen, 1998). On the other hand, SDFs are highly fermentable, producing a large amount of SCFA and inducing lowering of intraluminal colonic pH, which is proven to be favorable for intestinal health. SCFAs are the functional end products of most DFs fermentation. The increasing amount of SCFA is considered to be significant for the health of the colonic epithelium (Tuncil, Thakkar, Marcia, Hamaker, & Lindemann, 2018). As one of the well-known SCFAs, butyrate is a significant energy source for intestinal epithelial cells, and it plays an important role in the maintenance of colonic homeostasis through acting on various colonic mucosal functions. Besides, butyrate is also reported to promote satiety (Hamer et al., 2008).

DFs, such as RS, pectin, (hemi)cellulose and inulin, are important factors that have potential effect on the gut microbiota (Hamaker & Tuncil, 2014). Several investigations indicated that some DFs-enriched bakery products, such as chitosan-enriched biscuits (Terada et al., 1995) and inulin-enriched bread (Gråsten et al., 2003), may result in reductions of potentially harmful microfloral metabolites. Le Gall, Serena, Jorgensen, Theil, and Bach Knudsen (2009) found that the fermentation of the AX of the bread made from wheat aleurone flour has been reported to enhance the butyrate production in the large intestine. Besides, the beneficial effects of bread enriched with wholegrain rye flour over white wheat bread on the production of SCFAs were confirmed by Lappi et al. (2014). Similar, by consuming bread enriched with wheat pentosan and inulin for 3-weeks, Gråsten et al. (2003) demonstrated that the pentosan appeared to be more easily fermented than inulin by producing more total SCFAs and butyrate in the fecal, which may have favorable effects on colonic health. Notably, the DFs composition in foods has been shown to have an impact on the site for their degradation in the large intestine. By using a model experiment with pigs, Le Gall et al. (2009) indicated that AXs of the bread made from rye aleurone flour were mainly fermented in the caecum, AXs of the bread made from wheat aleurone flour as much in the colon as in the caecum, while the fibers in the bread made from white wheat flour, rich in cellulose, were fermented in the more distal locations of the large intestine.

6.5 Other effects

As aforementioned, the "associated compounds" are included in an official definition of DFs, and many cereals polysaccharide fraction (e.g., NSP) is associated with a substantial amount of phenolic compounds (e.g., ferulic acid). Therefore, the beneficial effects of the cereal DFs are reflected not only by the polysaccharide moiety but also by the associated phenolic compounds. By reviewing many intervention studies conducted using cereal bran and whole grains based foods, Vitaglione et al. (2008) concluded that cereal DFs are regarded as natural functional ingredients to deliver phenolic compounds into the gut. They illustrated a hypothesis that the DF (both SDF and IDF)-phenolic compounds may release in the gut slowly and continuously, which is supposed to show health benefits such as quenching the radicals and reduce oxidative stress (Rave, Roggen, Dellweg, Heise, & tom Dieck, H., 2007). Various baking practices, such as different ingredients and temperatures, may induce different changes in phenolic compounds among the bakery products. For instance, an overall significant reduction of total phenolic content was observed during the bread-making process (Blandino et al., 2013; Menga, Fares, Troccoli, Cattivelli, & Baiano, 2010; Yu, Nanguet, & Beta, 2013), however, a general increase of total phenolics was found in muffin products (Li, Pickard, & Beta, 2007). Besides, an increase in the bound phenolic acids content was found in cooked pasta due to that cooking in boiling water probably enhance the extraction of bound phenolics from pasta, especially ferulic acid ester-linked to cell walls (Fares, Platani, Baiano, & Menga, 2010).

The addition of the fibers from fruits and vegetables has been reported to have various surprising physiological functions which are involved with both the DF and phenolic compounds (Betoret & Rosell, 2019; Garcia-Amezquita et al., 2018). Al-Sayed and Ahmed (2013) found that the addition of watermelon or sharlyn melon rinds DF (5–10%) in cakes reduced the lipid oxidation and free fatty acids formation due to the presence of antioxidant compounds in DF. In addition, guava pulp powder was successfully used to complement the DF content of bread along with improvement in antioxidant potential without adversely affecting the organoleptic properties (Arslan et al., 2017). In addition, the supplementation of vegetable flour, like artichoke, cauliflower, broccoli, fennel and zucchini, into bread has been reported to improve the antioxidant content and reduce the GI of bread (Saccotelli et al., 2018). Certainly, more human studies are necessary to validate the hypothesis about the role of DF-phenolic compounds deriving from regular consumption of cereal DF-rich foods.

Besides, the hypocholesterolemic properties of fiber-enriched bakery products, which can be considered as an efficient antiatherosclerotic diet, have also been studied. Previously, clinical trials suggest that chitosan could reduce the total serum cholesterol and increase the serum HDL-cholesterol when ingested in fortified biscuits at a dose level of 6 g/day by healthy adult males (Maezaki et al., 1993). More recently, Ausar et al. (2003) found that compared to the control group, the patients consuming chitosan-containing bread (2% chitosan) had significantly lower mean levels of LDL cholesterol and higher mean levels of high-density lipoprotein (HDL) cholesterol at the end of the study.

7. Improving the compatibility of DFs in bakery products

The supplementation of DFs in bakery foods is prevalent, aiming to reach a positive effect on people's health. Regardless of the improvement in the nutritional value and the associated health benefits, their acceptability is relatively low. DFs can affect dough and finally show a negative effect on the quality of end-products, which makes it less appealing to consumers. As abovementioned, the adverse effects of DFs on the quality of bakery products have been well studied. Thus, various ways are being investigated to reduce this adverse effect during dough processing so that the final bakery products can not only have improved nutritional quality but also have acceptable sensory quality. The approaches are mainly through adding additives, and using food processing (e.g., grinding, heating, and extrusion) or bioprocessing (e.g., enzymatic treatment, fermentation, and sourdough) techniques (Prückler et al., 2014; Salmenkallio-Marttila et al., 2001).

7.1 Grinding

Grinding is likely to be the easiest way to enlarge the surface area of bran, which is expected to increase the bioaccessibility of nutrients and soluble fiber. Recently, the utilization of micronization in the food field indicates that the structure, surface area, and functional properties of the particles of various fiber-rich plant materials are altered after milling. The influence of particle size reduction on the properties of wheat bran is studied. Intentionally or not, the reduction of particle size changes its chemical, physical properties, and physiological effects. Studies have shown that finer wheat bran induced more short-chain fatty acids than large-particle bran due to the enlarged accessible surface area (Jenkins et al., 1999; Stewart & Slavin, 2009). By reducing the particle size of wheat bran, the water-binding ability of wheat bran decreased (Noort et al., 2010). Van Craeyveld et al. (2009) enhanced the content of water-soluble arabinoxylan of wheat bran from 4% to 61% by ball milling for 120 h. A combination of cryogenic grinding and electrostatic separation can enrich the aleurone layer with increasing minerals, vitamins, phenolic antioxidants, and lignans, and improve the bioaccessible phenolic acids of wheat bran (Hemery et al., 2009; Hemery et al., 2010; Hemery, Rouau, Lullien-Pellerin, Barron, & Abecassis, 2007). Therefore, decreasing the particle size by various methods including superfine grinding is an alternative way to improve the nutritional potential of wheat bran, but the subsequent effect on the quality of end-products is not always positive. For instance, Niu, Hou, Wang, and Chen (2014) found that superfine grinding of whole-what flour could improve the quality of whole-wheat noodles by significantly increasing the cooking yield. However, Xu, Xu, Wang, and Zhou (2018) demonstrated that the superfine grinding of wheat bran (from ~433 to ~39 mm) had a detrimental effect on the quality of steamed bread by decreasing its specific volume. Contrarily, Liu, Hou, Lee, Marquart, and Dubat (2016) indicated that fine grinding of whole wheat flour improved tortilla baking performance by increasing the breaking force and extensibility without affecting the diameter and thickness. They demonstrated that the finer bran particles might have a less destructive influence on gluten network formation in the dough. Furthermore, the particle size of whole wheat flour or wheat bran has been found to play an important role in determining both quality and digestibility of bakery products (Jin et al., 2021; Lin et al., 2020).

7.2 Extrusion

Extrusion using very high thermal and shear energy seems to be a state-of-the-art technology to change the texture and physicochemical properties, such as increasing the DF solubility of wheat bran. Wang, Klopfenstein, and Ponte (1993) indicated that the maximum fiber solubility increased about 2.5-fold after processing wheat bran at high moisture (28.8%) and high screw speed (400 rpm). Similarly, other researchers also found an increase (from 2.1% to 3.1%) of SDF in wheat bran after utilizing a twin-screw extruder conducted at 400 rpm and barrel temperatures of 80 to 130 °C (Kahlon, de Berrios, Smith, & Pan, 2006). The extruded bran is further reported to show an impact on the properties of dough and end-products. For instance, Gómez, Jiménez, Ruiz, and Oliete (2011) found that the extruded bran increased the dough development time and tenacity, while reducing the loss of dough height during fermentation to a greater extent than untreated bran. Extruded bran thus induced greater gas production than untreated bran. Therefore, bread with extruded bran exhibited a higher volume and lower hardness than that with untreated bran when improver added, but extruded bran had no significant effect on the organoleptic evaluation of bread. Furthermore, Wójtowicz and Moscicki (2011) found that higher screw speed improved the sensory attributes and the firmness of extruded pasta-like products. As for its physiological properties, the cholesterol-lowering potential of wheat bran fed to hamsters was enhanced by a proper extrusion process (Kahlon et al., 2006). The DFs in wheat bran, principally NSP, is mostly water-unextractable and is poorly utilized by human gut microbiota. Extrusion was reported to increase the content of WE-NSP of wheat bran, which enabled greater fermentability by human fecal microbiota. However, once extruded bran was used in a whole wheat bread formulation, the changes in fermentation outcomes were no longer evident (Arcila, Weier, & Rose, 2015).

7.3 Thermal treatment

Thermal treatments are usually applied to inactivate the enzymes in cereal brans and thus improve the shelf life of bran when used in food products. The chemical composition and physical properties of the bran have been changed after heat treatment due to the interaction between bran, proteins and water (Caprez, Arrigoni, Amado, & Neukom, 1986; Wang, Hamer, et al., 2002). Also, thermal treatments like roasting, steaming, cooking under forced air or microwave radiation have been reported to show an effect on the dough properties and end-product. For instance, the addition of autoclaved bran improved the texture and crumb structure of bread (Zhang, Gao, & He, 2019). Nandeesh, Jyotsna, and Venkateswara Rao (2011) investigated the effect of various thermally treated wheat bran on the rheological and quality characteristics of the soft dough biscuits, and found that the dough stability and Extensograph resistance to extension increased with the addition of wheat bran. Besides, Aktas-Akyildiz et al. (2017) found that the steam explosion treatment (120–160 °C, 5–10 min) increased the WE-AX content and reducing sugars in wheat bran, and they attributed this to the fact that cell wall structure of wheat bran was destroyed by the severe steam explosion conditions. Subsequently, a 20% addition of heat-treated bran reduced the bread quality. Nevertheless, heat-treated bran followed by enzymatic treatment enhanced bread quality with increased specific volume and decreased crumb hardness. Moreover, Sui, Xie, Liu, Wu, and Zhang (2018) found that steam explosion bran could improve the stability and elasticity of dough by increasing the storage modulus

(G') and loss modulus (G''), and they attributed to its ability of competitive water by promoting the partial dehydration of gluten and caused protein conformational changes of the collapse of β-turns into β-sheet structures. Similarly, flour with heated bran (190 °C, 30 s) was more stable and showed less softening compared to that with unheated bran (Wang, Liu, & Wen, 2017). In addition, the treatment of steam explosion under high temperature and high-pressure conditions is also reported to be a good method to release the phenolic acids in the wheat bran and to obtain a high yield of natural antioxidants (Zheng, Zhang, & Liu, 2015). Furthermore, heat treatment was found to improve the bioaccessibility and bioavailability of phenolic compounds in wheat bran (Wang, He, & Chen, 2014).

7.4 Enzymatic treatment

The use of bioprocessing techniques is also a good approach to improve the bioaccessibility of health-promoting compounds in cereal brans. Of which, enzymatic treatment and fermentation are particularly popular since they provide various tools to modify the matrix of whole-grain constitutes.

Proper enzymatic treatment can modify the complex cell wall structure of plant bran, especially cereal brans. Expensive purified enzymes are usually unfeasible, therefore breeding microorganisms that can produce the desired enzyme provide possible and technologically feasible methods. To date, xylanases are usually used in bread-making to improve dough processability and bread quality. Uysal et al. (2007) used xylanase to improve the properties of cookies supplemented with fibers from lemon, apple, and wheat and wheat bran. They found that xylanase (0.4%) decreased the cellulose content in the control sample (without fiber) whereas showed no significant impact in those samples fortified with high fiber levels (15%, 20%, and 30%). The addition of xylanase reduced the hardness and increased the spread values as compared to the control without enzyme, but showed an insignificant effect on the sensory score. Also, Shah, Shah, and Madamwar (2006) found that the quality of whole wheat bread improved by supplementation of xylanase (*Aspergillus foetidus*) as suggested by an increasing specific volume and decreasing firmness that are desirable for bread. Dornez, Verjans, Arnaut, Delcour, and Courtin (2011) indicated that (hyper-)thermophilic xylanases may help to produce arabinoxylan oligosaccharides during bread making through extensively degrading parts of the AX fraction. Besides, to improve the quality of fiber-enriched wheat breads, Laurikainen, Härkönen, Autio, and Poutanen (1998) investigated the impact of three enzymes (hemicellulolytic culture filtrate, xylanase and α-amylase) on bread containing rye bran. The team found that the addition of the hemicellulolytic culture filtrate and the α-amylase increased the specific volume of bread supplemented with rye by almost 20%, decreased the firmness by 54% and 37% respectively and reduced the staling rate. They attributed these improvements of bread quality to the fact that cell wall polysaccharides were changed by the enzyme, specifically, the area of endosperm cell walls and the total DF were all reduced, while the number of extractable pentosans increased for all the enzymatic treatments.

A combination of different enzymes is also a promising method. For instance, Flander et al. (2008) investigated the effects of laccase and xylanase and their combination on oat, wheat, and mixed oat-wheat doughs as well as the quality of corresponding breads. They indicated that xylanase increased the contents of WE-AX and water-soluble polysaccharides in the oat-wheat dough, which increased the softness of the dough, as well as the specific volume and softness of the bread. Moreover, the combination of laccase and xylanase produced slight hydrolysis of β-glucan by the β-glucanase side activity of laccase and enhanced the availability of AXs for xylanase with concomitant reduction of the amount and molar mass of water-soluble polysaccharides. Subsequently, the volume of oat bread was increased.

7.5 Fermentation

During fermentation, either the endogenous or the bacterial enzymes can modify the structure, bioactivity, and bioavailability of grain components (Hole et al., 2012). Fermentation seems to exhibit a positive impact on the nutritional values and health properties of whole grain bakery products. Yeast fermentation promoted the bioactivity and baking properties of wheat bran, inducing solubilization of AXs (Katina et al., 2012). Lactic acid bacteria fermentation with proper strains is also a useful tool to obtain specific results. For example, fermentation with lactic acid bacteria has a positive effect in promoting the nutritional quality, textural property and palatability of fiber-enrich bakery products (Poutanen, Flander, & Katina, 2009; Rizzello, Cassone, Coda, & Gobbetti, 2011; Salmenkallio-Marttila et al., 2001). Further, this fermentation can also enhance the amount of bioactive peptides, the solubility of DF (e.g., AXs) and bioavailability of minerals, as well as reduce the GI of the food (Rizzello, Cassone, Di Cagno, & Gobbetti, 2008).

In addition, sourdough fermentation is a common practice, especially in rye bread-making. It has been reported to promote the texture, sensory impression and shelf life of bread probably through modifying the dough components by acidification, as well as changing the proteolysis of gluten and starch hydrolysis (Coda, Rizzello, Curiel, Poutanen, & Katina, 2014; Poutanen et al., 2009). Salmenkallio-Marttila et al. (2001) investigated the impact of wheat bran fermentation on bread-making and concluded that fermentation with yeast and lactic acid bacteria enhanced the specific volume, firmness and shelf life of bread with 20% addition of wheat bran. The use of this pre-fermented bran improved the rheological characteristics of dough and the quality of bread having higher moisture retention and a better shelf-life (Messia et al., 2016).

Fermentation and enzymatic treatment are also combined to induce additional effects. Katina et al. (2006) investigated the combined effects of sourdough fermentation and enzyme mixture of α-amylase, xylanase, and lipase on bread containing 20% wheat bran. Individually, both treatments were observed to increase the loaf volume (from 3.6 to 4.1 mL/g), but the greatest increase was found in the combined group (4.5 mL/g). Moreover, the combination treatment produced a similar crumb textural feature as the control bread, and decreased the staling rate of bread, mainly due to the degradation of cell wall components, the improvement of the gluten network, the changes of water migration during staling, and the reduction of starch retrogradation. Further, the fortification of pre-fermented bran (lactic acid bacteria fermentation) and enzymatic treatment (xylanase, amylase, cellulase) increased the AXs solubility in the dough, improving the rheological characteristics of dough and the quality of bread (e.g., higher moisture retention and better shelf-life) (Messia et al., 2016).

7.6 Other methods

The positive impacts of gluten addition on the quality of fiber-enrich bread have been studied (Flander, Salmenkallio-Marttila, Suortti, & Autio, 2007; Ma & Baik, 2017; Salmenkallio-Marttila, Roininen, & Autio, 2004). An optimized addition of water and gluten in oat flour was reported to improve the volume and texture of bread (Flander et al., 2007). They also demonstrated that the temperature and time of the proofing process show a profound impact on bread quality, including the sensory properties of bread such as flavor and crispiness of the crust. Longer proofing time at a high temperature (39–40 °C) produced bread with a higher loaf volume and softer crumb texture (Flander et al., 2008). In addition, the prehydration of bran before being added into refined flour for dough preparation was found to improve bread quality (Park, Fuerst, & Baik, 2017). They attributed to the fact that prehydration can alter bran composition, allowing enough water to be absorbed by fibrous materials of bran and preventing water competition among dough constituents.

8. Conclusions

Bakery products are prevalently consumed foods in the world, and they have been regarded as convenient dietary vehicles for delivering nutritive ingredients into people's diet, of which, dietary fiber (DF) is one of the most popular items. The food industry attempts to produce fiber-enriched bakery products with both increasing nutritional value and appealing palatability. The trends of fortifying DFs into bakery products have tend to various sources other than cereals. This chapter has reviewed the recent advances in the diversity of DFs sources, such as pseudo-cereals, fruits and vegetables, legumes, commercially hydrocolloids, RS and bacterial cellulose. As many new sources of DFs become available, the impact mechanism of various DFs on dough processing and the final bakery products are varied, although many studies have indicated the impact of some DFs on the gluten formation, more investigations are needed due to the new emerging DFs sources. Besides, the nutrition value of DF is commonly accepted, and many investigations have also revealed the physiological functionality of DF-enriched bakery products, including effect on swallowing and the oesophagus, gastric emptying and satiety, digestion and absorption, and colonic microflora. In addition, methods of improving the compatibility of DFs in bakery products and the quality of end-product have been proposed by many researchers, and they are mainly concentrated on manipulating the processing ways of DF, such as grinding, extrusion, thermal treatment, enzymatic treatment and fermentation. Thus, with the increase of DFs sources and consumer awareness for healthier diets, researchers and manufactures should put more effort in understanding the ways in which DFs may be used to improve the quality, acceptability and nutritional properties of DF-enriched bakery products.

References

Ajila, C. M., Leelavathi, K., & Prasada Rao, U. J. S. (2008). Improvement of dietary fibre content and antioxidant properties in soft dough biscuits with the incorporation of mango peel powder. *Journal of Cereal Science*, *48*(2), 319–326.

Aktas-Akyildiz, E., Mattila, O., Sozer, N., Poutanen, K., Koksel, H., & Nordlund, E. (2017). Effect of steam explosion on enzymatic hydrolysis and baking quality of wheat bran. *Journal of Cereal Science, 78*, 25–32.

Al-Sayed, H. M. A., & Ahmed, A. R. (2013). Utilization of watermelon rinds and sharlyn melon peels as a natural source of dietary fibre and antioxidants in cake. *Annals of Agricultural Science, 58*(1), 83–95.

Alba, K., Campbell, G. M., & Kontogiorgos, V. (2019). Dietary fibre from berry-processing waste and its impact on bread structure: a review. *Journal of the Science of Food and Agriculture, 99*(9), 4189–4199.

Alvarez-Jubete, L., Arendt, E. K., & Gallagher, E. (2009). Nutritive value and chemical composition of pseudocereals as gluten-free ingredients. *International Journal of Food Sciences and Nutrition, 60*(supp. 4), 240–257.

Åman, P., Andersson, A., Rakha, A., & Andersson, R. (2010). Rye, a healthy cereal full of dietary fibre. *Cereal Foods World, 55*(5), 231–234.

Amoah, I., Cairncross, C., & Rush, E. (2021). Swallowing and liking of vegetable-enriched bread compared with commercial breads as evaluated by older adults. *Frontiers in Nutrition, 7*, 599737.

Andersson, A. A. M., Armö, E., Grangeon, E., Fredriksson, H., Andersson, R., & Åman, P. (2004). Molecular weight and structure units of (1→3, 1→4)-β-glucans in dough and bread made from hull-less barley milling fractions. *Journal of Cereal Science, 40*(3), 195–204.

Andersson, R., Fransson, G., Tietjen, M., & Åman, P. (2009). Content and molecular-weight distribution of dietary fibre components in whole-grain rye flour and bread. *Journal of Agricultural and Food Chemistry, 57*(5), 2004–2008.

Angioloni, A., & Collar, C. (2011). Physicochemical and nutritional properties of reduced-caloric density high-fibre breads. *LWT - Food Science and Technology, 44*(3), 747–758.

Anil, M. (2007). Using of hazelnut testa as a source of dietary fibre in breadmaking. *Journal of Food Engineering, 80*(1), 61–67.

Anton, A. A., & Artfield, S. D. (2008). Hydrocolloids in gluten-free breads: A review. *International Journal of Food Sciences and Nutrition, 59*(1), 11–23.

Apolinário, A. C, de Lima Damasceno, B. P. G., de Macêdo Beltrão, N. E., Pessoa, A., Converti, A., & da Silva, J. A. (2014). Inulin-type fructans: A review on different aspects of biochemical and pharmaceutical technology. *Carbohydrate Polymers, 101*, 368–378.

Arcila, J. A., Weier, S. A., & Rose, D. J. (2015). Changes in dietary fibre fractions and gut microbial fermentation properties of wheat bran after extrusion and bread making. *Food Research International, 74*, 217–223.

Arslan, M., Rakha, A., Khan, M. R., & Zou, X. (2017). Complementing the dietary fibre and antioxidant potential of gluten free bread with guava pulp powder. *Journal of Food Measurement and Characterization, 11*(4), 1959–1968.

Arslan, M., Rakha, A., Xiaobo, Z., & Mahmood, M. A. (2019). Complimenting gluten free bakery products with dietary fibre: Opportunities and constraints. *Trends in Food Science & Technology, 83*, 194–202.

Ausar, S. F., Morcillo, M., León, A. E., Ribotta, P. D., Masih, R., Vilaro Mainero, M., et al. (2003). Improvement of HDL- and LDL-cholesterol levels in diabetic subjects by feeding bread containing chitosan. *Journal of Medicinal Food, 6*(4), 397–399.

Ayadi, M. A., Abdelmaksoud, W., Ennouri, M., & Attia, H. (2009). Cladodes from Opuntia ficus indica as a source of dietary fibre: Effect on dough characteristics and cake making. *Industrial Crops and Products, 30*(1), 40–47.

Bae, I. Y., Lee, H. I., Ko, A., & Lee, H. G. (2013). Substituting whole grain flour for wheat flour: Impact on cake quality and glycemic index. *Food Science and Biotechnology, 22*(5), 1–7.

Baixauli, R., Sanz, T., Salvador, A., & Fiszman, S. M. (2008). Muffins with resistant starch: Baking performance in relation to the rheological properties of the batter. *Journal of Cereal Science*, *47*(3), 502–509.

Bandyopadhyay, K., Chakraborty, C., & Bhattacharyya, S. (2014). Fortification of mango peel and kernel powder in cookies formulation. *Journal of Academia Industrial Research*, *2*(12), 661–664.

Barros, F., Alviola, J. N., & Rooney, L. W. (2010). Comparison of quality of refined and whole wheat tortillas. *Journal of Cereal Science*, *51*(1), 50–56.

Behall, K. M., Scholfield, D. J., & Hallfrisch, J. G. (2006). Barley β-glucan reduces plasma glucose and insulin responses compared with resistant starch in men. *Nutrition Research*, *26*(12), 644–650.

Ben Jeddou, K., Bouaziz, F., Zouari-Ellouzi, S., Chaari, F., Ellouz-Chaabouni, S., Ellouz-Ghorbel, R., et al. (2017). Improvement of texture and sensory properties of cakes by addition of potato peel powder with high level of dietary fibre and protein. *Food Chemistry*, *217*, 668–677.

Benítez, V., Cantera, S., Aguilera, Y., Mollá, E., Esteban, R. M., Díaz, M. F., et al. (2013). Impact of germination on starch, dietary fibre and physicochemical properties in non-conventional legumes. *Food Research International*, *50*(1), 64–69.

Benítez, V., Esteban, R. M., Moniz, E., Casado, N., Aguilera, Y., & Mollá, E. (2018). Breads fortified with wholegrain cereals and seeds as source of antioxidant dietary fibre and other bioactive compounds. *Journal of Cereal Science*, *82*, 113–120.

Betoret, E., & Rosell, C. M. (2019). Enrichment of bread with fruits and vegetables: Trends and strategies to increase functionality. *Cereal Chemistry*.

Bhushan, S., Kalia, K., Sharma, M., Singh, B., & Ahuja, P. S. (2008). Processing of apple pomace for bioactive molecules. *Critical Reviews in Biotechnology*, *28*(4), 285–296.

Bharath Kumar, S., & Prabhasankar, P. (2015). A study on noodle dough rheology and product quality characteristics of fresh and dried noodles as influenced by low glycemic index ingredient. *Journal of Food Science and Technology*, *52*(3), 1404–1413.

Blandino, M., Sovrani, V., Marinaccio, F., Reyneri, A., Rolle, L., Giacosa, S., et al. (2013). Nutritional and technological quality of bread enriched with an intermediated pearled wheat fraction. *Food Chemistry*, *141*(3), 2549–2557.

Bosmans, G. M., Lagrain, B., Deleu, L. J., Fierens, E., Hills, B. P., & Delcour, J. A. (2012). Assignments of proton populations in dough and bread using NMR relaxometry of starch, gluten, and flour model systems. *Journal of Agricultural and Food Chemistry*, *60*(21), 5461–5470.

Boukid, F., Zannini, E., Carini, E., & Vittadini, E. (2019). Pulses for bread fortification: A necessity or a choice? *Trends in Food Science & Technology*, *88*, 416–428.

Brennan, C. S. (2005). Dietary fibre, glycaemic response, and diabetes. *Molecular Nutrition & Food Research*, *49*(6), 560–570.

Brennan, C. S., Blake, D. E., Ellis, P. R., & Schofield, J. D. (1996). Effects of guar galactomannan on wheat bread microstructure and on the in vitro and in vivo digestibility of starch in bread. *Journal of Cereal Science*, *24*(2), 151–160.

Brennan, C. S., Kuri, V., & Tudorica, C. M. (2004). Inulin-enriched pasta: effects on textural properties and starch degradation. *Food Chemistry*, *86*(2), 189–193.

Brennan, C. S., & Samyue, E. (2004). Evaluation of starch degradation and textural characteristics of dietary fibre enriched biscuits. *International Journal of Food Properties*, 7(3), 647–657.

Brites, C. M., Trigo, M. J., Carrapiço, B., Alviña, M., & Bessa, R. J. (2011). Maize and resistant starch enriched breads reduce postprandial glycemic responses in rats. *Nutrition Research*, *31*(4), 302–308.

Brownlee, I. (2014). The impact of dietary fibre intake on the physiology and health of the stomach and upper gastrointestinal tract. *Bioactive Carbohydrates and Dietary Fibre*, *4*(2), 155–169.

Brownlee, I. A. (2011). The physiological roles of dietary fibre. *Food Hydrocolloids*, *25*(2), 238–250.

Campo, V. L., Kawano, D. F., Silva, D. B.d., & Carvalho, I. (2009). Carrageenans: Biological properties, chemical modifications and structural analysis – A review. *Carbohydrate Polymers*, 77(2), 167–180.

Capita, R., & Alonso-calleja, C. (2003). Intake of nutrients associated with an increased risk of cardiovascular disease in a Spanish population. *International Journal of Food Sciences and Nutrition*, *54*(1), 57–75.

Caprez, A., Arrigoni, E., Amado, R., & Neukom, A. (1986). Influence of different types of thermal treatment on the chemical composition and physical properties of wheat bran. *Journal of Cereal Science*, *4*.

Capriles, V. D., & Arêas, J. A. G. (2014). Novel approaches in gluten-free breadmaking: interface between food science, nutrition, and health. *Comprehensive Reviews in Food Science and Food Safety*, *13*(5), 871–890.

Cauvain, S. P., & Young, L. S. (2008). Current approaches to the classification of bakery products. In S. P. Cauvain, & L. S. Young (Eds.), *Baked products: science, technology practice* (1st ed., pp. 1–13). John Wiley & Sons.

Champ, M., Langkilde, A. M., Brouns, F., Kettlitz, B., & Collet, Y. L. B. (2003). Advances in dietary fibre characterisation. 1. Definition of dietary fibre, physiological relevance, health benefits and analytical aspects. *Nutrition Research Reviews*, *16*(1), 71–82.

Chaturvedi, N., Sharma, P., Shukla, K., Singh, R., & Yadav, S. (2011). Cereals nutraceuticals, health enoblement and diseases obviation: A comprehensive review. *Journal of Applied Pharmaceutical Science*, *1*(7), 6–12.

Chau, C.-F., & Huang, Y.-L. (2003). Comparison of the chemical composition and physicochemical properties of different fibres prepared from the peel of citrus sinensis L. Cv. Liucheng. *Journal of Agricultural and Food Chemistry*, *51*(9), 2615–2618.

Chen, H., Rubenthaler, G., Leung, H., & Baranowski, J. (1988). Chemical, physical, and baking properties of apple fibre compared with wheat and oat bran. *Cereal Chemistry*, *65*(3), 244–247.

Choo, C. L., & Aziz, N. A. A. (2010). Effects of banana flour and β-glucan on the nutritional and sensory evaluation of noodles. *Food Chemistry*, *119*(1), 34–40.

Chuang, S.-C., Norat, T., Murphy, N., Olsen, A., Tjønneland, A., Overvad, K., et al. (2012). Fibre intake and total and cause-specific mortality in the european prospective investigation into cancer and nutrition cohort. *The American Journal of Clinical Nutrition*, *96*(1), 164–174.

Chung, H. J., Donner, E., & Liu, Q. (2011). Resistant starches in foods. In M. Moo-Young (Ed.), *Comprehensive Biotechnology* (2nd ed., pp. 527–534). Burlington: Academic Press.

Cichero, J. A. Y. (2013). Thickening agents used for dysphagia management: effect on bioavailability of water, medication and feelings of satiety. *Nutrition Journal*, *12*(1), 54.

Coda, R., Rizzello, C. G., Curiel, J. A., Poutanen, K., & Katina, K. (2014). Effect of bioprocessing and particle size on the nutritional properties of wheat bran fractions. *Innovative Food Science & Emerging Technologies*, *25*, 19–27.

Codex, A. (2009). *Report of the 31st session of the codex committee on nutrition and foods for special dietary uses*. Düsseldorf, Germany, 2–6 November, 2009. In ALINORM 10/33/26.

Collar, C., & Angioloni, A. (2014). Pseudocereals and teff in complex breadmaking matrices: Impact on lipid dynamics. *Journal of Cereal Science*, *59*(2), 145–154.

Corral, M. L., Cerrutti, P., Vázquez, A., & Califano, A. (2017). Bacterial nanocellulose as a potential additive for wheat bread. *Food Hydrocolloids*, *67*, 189–196.

Dahm, C. C., Keogh, R. H., Spencer, E. A., Greenwood, D. C., Key, T. J., Fentiman, I. S., et al. (2010). Dietary fibre and colorectal cancer risk: A Nested Case–Control Study Using Food Diaries. *JNCI: Journal of the National Cancer Institute, 102*(9), 614–626.

Daou, C., & Zhang, H. (2012). Oat beta-glucan: Its role in health promotion and prevention of diseases. *Comprehensive Reviews in Food Science and Food Safety, 11*(4), 355–365.

Das, L., Raychaudhuri, U., & Chakraborty, R. (2015). Effects of hydrocolloids as texture improver in coriander bread. *Journal of Food Science and Technology, 52*(6), 3671–3680.

Demirkesen, I., Kelkar, S., Campanella, O. H., Sumnu, G., Sahin, S., & Okos, M. (2014). Characterization of structure of gluten-free breads by using X-ray microtomography. *Food Hydrocolloids, 36*, 37–44.

Dobraszczyk, B. J., & Salmanowicz, B. P. (2008). Comparison of predictions of baking volume using large deformation rheological properties. *Journal of Cereal Science, 47*(2), 292–301.

Dogan, M., Kayacier, A., & Ic, E. (2007). Rheological characteristics of some food hydrocolloids processed with gamma irradiation. *Food Hydrocolloids, 21*(3), 392–396.

Dornez, E., Verjans, P., Arnaut, F., Delcour, J. A., & Courtin, C. M. (2011). Use of psychrophilic xylanases provides insight into the xylanase functionality in bread making. *Journal of Agricultural and Food Chemistry, 59*(17), 9553–9562.

Drzikova, B., Dongowski, G., Gebhardt, E., & Habel, A. (2005). The composition of dietary fibre-rich extrudates from oat affects bile acid binding and fermentation in vitro. *Food Chemistry, 90*(1), 181–192.

Du, H., Daphne, A.v.d. L., Boshuizen, H. C., Forouhi, N. G., Wareham, N. J., Halkjær, J., et al. (2010). Dietary fibre and subsequent changes in body weight and waist circumference in European men and women. *The American Journal of Clinical Nutrition, 91*(2), 329–336.

Eastwood, M. A., & Morris, E. R. (1992). Physical properties of dietary fibre that influence physiological function: a model for polymers along the gastrointestinal tract. *The American Journal of Clinical Nutrition, 55*(2), 436–442.

Ebere, C., Emelike, N., & Kiin-Kabari, D. (2015). Physico-chemical and sensory properties of cookies prepared from wheat flour and cashew-apple residue as a source of fibre. *Asian Journal of Agriculture Food Sciences, 3*(2).

EFSA Panel on Dietetic Products, Nutrition, & Allergies. (2010). Scientific opinion on dietary reference values for carbohydrates and dietary fibre. *EFSA Journal, 8*(3), 1462.

Ekström, L. M. N. K., Henningsson Bok, E. A. E., Sjöö, M. E., & Östman, E. M. (2017). Oat β-glucan containing bread increases the glycaemic profile. *Journal of Functional Foods, 32*, 106–111.

Elleuch, M., Bedigian, D., Roiseux, O., Besbes, S., Blecker, C., & Attia, H. (2011). Dietary fibre and fibre-rich by-products of food processing: Characterisation, technological functionality and commercial applications: A review. *Food Chemistry, 124*(2), 411–421.

Englyst, K. N., Vinoy, S., Englyst, H. N., & Lang, V. (2003). Glycaemic index of cereal products explained by their content of rapidly and slowly available glucose. *British Journal of Nutrition, 89*(3), 329–340.

Faccin, G. L., Miotto, L. A., Vieira, L.d. N., Barreto, P. L. M., & Amante, E. R. (2009). Chemical, sensorial and rheological properties of a new organic rice bran beverage. *Rice Science, 16*(3), 226–234.

Fares, C., Platani, C., Baiano, A., & Menga, V. (2010). Effect of processing and cooking on phenolic acid profile and antioxidant capacity of durum wheat pasta enriched with debranning fractions of wheat. *Food Chemistry, 119*(3), 1023–1029.

Faris, M.e. A.-I. E., Takruri, H. R., & Issa, A. Y. (2013). Role of lentils (Lens culinaris L.) in human health and nutrition: a review. *Mediterranean Journal of Nutrition and Metabolism, 6*, 3–16.

Filipović, J., Popov, S., & Filipović, N. (2008). The behavior of different fibres at bread dough freezing. *Chemical Industry and Chemical Engineering Quarterly*, *14*(4), 257–259.

Finocchiaro, F., Ferrari, B., Gianinetti, A., Scazzina, F., Pellegrini, N., Caramanico, R., et al. (2012). Effects of barley β-glucan-enriched flour fractions on the glycaemic index of bread. *International Journal of Food Sciences and Nutrition*, *63*(1), 23–29.

Flamm, G., Glinsmann, W., Kritchevsky, D., Prosky, L., & Roberfroid, M. (2001). Inulin and oligofructose as dietary fibre: a review of the evidence. *Critical Reviews in Food Science and Nutrition*, *41*(5), 353–362.

Flander, L., Rouau, X., Morel, M.-H., Autio, K., Seppänen-Laakso, T., Kruus, K., et al. (2008). Effects of laccase and xylanase on the chemical and rheological properties of oat and wheat doughs. *Journal of Agricultural and Food Chemistry*, *56*(14), 5732–5742.

Flander, L., Salmenkallio-Marttila, M., Suortti, T., & Autio, K. (2007). Optimization of ingredients and baking process for improved wholemeal oat bread quality. *LWT - Food Science and Technology*, *40*(5), 860–870.

Foschia, M., Horstmann, S. W., Arendt, E. K., & Zannini, E. (2017). Legumes as functional ingredients in gluten-free bakery and pasta products. *Annual Review of Food Science and Technology*, *8*(1), 75–96.

Frost, G., Sleeth, M. L., Sahuri-Arisoylu, M., Lizarbe, B., Cerdan, S., Brody, L., et al. (2014). The short-chain fatty acid acetate reduces appetite via a central homeostatic mechanism. *Nature Communications*, *5*(1), 3611.

Fuentes-Zaragoza, E., Riquelme-Navarrete, M. J., Sánchez-Zapata, E., & Pérez-Álvarez, J. A. (2010). Resistant starch as functional ingredient: A review. *Food Research International*, *43*(4), 931–942.

Gao, J., Brennan, M. A., Mason, S. L., & Brennan, C. S. (2016). Effect of sugar replacement with stevianna and inulin on the texture and predictive glycaemic response of muffins. *International Journal of Food Science & Technology*, *51*(9), 1979–1987.

Garcia-Amezquita, L. E., Tejada-Ortigoza, V., Serna-Saldivar, S. O., & Welti-Chanes, J. (2018). Dietary fibre concentrates from fruit and vegetable by-products: processing, modification, and application as functional ingredients. *Food and Bioprocess Technology*, *11*(8), 1439–1463.

Garcı´a-Ochoa, F., Santos, V. E., Casas, J. A., & Gómez, E. (2000). Xanthan gum: production, recovery, and properties. *Biotechnology Advances*, *18*(7), 549–579.

Gebruers, K., Dornez, E., Boros, D., Fraś, A., Dynkowska, W., Bedő, Z., et al. (2008). Variation in the content of dietary fibre and components thereof in wheats in the HEALTHGRAIN diversity screen. *Journal of Agricultural and Food Chemistry*, *56*(21), 9740–9749.

Gelencsér, T., Gál, V., Hódsági, M., & Salgó, A. (2008). Evaluation of quality and digestibility characteristics of resistant starch-enriched pasta. *Food and Bioprocess Technology*, *1*(2), 171–179.

Ghoshal, G., & Mehta, S. (2018). Effect of chitosan on physicochemical and rheological attributes of bread. *Food Science and Technology International*, *25*(3), 198–211.

Glitsø, L. V., Brunsgaard, G., Højsgaard, S., Sandström, B., & Bach Knudsen, K. E. (1998). Intestinal degradation in pigs of rye dietary fibre with different structural characteristics. *British Journal of Nutrition*, *80*(5), 457–468.

Gómez, M., Jiménez, S., Ruiz, E., & Oliete, B. (2011). Effect of extruded wheat bran on dough rheology and bread quality. *LWT - Food Science and Technology*, *44*(10), 2231–2237.

Gómez, M., Moraleja, A., Oliete, B., Ruiz, E., & Caballero, P. A. (2010). Effect of fibre size on the quality of fibre-enriched layer cakes. *LWT - Food Science and Technology*, *43*(1), 33–38.

Gómez, M., Ronda, F., Blanco, C. A., Caballero, P. A., & Apesteguía, A. (2003). Effect of dietary fibre on dough rheology and bread quality. *European Food Research and Technology*, *216*(1), 51–56.

Gómez, M., Ronda, F., Caballero, P. A., Blanco, C. A., & Rosell, C. M. (2007). Functionality of different hydrocolloids on the quality and shelf-life of yellow layer cakes. *Food Hydrocolloids*, *21*(2), 167–173.
Goñi, I., & Valentı´n-Gamazo, C. (2003). Chickpea flour ingredient slows glycemic response to pasta in healthy volunteers. *Food Chemistry*, *81*(4), 511–515.
Goulas, A., Tzortzis, G., & Gibson, G. R. (2007). Development of a process for the production and purification of α- and β-galactooligosaccharides from Bifidobacterium bifidum NCIMB 41171. *International Dairy Journal*, *17*(6), 648–656.
Gråsten, S., Liukkonen, K.-H., Chrevatidis, A., El-Nezami, H., Poutanen, K., & Mykkänen, H. (2003). Effects of wheat pentosan and inulin on the metabolic activity of faecal microbiota and on bowel function in healthy humans. *Nutrition Research*, *23*(11), 1503–1514.
Grigelmo-Miguel, N., Carreras-Boladeras, E., & Martín-Belloso, O. (2001). Influence of the addition of peach dietary fibre in composition, physical properties and acceptability of reduced-fat muffins. *Food Science and Technology International*, 7(5), 425–431.
Guarda, A., Rosell, C. M., Benedito, C., & Galotto, M. J. (2004). Different hydrocolloids as bread improvers and antistaling agents. *Food Hydrocolloids*, *18*(2), 241–247.
Guillon, F., & Champ, M. (2000). Structural and physical properties of dietary fibres, and consequences of processing on human physiology. *Food Research International*, *33*(3), 233–245.
Gularte, M. A., de la Hera, E., Gómez, M., & Rosell, C. M. (2012). Effect of different fibres on batter and gluten-free layer cake properties. *LWT - Food Science and Technology*, *48*(2), 209–214.
Gupta, S., Saurabh, C. K., Variyar, P. S., & Sharma, A. (2015). Comparative analysis of dietary fibre activities of enzymatic and gamma depolymerized guar gum. *Food Hydrocolloids*, *48*, 149–154.
Hager, A.-S., Ryan, L. A. M., Schwab, C., Gänzle, M. G., O'Doherty, J. V., & Arendt, E. K. (2011). Influence of the soluble fibres inulin and oat β-glucan on quality of dough and bread. *European Food Research and Technology*, *232*(3), 405–413.
Hamed, I., Özogul, F., & Regenstein, J. M. (2016). Industrial applications of crustacean by-products (chitin, chitosan, and chitooligosaccharides): A review. *Trends in Food Science & Technology*, *48*, 40–50.
Hamer, H. M., Jonkers, D., Venema, K., Vanhoutvin, S., Troost, F. J., & Brummer, R. J. (2008). Review article: the role of butyrate on colonic function. *Alimentary Pharmacology & Therapeutics*, *27*(2), 104–119.
Hamaker, B. R., & Tuncil, Y. E. (2014). A perspective on the complexity of dietary fiber structures and their potential effect on the gut microbiota. *Journal of Molecular Biology*, *426*(23), 3838–3850.
Han, J., Janz, J. A. M., & Gerlat, M. (2010). Development of gluten-free cracker snacks using pulse flours and fractions. *Food Research International*, *43*(2), 627–633.
Hansen, H. B., Rasmussen, C. V., Bach Knudsen, K. E., & Hansen, Å. (2003). Effects of genotype and harvest year on content and composition of dietary fibre in rye (Secale cereale L) grain. *Journal of the Science of Food and Agriculture*, *83*(1), 76–85.
Hashemi, J. (2015). Biomedical effects of barley-a review. *New York Science Journal*, *8*(3), 52–55.
Hejrani, T., Sheikholeslami, Z., Mortazavi, A., & Davoodi, M. G. (2017). The properties of part baked frozen bread with guar and xanthan gums. *Food Hydrocolloids*, *71*, 252–257.
Hemery, Y., Rouau, X., Dragan, C., Bilici, M., Beleca, R., & Dascalescu, L. (2009). Electrostatic properties of wheat bran and its constitutive layers: Influence of particle size, composition, and moisture content. *Journal of Food Engineering*, *93*(1), 114–124.
Hemery, Y., Rouau, X., Lullien-Pellerin, V., Barron, C., & Abecassis, J. (2007). Dry processes to develop wheat fractions and products with enhanced nutritional quality. *Journal of Cereal Science*, *46*(3), 327–347.

Hemery, Y. M., Anson, N. M., Havenaar, R., Haenen, G. R., Noort, M. W., & Rouau, X. (2010). Dry-fractionation of wheat bran increases the bioaccessibility of phenolic acids in breads made from processed bran fractions. *Food Research International, 43*(5), 1429–1438.

Hipsley, E. H. (1953). Dietary "fibre" and pregnancy toxaemia. *British Medical Journal, 2*(4833), 420–422.

Hlebowicz, J. (2009). Postprandial blood glucose response in relation to gastric emptying and satiety in healthy subjects. *Appetite, 53*(2), 249–252.

Ho, L.-H., Tan, T.-C., Abdul Aziz, N. A., & Bhat, R. (2015). In vitro starch digestibility of bread with banana (Musa acuminata X balbisiana ABB cv. Awak) pseudo-stem flour and hydrocolloids. *Food Bioscience, 12*, 10–17.

Hole, A. S., Rud, I., Grimmer, S., Sigl, S., Narvhus, J., & Sahlstrøm, S. (2012). Improved bioavailability of dietary phenolic acids in whole grain barley and oat groat following fermentation with probiotic Lactobacillus acidophilus, Lactobacillus johnsonii, and Lactobacillus reuteri. *Journal of Agricultural and Food Chemistry, 60*(25), 6369–6375.

Houghton, D., Wilcox, M. D., Brownlee, I. A., Chater, P., Seal, C. J., & Pearson, J. P. (2014). Method for quantifying alginate and determining release from a food vehicle in gastrointestinal digesta. *Food Chemistry, 151*, 352–357.

Huang, S. C., Shiau, C. Y., Liu, T. E., Chu, C. L., & Hwang, D. F. (2005). Effects of rice bran on sensory and physico-chemical properties of emulsified pork meatballs. *Meat Science, 70*(4), 613–619.

Inglett, G. E., & Chen, D. (2011). Antioxidant activity and phenolic content of air-classified corn bran. *Cereal Chemistry, 88*(1), 36–40.

Izydorczyk, M. S., & Dexter, J. E. (2008). Barley β-glucans and arabinoxylans: Molecular structure, physicochemical properties, and uses in food products–A review. *Food Research International, 41*(9), 850–868.

Izydorczyk, M. S., Hussain, A., & MacGregor, A. W. (2001). Effect of barley and barley components on rheological properties of wheat dough. *Journal of Cereal Science, 34*(3), 251–260.

Jenkins, D. J., Kendall, C. W., Vuksan, V., Augustin, L. S., Li, Y.-M., Lee, B., et al. (1999). The effect of wheat bran particle size on laxation and colonic fermentation. *Journal of the American College of Nutrition, 18*(4), 339–345.

Jenkins, D. J. A., Kendall, C. W. C., & Ransom, T. P. P. (1998). Dietary fibre, the evolution of the human diet and coronary heart disease. *Nutrition Research, 18*(4), 633–652.

Jiménez-Escrig, A., & Sánchez-Muniz, F. J. (2000). Dietary fibre from edible seaweeds: Chemical structure, physicochemical properties and effects on cholesterol metabolism. *Nutrition Research, 20*(4), 585–598.

Jin, X., Lin, S., Gao, J., Wang, Y., Ying, J., Dong, Z., et al. (2021). Effect of coarse and superfine-ground wheat brans on the microstructure and quality attributes of dried white noodle. *Food and Bioprocess Technology*.

Juarez-Garcia, E., Agama-Acevedo, E., Sáyago-Ayerdi, S. G., Rodríguez-Ambriz, S. L., & Bello-Pérez, L. A. (2006). Composition, digestibility and application in breadmaking of banana flour. *Plant Foods for Human Nutrition, 61*(3), 131.

Sohail, M., Rakha, A., Butt, M. S., Iqbal, M. J., & Rashid, S. (2017). Rice bran nutraceutics: A comprehensive review. *Critical Reviews in Food Science and Nutrition, 57*(17), 3771–3780.

Kahlon, T. S., de Berrios, J., Smith, G. E., & Pan, J. L. (2006). Extrusion conditions modify hypocholesterolemic properties of wheat bran fed to hamsters. *Cereal Chemistry, 83*(2), 152–156.

Kahraman, K., Aktas-Akyildiz, E., Ozturk, S., & Koksel, H. (2019). Effect of different resistant starch sources and wheat bran on dietary fibre content and in vitro glycaemic index values of cookies. *Journal of Cereal Science, 90*, 102851.

Kalinga, D., & Mishra, V. K. (2009). Rheological and physical properties of low fat cakes produced by addition of cereal β-glucan concentrates. *Journal of Food Processing and Preservation*, *33*(3), 384–400.

Katina, K. (2003). High-fibre baking. In *Bread making: Improving quality* (pp. 487–499). Cambridge: Woodhead Publishing Limited.

Katina, K., Juvonen, R., Laitila, A., Flander, L., Nordlund, E., Kariluoto, S., et al. (2012). Fermented wheat bran as a functional ingredient in baking. *Cereal Chemistry*, *89*(2), 126–134.

Katina, K., Salmenkallio-Marttila, M., Partanen, R., Forssell, P., & Autio, K. (2006). Effects of sourdough and enzymes on staling of high-fibre wheat bread. *LWT - Food Science and Technology*, *39*(5), 479–491.

Keenan, J. M., Goulson, M., Shamliyan, T., Knutson, N., Kolberg, L., & Curry, L. (2007). The effects of concentrated barley β-glucan on blood lipids in a population of hypercholesterolaemic men and women. *British Journal of Nutrition*, *97*(6), 1162–1168.

Kerch, G., Rustichelli, F., Ausili, P., Zicans, J., Meri, R. M., & Glonin, A. (2008). Effect of chitosan on physical and chemical processes during bread baking and staling. *European Food Research and Technology*, *226*(6), 1459–1464.

Khan, M. A., Nadeem, M., Rakha, A., Shakoor, S., Shehzad, A., & Khan, M. R. (2016). Structural characterization of oat bran (1 → 3), (1 → 4)-β-D-glucans by lichenase hydrolysis through high-performance anion exchange chromatography with pulsed amperometric detection. *International Journal of Food Properties*, *19*(4), 929–935.

Kian, L. K., Jawaid, M., Ariffin, H., & Karim, Z. (2018). Isolation and characterization of nanocrystalline cellulose from roselle-derived microcrystalline cellulose. *International Journal of Biological Macromolecules*, *114*, 54–63.

Kim, H.-W., Miller, D. K., Lee, Y. J., & Kim, Y. H. B. (2016). Effects of soy hull pectin and insoluble fibre on physicochemical and oxidative characteristics of fresh and frozen/thawed beef patties. *Meat Science*, *117*, 63–67.

Kiumarsi, M., Shahbazi, M., Yeganehzad, S., Majchrzak, D., Lieleg, O., & Winkeljann, B. (2019). Relation between structural, mechanical and sensory properties of gluten-free bread as affected by modified dietary fibres. *Food Chemistry*, *277*, 664–673.

Korus, J., Witczak, M., Ziobro, R., & Juszczak, L. (2009). The impact of resistant starch on characteristics of gluten-free dough and bread. *Food Hydrocolloids*, *23*(3), 988–995.

Kouris-Blazos, A., & Belski, R. (2016). Health benefits of legumes and pulses with a focus on Australian sweet lupins. *Asia Pacific Journal of Clinical Nutrition*, *25*(1), 1–17.

Krishnan, M., & Prabhasankar, P. (2010). Studies on pasting, microstructure, sensory, and nutritional profile of pasta influenced by sprouted finger millet (eleucina coracana) and green banana (musa paradisiaca) flours. *Journal of Texture Studies*, *41*(6), 825–841.

Kristensen, M., & Bugel, S. (2011). A diet rich in oat bran improves blood lipids and hemostatic factors, and reduces apparent energy digestibility in young healthy volunteers. *European Journal of Clinical Nutrition*, *65*(9), 1053–1058.

Ktenioudaki, A., Butler, F., & Gallagher, E. (2011). Dough characteristics of Irish wheat varieties I. Rheological properties and prediction of baking volume. *LWT - Food Science and Technology*, *44*(3), 594–601.

Ktenioudaki, A., & Gallagher, E. (2012). Recent advances in the development of high-fibre baked products. *Trends in Food Science & Technology*, *28*(1), 4–14.

Kumar, M. N. V. R., Muzzarelli, R. A. A., Muzzarelli, C., Sashiwa, H., & Domb, A. J. (2004). Chitosan chemistry and pharmaceutical perspectives. *Chemical Reviews*, *104*(12), 6017–6084.

Laguna, L., Salvador, A., Sanz, T., & Fiszman, S. M. (2011). Performance of a resistant starch rich ingredient in the baking and eating quality of short-dough biscuits. *LWT - Food Science and Technology*, *44*(3), 737–746.

Lahouar, L., Ghrairi, F., Arem, A. E., Sghaeir, W., Felah, M. E., Salem, H. B., et al. (2014). Attenuation of histopathological alterations of colon, liver and lung by dietary fibre of barley Rihane in azoxymethane-treated rats. *Food Chemistry*, *149*, 271–276.

Lai, V. M. F., Lu, S., He, W. H., & Chen, H. H. (2007). Non-starch polysaccharide compositions of rice grains with respect to rice variety and degree of milling. *Food Chemistry*, *101*(3), 1205–1210.

Lappi, J., Mykkänen, H., Knudsen, K. E. B., Kirjavainen, P., Katina, K., Pihlajamäki, J., et al. (2014). Postprandial glucose metabolism and SCFA after consuming wholegrain rye bread and wheat bread enriched with bioprocessed rye bran in individuals with mild gastrointestinal symptoms. *Nutrition Journal*, *13*(1), 1–9.

Larrea, M. A., Chang, Y. K., & Martinez-Bustos, F. (2005). Some functional properties of extruded orange pulp and its effect on the quality of cookies. *LWT - Food Science and Technology*, *38*(3), 213–220.

Laurikainen, T., Härkönen, H., Autio, K., & Poutanen, K. (1998). Effects of enzymes in fibre-enriched baking. *Journal of the Science of Food and Agriculture*, *76*(2), 239–249.

Le Bleis, F., Chaunier, L., Chiron, H., Della Valle, G., & Saulnier, L. (2015). Rheological properties of wheat flour dough and French bread enriched with wheat bran. *Journal of Cereal Science*, *65*, 167–174.

Le Bleis, F., Chaunier, L., Montigaud, P., & Della Valle, G. (2016). Destructuration mechanisms of bread enriched with fibers during mastication. *Food Research International*, *80*, 1–11.

Le Gall, M., Serena, A., Jorgensen, H., Theil, P. K., & Bach Knudsen, K. E. (2009). The role of whole-wheat grain and wheat and rye ingredients on the digestion and fermentation processes in the gut—a model experiment with pigs. *British Journal of Nutrition*, *102*(11), 1590–1600.

Leathers, T. D. (2003). Biotechnological production and applications of pullulan. *Applied Microbiology and Biotechnology*, *62*(5), 468–473.

Lee, S., Warner, K., & Inglett, G. E. (2005). Rheological properties and baking performance of new oat β-glucan-rich hydrocolloids. *Journal of Agricultural and Food Chemistry*, *53*(25), 9805–9809.

Lee, Y.-K., Kim, M.-J., Lee, S.-B., & Kim, S.-D. (2005). Quality characteristics of kipfel cookie prepared with chitosan-chungkukjang. *Journal of the East Asian Society of Dietary Life*, *15*(4), 437–443.

Li, W., Pickard, M. D., & Beta, T. (2007). Effect of thermal processing on antioxidant properties of purple wheat bran. *Food Chemistry*, *104*(3), 1080–1086.

Li, X., Li, M., Pu, Y., Ragauskas, A. J., Klett, A. S., Thies, M., et al. (2018). Inhibitory effects of lignin on enzymatic hydrolysis: The role of lignin chemistry and molecular weight. *Renewable Energy*, *123*, 664–674.

Lin, S., Gao, J., Jin, X., Wang, Y., Dong, Z., Ying, J., et al. (2020). Whole-wheat flour particle size influences dough properties, bread structure and in vitro starch digestibility. *Food & Function*, *11*(4), 3610–3620.

Lin, S., Jin, X., Gao, J., Qiu, Z., Ying, J., Wang, Y., et al. (2021). Impact of wheat bran micronization on dough properties and bread quality: Part I–Bran functionality and dough properties. *Food Chemistry*, *353*, 129407.

Lin, S., Jin, X., Gao, J., Kim, E. H. J., Morgenstern, M. P., Dong, Z., et al. (2021). Bread breakdown pathways during mastication: impact of wheat bran fortification. *Food & Function*. https://doi.org/10.1039/D1FO02057A.

Lin, Y., Huybrechts, I., Vereecken, C., Mouratidou, T., Valtueña, J., Kersting, M., et al. (2015). Dietary fibre intake and its association with indicators of adiposity and serum biomarkers in European adolescents: the HELENA study. *European Journal of Nutrition*, *54*(5), 771–782.

Liu, T., Hou, G. G., Lee, B., Marquart, L., & Dubat, A. (2016). Effects of particle size on the quality attributes of reconstituted whole-wheat flour and tortillas made from it. *Journal of Cereal Science*, *71*, 145–152.

Ma, F., & Baik, B.-K. (2017). Qualitative effect of added gluten on dough properties and quality of chinese steamed bread. *Cereal Chemistry*, *94*(5), 827–833.

Madhujith, T., Izydorczyk, M., & Shahidi, F. (2006). Antioxidant properties of pearled barley fractions. *Journal of Agricultural and Food Chemistry*, *54*(9), 3283–3289.

Maes, C., & Delcour, J. A. (2002). Structural characterisation of water-extractable and water-unextractable arabinoxylans in wheat bran. *Journal of Cereal Science*, *35*(3), 315–326.

Madhumitha, S., & Prabhasankar, P. (2011). Influence of additives on functional and nutritional quality characteristics of black gram flour incorporated pasta. *Journal of Texture Studies*, *42*(6), 441–450.

Maezaki, Y., Tsuji, K., Nakagawa, Y., Kawai, Y., Akimoto, M., Tsugita, T., et al. (1993). Hypocholesterolemic effect ofchitosan in adult males. *Bioscience Biotechnology and Biochemistry*, *57*(9), 1439–1444.

Majzoobi, M., Farahnaky, A., Nematolahi, Z., Mohammadi, H. M., & Taghipour, A. M. J. (2013). Effect of different levels and particle sizes of wheat bran on the quality of flat bread. *Journal of Agricultural Science and Technology*, *15*, 115–123.

Makeri, M., Nkama, I., & Badau, M. (2013). Physico-chemical, malting and biochemical properties of some improved Nigerian barley cultivars and their malts. *International Food Research Journal*, *20*(4).

Maki, K. C., Beiseigel, J. M., Jonnalagadda, S. S., Gugger, C. K., Reeves, M. S., Farmer, M. V., et al. (2010). Whole-grain ready-to-eat oat cereal, as part of a dietary program for weight loss, reduces low-density lipoprotein cholesterol in adults with overweight and obesity more than a dietary program including low-fibre control foods. *Journal of the American Dietetic Association*, *110*(2), 205–214.

Malunga, L. N., Izydorczyk, M., & Beta, T. (2017). Antiglycemic effect of water extractable arabinoxylan from wheat aleurone and bran. *Journal of Nutrition and Metabolism*, *2017*, 5784759.

Maphosa, Y., & Jideani, V. A. (2015). Dietary fibre extraction for human nutrition—A review. *Food Reviews International*, *32*(1), 98–115.

Marangoni, F., & Poli, A. (2008). The glycemic index of bread and biscuits is markedly reduced by the addition of a proprietary fibre mixture to the ingredients. *Nutrition, Metabolism, and Cardiovascular Diseases*, *18*(9), 602–605.

Marciani, L., Pritchard, S. E., Hellier-Woods, C., Costigan, C., Hoad, C. L., Gowland, P. A., et al. (2013). Delayed gastric emptying and reduced postprandial small bowel water content of equicaloric whole meal bread versus rice meals in healthy subjects: novel MRI insights. *European Journal of Clinical Nutrition*, *67*(7), 754–758.

Martínez-Cervera, S., Salvador, A., Muguerza, B., Moulay, L., & Fiszman, S. M. (2011). Cocoa fibre and its application as a fat replacer in chocolate muffins. *LWT - Food Science and Technology*, *44*(3), 729–736.

McKee, L. H., & Latner, T. A. (2000). Underutilized sources of dietary fiber: A review. *Plant Foods for Human Nutrition*, *55*(4), 285–304.

Melini, F., Melini, V., Luziatelli, F., & Ruzzi, M. (2017). Current and forward-looking approaches to technological and nutritional improvements of gluten-free bread with legume flours: A critical review. *Comprehensive Reviews in Food Science and Food Safety*, *16*(5), 1101–1122.

Menga, V., Fares, C., Troccoli, A., Cattivelli, L., & Baiano, A. (2010). Effects of genotype, location and baking on the phenolic content and some antioxidant properties of cereal species. *International Journal of Food Science & Technology*, *45*(1), 7–16.

Messia, M. C., Reale, A., Maiuro, L., Candigliota, T., Sorrentino, E., & Marconi, E. (2016). Effects of pre-fermented wheat bran on dough and bread characteristics. *Journal of Cereal Science, 69*, 138–144.

Miller-Jones, J. (2004). Dietary fibre intake, disease prevention, and health promotion: an overview with emphasis on evidence from epidemiology. *Bioactive Carbohydrates for Food Feed*, 143–164.

Miñarro, B., Albanell, E., Aguilar, N., Guamis, B., & Capellas, M. (2012). Effect of legume flours on baking characteristics of gluten-free bread. *Journal of Cereal Science, 56*(2), 476–481.

Mir, N. A., Riar, C. S., & Singh, S. (2018). Nutritional constituents of pseudo cereals and their potential use in food systems: A review. *Trends in Food Science & Technology, 75*, 170–180.

Miyazaki, M., Van Hung, P., Maeda, T., & Morita, N. (2006). Recent advances in application of modified starches for breadmaking. *Trends in Food Fcience & Technology, 17*(11), 591–599.

Mohammed, I., Ahmed, A. R., & Senge, B. (2012). Dough rheology and bread quality of wheat–chickpea flour blends. *Industrial Crops and Products, 36*(1), 196–202.

Moure, A., Gullón, P., Domínguez, H., & Parajó, J. C. (2006). Advances in the manufacture, purification and applications of xylo-oligosaccharides as food additives and nutraceuticals. *Process Biochemistry, 41*(9), 1913–1923.

Mudgil, D. (2017). Chapter 3 - The interaction between insoluble and soluble fibre. In R. A. Samaan (Ed.), *Dietary Fibre for the prevention of cardiovascular disease* (pp. 35–59). Academic Press.

Mudgil, D., & Barak, S. (2013). Composition, properties and health benefits of indigestible carbohydrate polymers as dietary fibre: a review. *International Journal of Biological Macromolecules, 61*, 1–6.

Mudgil, D., & Barak, S. (2019). Chapter 2 - Classification, technological properties, and sustainable sources. In C. M. Galanakis (Ed.), *Dietary Fibre: Properties, Recovery, and Applications* (pp. 27–58). Academic Press.

Mudgil, D., Barak, S., Patel, A., & Shah, N. (2018). Partially hydrolyzed guar gum as a potential prebiotic source. *International Journal of Biological Macromolecules, 112*, 207–210.

Nandeesh, K., Jyotsna, R., & Venkateswara Rao, G. (2011). Effect of differently treated wheat bran on rheology, microstructure and quality characteristics of soft dough biscuits. *Journal of Food Processing and Preservation, 35*(2), 179–200.

Nelson, A. (2001). Properties of high-fibre ingredients. *Cereal Foods World, 46*(3), 93–97.

Newman, R. K., & Newman, C. W. (2008). *Barley for food and health: Science, technology, and products*. John Wiley & Sons.

Nieman, K. M., Rains, T. M., Dicklin, M. R., & Maki, K. C. (2014). Chapter 5 - Wheat fibre in postprandial metabolic profile and health. In R. R. Watson, V. R. Preedy, & S. Zibadi (Eds.), *Wheat and Rice in Disease Prevention and Health* (pp. 59–66). San Diego: Academic Press.

Nilsson, A. C., OÖstman, E. M., Granfeldt, Y., & Björck, I. M. E. (2008). Effect of cereal test breakfasts differing in glycemic index and content of indigestible carbohydrates on daylong glucose tolerance in healthy subjects. *The American Journal of Clinical Nutrition, 87*(3), 645–654.

Niu, M., Hou, G. G., Wang, L., & Chen, Z. (2014). Effects of superfine grinding on the quality characteristics of whole-wheat flour and its raw noodle product. *Journal of Cereal Science, 60*(2), 382–388.

Noort, M. W. J., Haaster, D. V., Hemery, Y., Schols, H. A., & Hamer, R. J. (2010). The effect of particle size of wheat bran fractions on bread quality – Evidence for fibre–protein interactions. *Journal of Cereal Science, 52*(1), 59–64.

Nordlund, E., Katina, K., Aura, A.-M., & Poutanen, K. (2013). Changes in bran structure by bioprocessing with enzymes and yeast modifies the in vitro digestibility and fermentability of bran protein and dietary fibre complex. *Journal of Cereal Science*, *58*(1), 200–208.

Novotni, D., Ćurić, D., Bituh, M., Colić Barić, I., Škevin, D., & Čukelj, N. (2011). Glycemic index and phenolics of partially-baked frozen bread with sourdough. *International Journal of Food Sciences and Nutrition*, *62*(1), 26–33.

Oh, I. K., Bae, I. Y., & Lee, H. G. (2014). In vitro starch digestion and cake quality: impact of the ratio of soluble and insoluble dietary fibre. *International Journal of Biological Macromolecules*, *63*, 98–103.

Okada, H., Fukushi, E., Yamamori, A., Kawazoe, N., Onodera, S., Kawabata, J., et al. (2010). Novel fructopyranose oligosaccharides isolated from fermented beverage of plant extract. *Carbohydrate Research*, *345*(3), 414–418.

Onipe, O. O., Jideani, A. I. O., & Beswa, D. (2015). Composition and functionality of wheat bran and its application in some cereal food products. *International Journal of Food Science & Technology*, *50*(12), 2509–2518.

Ortiz de Erive, M., Wang, T., He, F., & Chen, G. (2020). Development of high-fibre wheat bread using microfluidized corn bran. *Food Chemistry*, *310*, 125921.

Ou, J., & Sun, Z. (2014). Feruloylated oligosaccharides: Structure, metabolism and function. *Journal of Functional Foods*, 7, 90–100.

Özkaya, H., Özkaya, B., Duman, B., & Turksoy, S. (2017). Effect of dephytinization by fermentation and hydrothermal autoclaving treatments on the antioxidant activity, dietary fiber, and phenolic content of oat bran. *Journal of Agricultural and Food Chemistry*, *65*(28), 5713–5719.

Park, E. Y., Fuerst, E. P., & Baik, B. K. (2017). Effects of bran prehydration on functional characteristics and bread-baking quality of bran and flour blends. *Cereal Chemistry*, *94*(5).

Peng, B., Li, Y., Ding, S., & Yang, J. (2017). Characterization of textural, rheological, thermal, microstructural, and water mobility in wheat flour dough and bread affected by trehalose. *Food Chemistry*, *233*, 369–377.

Peressini, D., & Sensidoni, A. (2009). Effect of soluble dietary fibre addition on rheological and breadmaking properties of wheat doughs. *Journal of Cereal Science*, *49*(2), 190–201.

Peters, U., Sinha, R., Chatterjee, N., Subar, A. F., Ziegler, R. G., Kulldorff, M., et al. (2003). Dietary fibre and colorectal adenoma in a colorectal cancer early detection programme. *The Lancet*, *361*(9368), 1491–1495.

Phimolsiripol, Y., Siripatrawan, U., Teekachunhatean, S., Wangtueai, S., Seesuriyachan, P., Surawang, S., et al. (2017). Technological properties, in vitro starch digestibility and in vivo glycaemic index of bread containing crude malva nut gum. *International Journal of Food Science & Technology*, *52*(4), 1035–1041.

Polaki, A., Xasapis, P., Fasseas, C., Yanniotis, S., & Mandala, I. (2010). Fibre and hydrocolloid content affect the microstructural and sensory characteristics of fresh and frozen stored bread. *Journal of Food Engineering*, *97*(1), 1–7.

Ponzio, N. R., Ferrero, C., & Puppo, M. C. (2013). Wheat varietal flours: influence of pectin and DATEM on dough and bread quality. *International Journal of Food Properties*, *16*(1), 33–44.

Poutanen, K., Flander, L., & Katina, K. (2009). Sourdough and cereal fermentation in a nutritional perspective. *Food Microbiology*, *26*(7), 693–699.

Poutanen, K. S., Fiszman, S., Marsaux, C. F. M., Pentikäinen, S. P., Steinert, R. E., & Mela, D. J. (2018). Recommendations for characterization and reporting of dietary fibres in nutrition research. *The American Journal of Clinical Nutrition*, *108*(3), 437–444.

Prückler, M., Siebenhandl-Ehn, S., Apprich, S., Höltinger, S., Haas, C., Schmid, E., et al. (2014). Wheat bran-based biorefinery 1: Composition of wheat bran and strategies of functionalization. *LWT - Food Science and Technology*, *56*(2), 211–221.

Qiang, X., YongLie, C., & QianBing, W. (2009). Health benefit application of functional oligosaccharides. *Carbohydrate Polymers*, 77(3), 435–441.

Ragaee, S., Abdelaal, E., & Noaman, M. (2006). Antioxidant activity and nutrient composition of selected cereals for food use. *Food Chemistry*, *98*(1), 32–38.

Rakha, A., Åman, P., & Andersson, R. (2010). Characterisation of dietary fibre components in rye products. *Food Chemistry*, *119*(3), 859–867.

Ramasamy, U. S., Venema, K., Gruppen, H., & Schols, H. A. (2014). The fate of chicory root pulp polysaccharides during fermentation in the TNO in vitro model of the colon (TIM-2). *Bioactive Carbohydrates and Dietary Fibre*, *4*(1), 48–57.

Rao, R. S. P., & Muralikrishna, G. (2004). Non-starch polysaccharide–phenolic acid complexes from native and germinated cereals and millet. *Food Chemistry*, *84*(4), 527–531.

Rao, R. S. P., & Muralikrishna, G. (2006). Water soluble feruloyl arabinoxylans from rice and ragi: Changes upon malting and their consequence on antioxidant activity. *Phytochemistry*, *67*(1), 91–99.

Rave, K., Roggen, K., Dellweg, S., Heise, T., & tom Dieck, H. (2007). Improvement of insulin resistance after diet with a whole-grain based dietary product: results of a randomized, controlled cross-over study in obese subjects with elevated fasting blood glucose. *British Journal of Nutrition*, *98*(5), 929–936.

Rehinan, Z.-U., Rashid, M., & Shah, W. H. (2004). Insoluble dietary fibre components of food legumes as affected by soaking and cooking processes. *Food Chemistry*, *85*(2), 245–249.

Reshmi, S. K., Sudha, M. L., & Shashirekha, M. N. (2017). Starch digestibility and predicted glycemic index in the bread fortified with pomelo (Citrus maxima) fruit segments. *Food Chemistry*, *237*, 957–965.

Rieder, A., Holtekjølen, A. K., Sahlstrøm, S., & Moldestad, A. (2012). Effect of barley and oat flour types and sourdoughs on dough rheology and bread quality of composite wheat bread. *Journal of Cereal Science*, *55*(1), 44–52.

Rizzello, C. G., Cassone, A., Coda, R., & Gobbetti, M. (2011). Antifungal activity of sourdough fermented wheat germ used as an ingredient for bread making. *Food Chemistry*, *127*(3), 952–959.

Rizzello, C. G., Cassone, A., Di Cagno, R., & Gobbetti, M. (2008). Synthesis of angiotensin I-converting enzyme (ACE)-inhibitory peptides and γ-aminobutyric acid (GABA) during sourdough fermentation by selected lactic acid bacteria. *Journal of Agricultural and Food Chemistry*, *56*(16), 6936–6943.

Rocha Parra, A. F., Ribotta, P. D., & Ferrero, C. (2015). Apple pomace in gluten-free formulations: effect on rheology and product quality. *International Journal of Food Science & Technology*, *50*(3), 682–690.

Rodríguez, R., Jiménez, A., Fernández-Bolaños, J., Guillén, R., & Heredia, A. (2006). Dietary fibre from vegetable products as source of functional ingredients. *Trends in Food Science & Technology*, *17*(1), 3–15.

Rose, D. J., Inglett, G. E., & Liu, S. X. (2010). Utilisation of corn (Zea mays) bran and corn fiber in the production of food components. *Journal of the Science of Food and Agriculture*, *90*(6), 915–924.

Rosell, C., Rojas, J., & De Barber, C. B. (2001). Influence of hydrocolloids on dough rheology and bread quality. *Food Hydrocolloids*, *15*(1), 75–81.

Rosell, C. M., Santos, E., & Collar, C. (2006). Mixing properties of fibre-enriched wheat bread doughs: A response surface methodology study. *European Food Research and Technology*, *223*(3), 333–340.

Rupasinghe, H. V., Wang, L., Pitts, N. L., & Astatkie, T. (2009). Baking and sensory characteristics of muffins incorporated with apple skin powder. *Journal of Food Quality*, *32*(6), 685–694.

Saccotelli, M. A., Spinelli, S., Conte, A., & Nobile, M. A. D. (2018). Gluten-free bread enriched with vegetable flours. *Food and Nutrition Sciences, 09No.04*, 13.

Saha, D., & Bhattacharya, S. (2010). Hydrocolloids as thickening and gelling agents in food: a critical review. *Journal of Food Science and Technology, 47*(6), 587–597.

Sajilata, M. G., Singhal, R. S., & Kulkarni, P. R. (2006). Resistant starch–A review. *Comprehensive Reviews in Food Science and Food Safety, 5*(1), 1–17.

Salehi, F. (2020). Effect of common and new gums on the quality, physical, and textural properties of bakery products: A review. *Journal of Texture Studies, 51*(2), 361–370.

Salmenkallio-Marttila, M., Katina, K., & Autio, K. (2001). Effects of bran fermentation on quality and microstructure of high-fibre wheat bread. *Cereal Chemistry, 78*(4), 429–435.

Salmenkallio-Marttila, M., Roininen, K., & Autio, K. (2004). Effects of gluten and transglutaminase on microstructure, sensory characteristics and instrumental texture of oat bread. *Agricultural and Food Science, 13*(1-2), 138–150.

Sanz Penella, J. M., Collar, C., & Haros, M. (2008). Effect of wheat bran and enzyme addition on dough functional performance and phytic acid levels in bread. *Journal of Cereal Science, 48*(3), 715–721.

Sanz, T., Salvador, A., & Fiszman, S. M. (2008). Resistant starch (RS) in battered fried products: Functionality and high-fibre benefit. *Food Hydrocolloids, 22*(4), 543–549.

Sarawong, C., Gutiérrez, Z. R., Berghofer, E., & Schoenlechner, R. (2014). Effect of green plantain flour addition to gluten-free bread on functional bread properties and resistant starch content. *International Journal of Food Science & Technology, 49*(8), 1825–1833.

Sasaki, T. (2018). Effects of xanthan and guar gums on starch digestibility and texture of rice flour blend bread. *Cereal Chemistry, 95*(1), 177–184.

Satija, A., & Hu, F. B. (2012). Cardiovascular benefits of dietary fibre. *Current Atherosclerosis Reports, 14*(6), 505–514.

Scazzina, F., Siebenhandl-Ehn, S., & Pellegrini, N. (2013). The effect of dietary fibre on reducing the glycaemic index of bread. *British Journal of Nutrition, 109*(7), 1163–1174.

Sera, N., Morita, K., Nagasoe, M., Tokieda, H., Kitaura, T., & Tokiwa, H. (2005). Binding effect of polychlorinated compounds and environmental carcinogens on rice bran fibre. *The Journal of Nutritional Biochemistry, 16*(1), 50–58.

Shah, A. R., Shah, R. K., & Madamwar, D. (2006). Improvement of the quality of whole wheat bread by supplementation of xylanasefrom Aspergillus foetidus. *Bioresource Technology, 97*, 2047–2053.

Shewry, P. R., Piironen, V., Lampi, A.-M., Nyström, L., Li, L., Rakszegi, M., et al. (2008). Phytochemical and fibre components in oat varieties in the HEALTHGRAIN diversity screen. *Journal of Agricultural and Food Chemistry, 56*(21), 9777–9784.

Shi, Z., Zhang, Y., Phillips, G. O., & Yang, G. (2014). Utilization of bacterial cellulose in food. *Food Hydrocolloids, 35*, 539–545.

Shoaib, M., Shehzad, A., Omar, M., Rakha, A., Raza, H., Sharif, H. R., et al. (2016). Inulin: Properties, health benefits and food applications. *Carbohydrate Polymers, 147*, 444–454.

Sivam, A. S., Sun-Waterhouse, D., Quek, S., & Perera, C. O. (2010). Properties of bread dough with added fibre polysaccharides and phenolic antioxidants: A review. *Journal of Food Science, 75*(8), R163–R174.

Slavin, J. (2013). Fibre and prebiotics: mechanisms and health benefits. *Nutrients, 5*(4), 1417–1435.

Slavin, J., Stewart, M., Timm, D., Grabitske, H., & Hospattankar, A. (2010). Fermentation patterns and short chain fatty acid profiles of wheat dextrin and other functional fibres. *Dietary Fibre: New Frontiers for Food and Health* (pp. 177–191). Wageningen Academic Publishers.

Smith, K. N., Queenan, K. M., Thomas, W., Fulcher, R. G., & Slavin, J. L. (2008). Physiological effects of concentrated barley β-glucan in mildly hypercholesterolemic adults. *Journal of the American College of Nutrition, 27*(3), 434–440.

Soler, M., Bosetti, C., Franceschi, S., Negri, E., Zambon, P., Talamini, R., et al. (2001). Fibre intake and the risk of oral, pharyngeal and esophageal cancer. *International Journal of Cancer, 91*(3), 283–287.

Stephen, A. M., Champ, M. M. J., Cloran, S. J., Fleith, M., van Lieshout, L., Mejborn, H., et al. (2017). Dietary fibre in Europe: current state of knowledge on definitions, sources, recommendations, intakes and relationships to health. *Nutrition Research Reviews, 30*(2), 149–190.

Stewart, M. L., & Slavin, J. L. (2009). Particle size and fraction of wheat bran influence short-chain fatty acid production in vitro. *British Journal of Nutrition, 102*(10), 1404–1407.

Stojceska, V., & Ainsworth, P. (2008). The effect of different enzymes on the quality of high-fibre enriched brewer's spent grain breads. *Food Chemistry, 110*(4), 865–872.

Stojceska, V., Butler, F., Gallagher, E., & Keehan, D. (2007). A comparison of the ability of several small and large deformation rheological measurements of wheat dough to predict baking behaviour. *Journal of Food Engineering, 83*(4), 475–482.

Sudha, M., Vetrimani, R., & Leelavathi, K. (2007). Influence of fibre from different cereals on the rheological characteristics of wheat flour dough and on biscuit quality. *Food Chemistry, 100*(4), 1365–1370.

Sudha, M. L., Baskaran, V., & Leelavathi, K. (2007). Apple pomace as a source of dietary fibre and polyphenols and its effect on the rheological characteristics and cake making. *Food Chemistry, 104*(2), 686–692.

Sui, W., Xie, X., Liu, R., Wu, T., & Zhang, M. (2018). Effect of wheat bran modification by steam explosion on structural characteristics and rheological properties of wheat flour dough. *Food Hydrocolloids, 84*, 571–580.

Sullivan, P., O'Flaherty, J., Brunton, N., Arendt, E., & Gallagher, E. (2010). Fundamental rheological and textural properties of doughs and breads produced from milled pearled barley flour. *European Food Research and Technology, 231*(3), 441–453.

Sworn, G., Kerdavid, E., & Fayos, J. (2008). The Role of Hydrocolloids in the Management of Dysphagia. In *14. Gums and Stabilisers for the Food Industry* (pp. 392–401). The Royal Society of Chemistry.

Terada, A., Hara, H., Sato, D., Higashi, T., Nakayama, S., Tsuji, K., et al. (1995). Effect of dietary chitosan on faecal microbiota and faecal metabolites of humans. *Microbial Ecology in Health and Disease, 8*(1), 15–21.

Terry, P., Giovannucci, E., Michels, K. B., Bergkvist, L., Hansen, H., Holmberg, L., et al. (2001). Fruit, vegetables, dietary fibre, and risk of colorectal cancer. *Journal of the National Cancer Institute, 93*(7), 525–533.

Touyarou, P., Sulmont-Rossé, C., Gagnaire, A., Issanchou, S., & Brondel, L. (2012). Monotonous consumption of fibre-enriched bread at breakfast increases satiety and influences subsequent food intake. *Appetite, 58*(2), 575–581.

Trowell, H. (1972). Crude fibre, dietary fibre and atherosclerosis. *Atherosclerosis, 16*(1), 138–140.

Trowell, H., Southgate, D. T., Wolever, T. S., Leeds, A., Gassull, M., & Jenkins, D. A. (1976). Dietary fibre redefined. *The Lancet, 307*(7966), 967.

Tudoricǎ, C. M., Kuri, V., & Brennan, C. S. (2002). Nutritional and physicochemical characteristics of dietary fibre enriched pasta. *Journal of Agricultural and Food Chemistry, 50*(2), 347–356.

Tuncil, Y. E., Thakkar, R. D., Marcia, A. D. R., Hamaker, B. R., & Lindemann, S. R. (2018). Divergent short-chain fatty acid production and succession of colonic microbiota arise in fermentation of variously-sized wheat bran fractions. *Scientific Reports, 8*(1), 16655.

Turfani, V., Narducci, V., Durazzo, A., Galli, V., & Carcea, M. (2017). Technological, nutritional and functional properties of wheat bread enriched with lentil or carob flours. *LWT - Food Science and Technology, 78*, 361–366.

Utrilla-Coello, R. G., Osorio-Díaz, P., & Bello-Pérez, L. A. (2007). Alternative use of chickpea flour in breadmaking: chemical composition and starch digestibility of bread. *Food Science and Technology International*, *13*(4), 323–327.

Uysal, H., Bilgicli, N., Elgun, A., Ibanoglu, S., Herken, E. N., & Demir, M. K. (2007). Effect of dietary fibre and xylanase enzymeaddition on the selected properties of wire-cut cookies. *Journal of Food Engineering*, *78*, 1074–1078.

Van Craeyveld, V., Holopainen, U., Selinheimo, E., Poutanen, K., Delcour, J. A., & Courtin, C. M. (2009). Extensive dry ball milling of wheat and rye bran leads to in situ production of arabinoxylan oligosaccharides through nanoscale fragmentation. *Journal of Agricultural and Food Chemistry*, *57*(18), 8467–8473.

Verónica, M., Montserrat, M., & Maria, P. V. (2012). Pharmacological applications of lignins and lignins related compounds: An overview. *Current Organic Chemistry*, *16*(16), 1863–1870.

Villarino, C. B., Jayasena, V., Coorey, R., Chakrabarti-Bell, S., & Johnson, S. K. (2016). Nutritional, health, and technological functionality of lupin flour addition to bread and other baked products: benefits and challenges. *Critical Reviews in Food Science and Nutrition*, *56*(5), 835–857.

Vitaglione, P., Napolitano, A., & Fogliano, V. (2008). Cereal dietary fibre: a natural functional ingredient to deliver phenolic compounds into the gut. *Trends in Food Science & Technology*, *19*(9), 451–463.

Vitali, D., Dragojević, I. V., & Šebečić, B. (2009). Effects of incorporation of integral raw materials and dietary fibre on the selected nutritional and functional properties of biscuits. *Food Chemistry*, *114*(4), 1462–1469.

Wang, H., Liu, C., & Wen, J. (2017). Comparison of unheated and heated bran on flour quality: effects of particle size and addition levels. *Journal of Food Processing and Preservation*, *41*(4), e12992.

Wang, L., Ye, F., Li, S., Wei, F., Chen, J., & Zhao, G. (2017). Wheat flour enriched with oat β -glucan: A study of hydration, rheological and fermentation properties of dough. *Journal of Cereal Science*, *75*, 143–150.

Wang, J., Rosell, C. M., & Barber, C. B. D. (2002). Effect of the addition of different fibres on wheat dough performance and bread quality. *Food Chemistry*, *79*(2), 221–226.

Wang, M., Hamer, R. J., van Vliet, T., & Oudgenoeg, G. (2002). Interaction of water extractable pentosans with gluten protein: effect on dough properties and gluten quality. *Journal of Cereal Science*, *36*(1), 25–37.

Wang, S., Li, C., Copeland, L., Niu, Q., & Wang, S. (2015). Starch retrogradation: A comprehensive review. *Comprehensive Reviews in Food Science and Food Safety*, *14*(5), 568–585.

Wang, T., He, F., & Chen, G. (2014). Improving bioaccessibility and bioavailability of phenolic compounds in cereal grains through processing technologies: A concise review. *Journal of Functional Foods*, 7, 101–111.

Wang, W. M., Klopfenstein, C. F., & Ponte, J. G. (1993). Effects of twin-screw extrusion on the physical properties of dietary fibre and other components of whole wheat and wheat bran and on the baking quality of the wheat bran. *Cereal Chemistry*, *70*(6), 707–711.

Wendin, K., Ekman, S., Bülow, M., Ekberg, O., Johansson, D., Rothenberg, E., et al. (2010). Objective and quantitative definitions of modified food textures based on sensory and rheological methodology. *Food & Nutrition Research*, *54*(1), 5134.

Westenbrink, S., Brunt, K., & van der Kamp, J. W. (2013). Dietary fibre: challenges in production and use of food composition data. *Food Chemistry*, *140*(3), 562–567.

Western, T. L., Skinner, D. J., & Haughn, G. W. (2000). Differentiation of mucilage secretory cells of the arabidopsis seed coat. *Plant Physiology*, *122*(2), 345–356.

Wijngaard, H. H., & Brunton, N. (2010). The optimisation of solid–liquid extraction of antioxidants from apple pomace by response surface methodology. *Journal of Food Engineering*, *96*(1), 134–140.

Wójtowicz, A., & Moscicki, L. (2011). Effect of wheat bran addition and screw speed on microstructure and textural characteristics of common wheat precooked pasta-like products. *Polish journal of food nutrition sciences*, *61*(2).
Wolfe, K. L., & Liu, R. H. (2003). Apple peels as a value-added food ingredient. *Journal of Agricultural and Food Chemistry*, *51*(6), 1676–1683.
Xia, W., Liu, P., Zhang, J., & Chen, J. (2011). Biological activities of chitosan and chitooligosaccharides. *Food Hydrocolloids*, *25*(2), 170–179.
Xu, X., Xu, Y., Wang, N., & Zhou, Y. (2018). Effects of superfine grinding of bran on the properties of dough and qualities of steamed bread. *Journal of Cereal Science*, *81*, 76–82.
Yang, Y., Yang, X., Zhai, F., & Cheng, Y. (2016). Dietary guidelines for Chinese (2016). *Journal of the Academy of Nutrition and Dietetics*, *116*(9), A37.
Yegin, S., Kopec, A., Kitts, D. D., & Zawistowski, J. (2020). Chapter 24 - Dietary fibre: a functional food ingredient with physiological benefits. In H. G. Preuss, & D. Bagchi (Eds.), *Dietary Sugar, Salt and Fat in Human Health* (pp. 531–555). Academic Press.
Yu, L., Nanguet, A. L., & Beta, T. (2013). Comparison of antioxidant properties of refined and whole wheat flour and bread. *Antioxidants (Basel)*, *2*(4), 370–383.
Zabidi, M. A., & Aziz, N. A. A. (2009). In vitro starch hydrolysis and estimated glycaemic index of bread substituted with different percentage of chempedak (Artocarpus integer) seed flour. *Food Chemistry*, *117*(1), 64–68.
Zhao, C., Wu, Y., Liu, X., Liu, B., Cao, H., Yu, H., et al. (2017). Functional properties, structural studies and chemo-enzymatic synthesis of oligosaccharides. *Trends in Food Science & Technology*, *66*, 135–145.
Zhang, Y., Gao, F., & He, Z. (2019). Effects of bran hydration and autoclaving on processing quality of Chinese steamed bread and noodles produced from whole grain wheat flour. *Cereal Chemistry*, *96*(1), 104–114.
Zheng, G. H., Rossnagel, B. G., Tyler, R. T., & Bhatty, R. S. (2000). Distribution of β-glucan in the grain of hull-less barley. *Cereal Chemistry*, *77*(2), 140–144.
Zheng, X., Zhang, R., & Liu, C. (2015). Extraction and antioxidant activity of phenolic compounds from wheat bran treated by steam explosion. *Tropical Journal of Pharmaceutical Research*, *14*(10), 1857–1863.
Zhou, Y., Dhital, S., Zhao, C., Ye, F., Chen, J., & Zhao, G. (2021). Dietary fiber-gluten protein interaction in wheat flour dough: Analysis, consequences and proposed mechanisms. *Food Hydrocolloids*, *111*, 106203.

CHAPTER THREE

The realm of plant proteins with focus on their application in developing new bakery products

Fatma Boukid*
Food Safety and Functionality Programme, Food Industry Area, Institute of Agriculture and Food Research and Technology (IRTA), Monells, Catalonia, Spain
*Corresponding author: e-mail address: fatma.boukid@irta.cat

Contents

Abstract

Plant proteins are spreading due to growing environmental, health and ethical concerns related to animal proteins. Proteins deriving from cereals, oilseeds, and pulses are witnessing a sharp growth showing a wide spectrum of applications from meat and fish analogues to infant formulations. Bakery products are one of the biggest markets of alternative protein applications for functional and nutritional motives. Fortifying bakery products with proteins can secure a better amino-acids profile and a higher protein intake. Conventional plant proteins (i.e., wheat and soy) dominate the bakery industry, but emerging sources (i.e., pea, chickpea, and faba) are also gaining traction. Each protein brings specific functional properties and nutritional value. Therefore, this chapter gives an overview of the main features of plant proteins and discusses their impact on the quality of bakery products.

Advances in Food and Nutrition Research, Volume 99
ISSN 1043-4526
https://doi.org/10.1016/bs.afnr.2021.11.001

1. Introduction

Due to growing consciousness over the relatedness between human health and nutrition, consumers are increasingly shifting towards foods with health-beneficial effects. Particular interests in protein-rich diets are rising due to their health-promoting effects, with particular attention to the source (Boukid, Rosell, Rosene, Bover-Cid, & Castellari, 2021). Plant-based diets have shown several health benefits, such as reducing body mass index, blood pressure, cholesterol levels, and some cancers (Gravely & Fraser, 2018; Tuso, Ismail, Ha, & Bartolotto, 2013). An estimated 2 billion people in the world do not have regular access to safe, nutritious, and sufficient food and one billion people in the world do not have access to a diet providing enough protein and energy. Furthermore, the global protein demand is expected to grow rapidly in the coming years due to an increasing world population expected to reach 10 billion people by 2050.

Food proteins can derive from the animal, plant, and microbial sources (Fig. 1). Consumers' appetite towards plant proteins is increasing for a multitude of motives including consciousness towards health, environment, and animal welfare. Environmentally, plant-based proteins are considered more sustainable than those animals due to their lower carbon footprint and land use (Harwatt, Sabaté, Eshel, Soret, & Ripple, 2017; Heusala, Sinkko, Mogensen, & Knudsen, 2020). Furthermore, plant proteins can be recovered as co- or by-products and then get back into the food chain as functional ingredients, thereby improving the sustainability of the process and contributing to circular economy (Boukid, 2021b; Brückner-Gühmann, Benthin, & Drusch, 2019; Garrido et al., 2014). Animal proteins can be

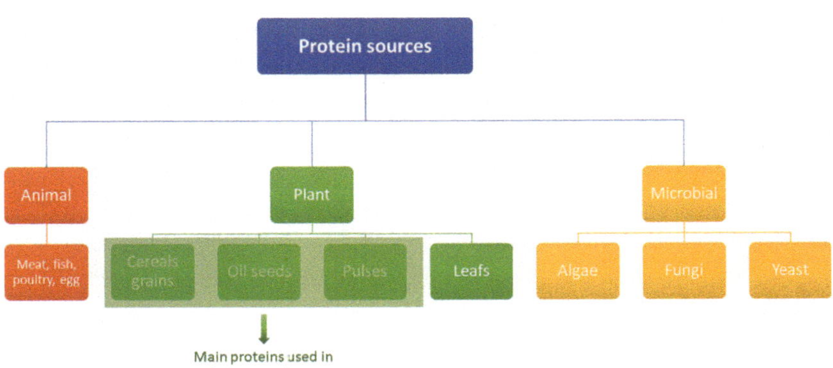

Fig. 1 Protein sources.

avoided for dietary restrictions such as lactose allergy or for the association of these proteins to health issues such as cardiovascular diseases. Although animal proteins have better amino acids profiles, blending several sources of plant proteins (e.g., cereals and pulses) can result in a balanced nutritional quality and high protein content (Boukid, Rosell, Rosene, et al., 2021; Grabowska, Tekidou, Boom, & van der Goot, 2014; Sá, Moreno, & Carciofi, 2020). Ethical choices (vegan, vegetarian, and flexitarian) are increasingly contributing to reshaping the protein market by favoring plant proteins over those deriving animals.

Plant proteins are applied across several food and beverages categories. Bakery products, including bread and baked goods, are staple foods that are widely used as vehicles for protein fortification with the aim to increase protein intake in the human diet and for the specific functionality of plant proteins (Hoehnel, Axel, Bez, Arendt, & Zannini, 2019; Liu et al., 2018). Dairy, soy, and egg have dominated the baking industry, but plant proteins like wheat, pea, pulses, potato, and other grains are also gaining a position in the market. Plant protein ingredients used in the bakery segment are expected to witness a compound annual growth rate of 5% over the forecast period 2018–2025 (GMInsights, 2018). Different forms and sources of plant proteins are currently available in the market differing in protein content and functional characteristics. Each type of protein brings specific attributes that can have a critical impact on product quality. The present chapter covers the main plant-based proteins used in bakery products with an emphasis on their impact on the nutritional value and technological quality of the reformulated products.

2. Realm of plant proteins

2.1 Main sources

2.1.1 Cereal grains

Cereals are staples foods rich in carbohydrates, protein, fats, and vitamins, minerals (Table 1). The main cereal grains used for human nutrition are wheat, barley, rice, maize, and sorghum. Wheat is one of the most grown cereals, with a global production volume of around 780 million tons in 2021. In the baking industry, wheat flour is the main ingredient for bread, crackers, cakes, cookies and pastries. Wheat species have a protein content ranging from 11% to 25%. Common wheat (*Triticum aestivum*) is recognized as gluten-containing cereal playing functional and nutritional roles in bakery products owing to gluten functionalities. Rice and maize are gluten-free cereals. Rice (*Oryza sativa* L.) includes several white and pigmented varieties

Table 1 Chemical composition of the main grains and oilseeds used for protein extraction (g/100 g).

	Cereals			Pulses			Oilseeds		
	Wheat	**Rice**	**Maize**	**Pea**	**Chickpea**	**Faba**	**Soy**	**Sunflower**	**Canola**
Carbohydrates	55–75	73–80	69–74	51–65	59–72	44–58	20–30	18–26	10–15
Fiber	12–15	1–8	2–7	17–27	3–7	11–12	5–12	36–40	10–12
Proteins	11–15	7–10	9–15	20–31	18–26	25–30	36–40	10–27	17–26
Fat	1–2	1–2	5–8	1–22	3–6	1–2	35–45	34–55	40–49
Ash	1–2	1–2	1–3	2–4	3–4	3–4	5–6	3–4	2–4
References	(Boukid, Folloni, Sforza, Vittadini, & Prandi, 2018)	(Beloshapka, Buff, Fahey, & Swanson, 2016; Verma & Srivastav, 2017)	(Mansilla, Nazar, & Pérez, 2020; Rodehutscord et al., 2016)	(Boukid, Rosell, & Castellari, 2021; Geerts, Mienis, Nikiforidis, van der Padt, & van der Goot, 2017)	(Zia-Ul-Haq et al., 2007)	(Gasim, Hamad, Abdelmula, & Mohamed Ahmed, 2015; Martinez et al., 2016)	(Silva et al., 2018; Wijewardana, Reddy, & Bellaloui, 2019)	(Alexandrino et al., 2017; González-Pérez & Vereijken, 2007)	(Anwar, Ali, & Nasr, 2015; Kotecka-Majchrzak, Sumara, Fornal, & Montowska, 2020)

having approximately 7% proteins (Pedro, Granato, & Rosso, 2016). Besides, pigmented varieties such as black rice are known as a rich source of polyphenols. Sorghum (*Sorghum bicolor* L.) contains around 70–75% starch and 11–14% protein (Xu, Wang, & Zhao, 2021). It is also recognized as a rich source of antioxidants, particularly phenolic compounds. Barley (*Hordeum vulgare* L.) flour contains around 54.2% starch, 2.4% crude fat, 2.9% total ash, 14.2% protein, 2.9% soluble dietary fiber, 10.2% insoluble dietary fiber and 13.1% of total dietary fiber (Huang, Jia, Zhang, Ma, & Ding, 2020). Compared to the other cereals, barley is known as a source of β-glucan (2.5–11.3%), and it is mostly used in brewing and to a less extent in bakery products.

Minor cereals such as oat, rye, and millet are also relevant sources of proteins. Oats (*Avena sativa* L.) are the sixth highest consumed type of cereal, with a global annual production of 23 million tons (FAO, 2019). Oat flour contains a high level of proteins (up to 20% protein) with a very balanced amino acid, dietary fiber and fat contents, dietary fiber and fats (Boukid, 2021b; Prosekov et al., 2018). The use of oats in the bakery industry is mainly in cakes and cookies. Oat flour is also gaining attention for partial substitution of wheat flour up to 30% in bread formulation to increase fiber, antioxidants, and proteins (Majzoobi, Raiss Jalali, & Farahnaky, 2016). Rye (*Secale cereale* L.) is a traditional raw material used for the production of bread in Northern and Eastern Europe (Stępniewska, Hassoon, Szafrańska, Cacak-Pietrzak, & Dziki, 2019). It is a good source of proteins (6–15%) and several bioactive compounds including fiber, polyphenols, minerals, and vitamins. Its bread, the partial substitution of wheat flour by rye flour was used to improve the viscosity of the dough owing to the swelling properties of pentosans. The addition of rye flour also improved the volume and texture of bread properties (Nordlund, Katina, Mykkänen, & Poutanen, 2016; Oest, Bindrich, Voß, Kaiser, & Rohn, 2020). Nevertheless, complete substitution cannot form a continuous network and elastic dough similar to wheat protein due to high total pentosans' content disturbing the dough network formation (Beck, Jekle, Selmair, Koehler, & Becker, 2011; Öhgren, Fabregat, & Langton, 2016).

2.1.2 Pulses

Pulses, the dry family of legumes, are staple crops worldwide and represent valuable protein sources for human nutrition and animal feed. Pulses have gained particular attention as a source of proteins due to their versatility, sustainability, high nutritional value, affordability and availability (Boukid et al., 2019; Heusala et al., 2020; Saldanha do Carmo et al., 2020).

Pulses are remarkably rich in protein (20–25%) with interesting nutritional and functional properties (e.g., solubility, emulsification capability and foaming) (Boukid, Zannini, Carini, & Vittadini, 2019). Currently, the main commercialized sources of proteins are from pea *(Pisum sativum* L.) and other emerging sources such as chickpea (*Cicer arietinum* L.), faba beans (*Vicia faba* L.), or lupine (*Lupinus luteus* L.) (Boukid, Rosell, & Castellari, 2021; Le Roux et al., 2020; Van Der Goot et al., 2016). Compared to cereals, pulses have a higher protein content (~20–40% protein) and can deliver distinct nutritional and functional properties depending on the type and extent of processing as well the interaction that might take place in each food matrix (Drulyte & Orlien, 2019; Felix, Lopez-Osorio, Romero, & Guerrero, 2018).

2.1.3 Oilseeds

Oilseeds are seeds grown primarily to produce edible oils, while the solvent-extracted meals and press cakes of oilseeds are rich sources of proteins (up to 50–70%) (Bernard, 2016). Oilseeds rich in proteins are soybeans, rapeseed/canola, cottonseed, peanut/groundnut, and sunflower seed, while sesame and flaxseed are minor sources (Riaz, 2011). Soy is the main oilseed used for protein production, where soy proteins represent around 40% of soybean seeds and up to 70% of the pressed cake or the soybean. Recently emerging sources are increasingly gaining attention, such as canola (*Brassica napus* L.) and sunflower (*Helianthus annuus* L.). These seeds are among the most important oily plants in the world. Their vegetable oils have a good nutritional value as they are rich composition in unsaturated fatty acids, omega-3 and omega-6 fatty acids and fat-soluble vitamins, and low in cholesterol (Kotecka-Majchrzak et al., 2020). These seeds have a high protein content ranging from 16% to 26% (Anwar et al., 2015; Kotecka-Majchrzak et al., 2020). Proteins are recovered after oil extraction, where the residue contains up to 50% protein. One of the key factors boosting the global oilseeds market is the easy availability of these crops and the high value of such by-products of oil extraction (ReportsandData, 2021).

2.2 Processing

Before the extraction, grains or seeds must go through pretreatments to improve the protein yield and quality. Dehulling and debranning are applied through abrasion and friction forces to remove the bran layers of grains (Fišteš, 2021), and consequently, make the endosperm more accessible for extraction. Defatting is necessary to remove the fat components (up to 4%) from oilseeds and it is performed using ethanol, hexane or supercritical

CO_2 extraction (Pelgrom, Wang, Boom, & Schutyser, 2015; Schutyser & van der Goot, 2011). Enzymes are also applied to break down polysaccharides and facilitate proteins extraction (Alrahmany, Avis, & Tsopmo, 2013; Prosekov et al., 2018). Afterwards, pretreated grains/ seeds go through the extraction process. The most used extraction methods are wet extraction, which enables the production of isolates, and dry fractionation, which produces protein concentrations. The pros and cons of both methods are summarized in Table 2.

2.2.1 *Alkaline extraction-isoelectric precipitation*

Alkaline extraction followed by isoelectric precipitation is the main process used for protein extraction at the industrial level. First, flours deriving from

Table 2 Advantages and drawbacks of extraction methods of chickpea proteins.

	Alkaline extraction	Dry extraction
Protein recovery	67–94%	31–66%
Purity	68–85%	45–55%
Extraction efficiency	55–77%	16–18%
Yield	18–25%	26–49%
Advantages	• High protein purity • Removal of antinutrients	• Intact structure • additives free • low use of energy • no use of chemical • no use of water • minimal processing
Drawbacks	– high use of energy – use of solvents – generation of aqueous wastes	– retain some antinutritional factors – low purity
References	(Boye et al., 2010; Boye, Zare, & Pletch, 2010; Espinosa-Ramírez & Serna-Saldívar, 2019; Paredes-López, Ordorica-Falomir, & Olivares-Vázquez, 1991; Pelgrom, Boom, & Schutyser, 2015; Eckert et al., 2019; Singhal, Stone, Vandenberg, Tyler, & Nickerson, 2016)	(Pelgrom, Boom, & Schutyser, 2015; Schutyser, Pelgrom, van der Goot, & Boom, 2015; Coda et al., 2015; Martinez et al., 2016; Saldanha do Carmo et al., 2020; Vogelsang-O'Dwyer et al., 2020).

cereals, pulses or oilseeds are solubilized under alkaline or acidic conditions and centrifuged to remove insoluble materials (such as starch and fiber. The recovered supernatant is subjected to isoelectric precipitation, and then the pellet is washed (several times) to remove soluble materials such as sugars, soluble fibers and fats and then centrifuged. The pellet is neutralized and dried. Depending on the sources, protein properties (i.e., structure, solubility, thermal stability and the isoelectric point) might differ and consequently, extraction conditions need to be adjusted to maximize the extraction yield (Jakobek, 2015; Ozdal, Capanoglu, & Altay, 2013). The wet process enables the production of protein isolates with high purity (90%) (Adenekan, Fadimu, Odunmbaku, & Oke, 2018). Nevertheless, this process can induce protein denaturation due to high drying temperatures (100–150 °C) that can negatively affect the integrity of protein structure and functionality (Jafari, Rajabzadeh, Tabtabaei, Marsolais, & Legge, 2016; Wu, Myers, & Johnson, 1997; Zhao, Zhang, Liu, Zhang, & Ao, 2018). This process also is characterized by the high use of water and energy and the high generation of industrial wastes (Chéreau et al., 2016; Ruiz, Xiao, Van Boekel, Minor, & Stieger, 2016).

A post-treatment is generally designed to enhance the functionality, increase the yield or/ and improve the aroma profile of plant proteins (Gao, Weng, Yan, Wang, & Wang, 2020). Ultrafiltration, electro-acidification and chromatography were reported to increase protein yield and the solubility of proteins (Giuseppin, Laus, & Schipper, 2014; Mondor, Ippersiel, Lamarche, & Boye, 2004). Ultrasound also increased soy protein isolates functionalities (i.e., solubility and emulsification) (Huang et al., 2020). Supercritical fluids enabled a better aroma profile and enhanced the yield of pea proteins (Vatansever & Hall, 2020; Vatansever, Rao, & Hall, 2020). Enzymatic treatment increased the yield, improved functionality and released peptides with health-beneficial biological activities (e.g., antioxidant, immunomodulatory, and antimicrobial and anti-inflammatory) (Mune Mune, Minka, & Henle, 2018; Roy, Boye, & Simpson, 2010).

2.2.2 Dry fractionation

Dry fractionation consists of two steps, namely size reduction (milling) and size separation (air classification or electrostatic separation) (Assatory, Vitelli, Rajabzadeh, & Legge, 2019). Air classification results in two fractions, i.e., the protein-rich fraction (fine particles) and the starch-rich fraction (coarse particles) (Boye, Zare, & Pletch, 2010; Schutyser & van der Goot, 2011). It is important to define the cut-point of protein-starch separation depending on the type of seeds (Boye, Aksay, et al., 2010; Boye, Zare, & Pletch, 2010; Schutyser & van der Goot, 2011). Milling followed by air classification produced protein concentrates with a purity ranging from 50% to 70% (Xing et al., 2020).

Electrostatic separation is performed based on the differences in dielectric properties between proteins and carbohydrates, unlike air classification based on the size and density of the particles (Aryee & Nickerson, 2012; Wang, Zhao, De Wit, Boom, & Schutyser, 2016). After milling, electrostatic separation is based on charging proteins to a higher extent compared to carbohydrates (Tabtabaei, Jafari, Rajabzadeh, & Legge, 2016). The efficiency of separation is closely related to the type of grains or seeds. For instance, protein content of lupin concentrates increased from 35% to 65% through electrostatic separation, while no effect was observed in peas (Pelgrom, Wang, et al., 2015).

2.3 Structural and nutritional characteristics

2.3.1 Cereal proteins

The most commercialized proteins deriving from cereals are wheat, rice, and maize. Cereal proteins consist of albumins, globulins, glutenins and gliadins based on their differential solubility in water, salt, and alcohol solutions. Amino acid composition of cereal proteins is reported in Table 3. Wheat protein is relatively rich in sulfur-containing amino acids but lacks lysine and threonine (Shewry, Tatham, Forde, Kreis, & Miflin, 1986). Gluten, wheat storage proteins, is rich in glutamine (30–35%) and proline (10–15%) but only contains small amounts of lysine, methionine, and threonine. Rice proteins are a good source of essential amino acids, particularly lysine (Amagliani, O'Regan, Kelly, & O'Mahony, 2016). Rice proteins are also adequate for infant diets and for people with allergy to soy/milk because it is generally recognized as hypoallergenic ingredients (Amagliani, O'Regan, Kelly, & O'Mahony, 2017; Wang, Xu, Chen, Zhou, & Wang, 2019; Wang, Xue, & Zhang, 2019). On the contrary, wheat proteins are known as triggers of several adverse reactions to gluten and to some other soluble proteins (e.g., amylase-trypsin inhibitors -ATIs related to baker's asthma (Boukid et al., 2017; Boukid, Mejri, Pellegrini, Sforza, & Prandi, 2017). Zein is the commercial form of maize protein (Cao, Liu, Zhang, Wang, & Ren, 2020). Zein is a rich source of glutamic acid (21–26%), leucine (20%), proline (10%), and alanine (10%), while is deficient in tryptophan and lysine (Dhillon, Kaur, Oberoi, Spier, & Brar, 2016). Cereals proteins from wheat rice and maize were reported to have specific bioactive peptides associated with

Table 3 Amino acid content of vegetable protein isolates (g per 100 g proteins).

	Wheat	Rice	Maize	Pea	Chickpea	Faba	Soy	Sunflower	Canola	FAO standard (adult)
Essential amino acids										
Isoleucine	2.0	2.0	1.7	2.3	4.8	4.8	1.9	6.7	5.2	1.3
Leucine	5.0	5.8	8.8	5.7	8.5	8.3	5.0	6.2	9.3	1.9
Threonine	1.8	2.0	1.8	2.5	3.0	4.1	2.3	3.4	5.3	0.9
Phenylalanine	3.7	3.7	3.4	3.7	5.3	5.0	3.2	12.7	5.1	1.9
Lysine	1.1	1.9	1.0	4.7	7.0	5.8	3.4	2.0	5.6	1.6
Histidine	1.4	1.5	1.1	1.6	3.2	2.5	1.5	3.1	3.2	1.6
Valine	2.3	2.8	2.1	2.7	4.4	5.9	2.2	3.6	5.9	1.3
Methionine	0.7	2.0	1.1	0.3	1.1	0.7	0.3	2.1	2.6	1.7
ΣEAA	18.0	21.7	21.0	23.5	37.3	37.1	19.8	39.7	42.1	12.2
Non-essential amino acids										
Serine	3.5	3.4	2.9	3.6	3.5	6.5	3.4	4.5	4.7	–
Glutamic acid	26.9	12.7	13.1	12.9	17.8	15.1	12.4	16.9	17.3	–
Glycine	2.4	3.4	1.6	2.8	3.6	7.2	2.7	4.9	5.1	–
Alanine	1.8	4.3	4.8	3.2	5.2	5.6	2.8	5.0	5.1	–
Cysteine	0.7	0.6	0.3	0.2	0.6	1.2	0.2	0.2	0.4	–

Arginine	2.4	5.4	1.7	5.9	8.5	7.8	4.8	9.8	7.7	–
Proline	8.8	3.4	5.2	3.1	4.1	5.5	3.3	5.1	4.3	–
Tyrosine	2.4	3.5	2.7	2.6	2.8	3.0	2.2	0.8	3.9	–
ΣNEAA	48.9	36.7	32.3	34.3	46.1	51.91	31.8	47.13	48.5	–
References	(Gorissen et al., 2018)	(Gorissen et al., 2018)	(Gorissen et al., 2018)	(Gorissen et al., 2018)	(Zia-Ul-Haq et al., 2007)	(Eckert et al., 2019; Singhal et al., 2016).	(Gorissen et al., 2018)	(González-Pérez & Vereijken, 2007; Oseyko, Romanovska, & Shevchyk, 2020)	(Tan, Mailer, Blanchard, & Agboola, 2011)	(WHO/FAO/UNU, 2007)

ΣEAA sum of all essential amino acids, ΣNEAA sum of all non-essential amino acids.

beneficial effects including anti-oxidative, anti-hypertensive, anti-cancer and anti-obesity activities (Karami, Peighambardoust, Hesari, Akbari-Adergani, & Andreu, 2019).

2.3.2 Pulses proteins

Pulses proteins consist of globulins and albumins, followed by prolamins and glutelins (Lan, Ohm, Chen, & Rao, 2019). Globulins are the major storage proteins (up to 60–80%), and they are classified into legumin (300–400 kDa, 11S), vicilin (150–170 kDa, 7S), and convicilin proteins (~70 kDa subunits, 7S) (Alavi, Chen, Wang, & Emam-Djomeh, 2021). Legumins have a hexametric structure with subunits containing α and β-chains, while vicilins have a trimeric structure (Warsame, Michael, O'Sullivan, & Tosi, 2020). Albumins (12–20%) contain metabolically active proteins including enzymes (Singh, Wani, Kaur, & Sogi, 2008; Warsame et al., 2020). Albumins are a rich source of essential amino acids, particularly sulfur-containing amino acids (tryptophan, threonine, and lysine) (Liu, Hung, & Bennett, 2008). Glutelins are present in pulses in low amounts with the exception of chickpea proteins having high content (around 18 g/100 g) (Chang, Alli, Konishi, & Ziomek, 2011). Prolamin was found in traces in pulse, unlike cereals (Singh et al., 2008).

Pulses proteins are rich sources of essential amino acids, including arginine, phenylalanine, leucine and isoleucine, and lysine (Gorissen et al., 2018; Millar, Barry-Ryan, Burke, McCarthy, & Gallagher, 2019). However, they are deficient in the sulfur-containing amino acids, mainly methionine and cysteine (Stone, Karalash, Tyler, Warkentin, & Nickerson, 2015). Protein digestibility-corrected amino acid score (PDCAAS) of pulses proteins isolates had lower values than concentrates [e.g., pea protein: 0.82 and 0.90; chickpea: 0.92 and 0.76, respectively] (Espinosa-Ramírez & Serna-Saldívar, 2019; Nosworthy et al., 2017). This can be attributed to changes in protein composition due to isolation process, that resulted in lowering PDCAAS value compared to protein concentrate (Mathai, Liu, & Stein, 2017). Indeed, not all the essential amino acid amounts meet the requirement of FAO/WHO for adults (WHO/FAO/UNU, 2007). In general, pulse proteins had lower PDCAAS values than whey proteins (1.0), casein (1.0) and soy protein (0.97–1) due to limiting sulfur-containing amino acids (Mathai et al., 2017; Rutherfurd, Fanning, Miller, & Moughan, 2015). Blending cereals and pulses proteins might overcome this deficiency. Notably, several studies reported the health-promoting properties (e.g., anti-hypertensive, antidiabetic and antioxidant) of pulses due to their rich composition in bioactive peptides (Chalamaiah, Yu, & Wu, 2018; Li & Aluko, 2010).

2.3.3 Oil seeds proteins

Soy proteins are mainly 7S globulin (conglycinin, 35%, ~150 and 200kDa) and 11S globulin (glycinin, 52%, ~300–380kDa) followed by 2S (8%) and 15S (5%) (Hsiao, Yu, Li, & Hsieh, 2015). Interestingly, soy proteins have an amino acid profile alongside with egg and milk protein. Indeed, all essential amino acids (isoleucine, leucine, lysine, methionine, phenylalanine, threonine, tryptophan, and valine) are present in soy proteins (Gorissen et al., 2018).

Like soy, sunflower proteins consist mainly of globulins (40–90%), albumins (10–30%) and traces of glutelins and prolamins. Storage proteins represent up to 85% of the total protein content and they are characterized by a high content of 11S helianthinin and 2S albumins. Sunflower proteins are rich in sulfur-containing amino acids but deficient in lysine, valine and isoleucine (González-Pérez & Vereijken, 2007).

Canola proteins were declared as a novel food ingredient since 2014 (European Parliament and of the Council, 2014). Canola proteins have higher amounts of metabolic proteins than storage proteins. They are characterized by a high level of albumins (45–60%) followed by globulins (20–25%) and traces of prolamins and glutamines. The two main types of storage proteins are 12S globulin (cruciferin) and 2S albumin (napin), representing 60% and 20% of the total protein content, respectively (Aider & Barbana, 2011; Wu & Muir, 2008). The amino acid profile of canola protein is characterized by high levels of methionine, lysine and arginine (Tan et al., 2011). PDCAAS of canola protein was found to be 0.86, slightly lower than soy protein isolate (Campbell, Rempel, & Wanasundara, 2016). Numerous peptides deriving from oil seeds were found to have anti-oxidative, anti-cancer, antimicrobial, and immunomodulatory effects (McGraw, Krul, Grunz-Borgmann, & Parrish, 2016; Padhi & Ramdath, 2017).

2.4 Adverse reactions to plant proteins

2.4.1 Wheat proteins

The consumption of wheat can trigger celiac disease, non-celiac gluten/wheat sensitivity, or wheat allergy. Gluten can trigger immune reactions, mainly celiac disease, which affects approximately 1% of people in the world (Boukid, Mejri, et al., 2017). The symptoms are mucosal inflammation, small bowel villous atrophy, increased intestinal permeability and malabsorption of macro- and micronutrients (de Punder & Pruimboom, 2013). Gliadins are considered the primary triggers of the abnormal immune response in celiac disease. Gluten elicits an adaptive Th1-mediated immune response in individuals carrying HLA-DQ2 or HLA-DQ8 as major genetic predisposition (Almeida et al., 2016; Catassi et al., 2010; Pisapia et al., 2016). Gluten degradation was tested using different methods such as thermal treatments, addition of polyphenols, germination, enzymes or fermentation. These methods showed different degrees of efficiency as intensively reviewed in previous works (Boukid, Mejri, et al., 2017; De Angelis et al., 2010). Amylase trypsin inhibitors are the main putative of wheat allergies and can cause inflammation and immune reactions (Boukid, Prandi, et al., 2017; Uvackova, Skultety, Bekesova, McClain, & Hajduch, 2013). Also, ω_5-gliadins can cause wheat-dependent exercise-induced anaphylaxis (Sapone et al., 2012). Gluten, amylase-trypsin inhibitors, FODMAPs, and agglutinin triggered non-celiac gluten/wheat sensitivity, which is frequently associated with other health issues such as irritable bowel syndrome (IBS) (Boukid, 2021a; Fasano, Sapone, Zevallos, & Schuppan, 2015).

2.4.2 Pulses

Allergic reactions to pulses are one of the most common food allergies. In most cases, cross-reactivity among lentil, chickpea and pea or/and cross-reactivity with peanut and soy were reported due to similarities protein structure (Mennini, Dahdah, Mazzina, & Fiocchi, 2016;Verma, Kumar, Das, & Dwivedi, 2013). Globulins were found the putative allergens in pulses allergies, where the main symptoms are respiratory, rhinitis and anaphylaxis (Patil, Niphadkar, & Bapat, 2001; Verma et al., 2012). In pea, two major allergens were identified Pis s1 (belongs to the vicilin) and Pis s2 (belongs to the convicilin) (Abrams & Gerstner, 2015). Eight peptides deriving from faba protein (70, 60, 48, 32, 23, 19, 15, and 10 kDa) showed IgE-binding capacity (Kumar et al., 2014). Chickpea allergy was mostly reported in India (up to 13% of the population) (Bar-El Dadon, Pascual, & Reifen, 2014; Wangorsch et al., 2020). Chickpea proteins (Cic PR-10, Cic a 1.01, 2S and Pa2) were also considered to evoke allergic reactions in chickpea-sensitive individuals (Bar-El Dadon et al., 2013; Kulkarni, Ananthanarayan, & Raman, 2013; Verma et al., 2016). Noteworthy, thermal processing was found efficient in allergens deactivation (Warnakulasuriya, Pillai, Stone, & Nickerson, 2018).

2.4.3 Oilseeds

Soy protein is one of the most common food allergens that can cause severe allergic reactions (e.g., hives, itching, swelling, eczema, vomiting, diarrhea and nausea) (Ito, 2015; Li & Damodaran, 2017). Glycinin and β-conglycinin are the major allergens of soy (Taylor, Remington, Panda, Goodman, & Baumert, 2015). Sunflower allergens have been reported in the allergen database (WHO/IUIS Allergen Nomenclature), including protein AP10

(Hel a 3, 9 kDa). Canola allergies have been previously described to be triggered by napins and cruciferin (Anzani, Boukid, Drummond, Mullen, & Álvarez, 2020; Wanasundara, Tan, Alashi, Pudel, & Blanchard, 2017). Several mitigation strategies (e.g., pulsed electrical fields, irradiation, ultrasound, genetic or chemical modifications) were investigated to reduce the allergenic potential of soy protein (Katz et al., 2014; Meinlschmidt, Sussmann, Schweiggert-Weisz, & Eisner, 2016). Identified allergens in sunflower proteins are Hel a 1 (34 kDa). Hel a 2 (4.7 kDa) and Hel a 3 (9 kDa). These proteins can cross react with members of the *Asteraceae* due to the presence of Hel a 2 (Ukleja-Sokołowska, Gawrońska-Ukleja, Zbikowska-Gotz, Bartuzi, & Sokołowski, 2016). Symptoms of an allergic reaction can include oral, respiratory and skin reactions, similarly to peanut allergy (Lavine & Ben-Shoshan, 2015; Murtagh et al., 2003). Canola protein was found to trigger immunogenic responses towards napin (most potent) and cruciferin (less potent) (Puumalainen et al., 2015). The napin isoform Bra n 1 was reported to cross react with mustard (Fiocchi, Dahdah, Riccardi, Mazzina, & Fierro, 2016). Allergenic potential of canola protein is not fully deciphered but suitable labelling of food containing these proteins is needed (Fiocchi et al., 2016; Puumalainen et al., 2015).

2.5 Functionalities of plant proteins

Gluten owes its functionality to glutenin conferring dough elasticity, and gliadin conferring viscosity and extensibility (Shewry, Halford, Belton, & Tatham, 2002). Vital gluten is crucial for baking quality thanks to its water absorption capacity, cohesiveness, viscosity, and elasticity (Bardini et al., 2018; Ortolan, Corrêa, da Cunha, & Steel, 2017). Gluten can aggregate to form protein network and thus plays a principal role in the final texture of bread or baked products such as cake (Delcour et al., 2012). The functional properties of gluten are related to the ratio of glutenins to gliadins. Gluten is heat stable and has the capacity to act as a binding and when added can improve texture, flavor, and moisture retention. Gluten play an important role in CO_2 entrapment during fermentation and quality of the textural quality of bread (Farbo et al., 2020; Martínez & Gómez, 2017).

Proteins deriving from gluten-free cereals do not possess the viscoelastic properties of gluten and cannot generate a gluten-like network (Boukid, Rosell, Rosene, et al., 2021). Rice proteins in their native form are characterized by low solubility, emulsification formation and stability (Amagliani et al., 2017; Rani, Pooja, & Pal, 2018). Treated rice proteins showed improved functionalities, and the level of improvement depends on the severity of the processing (Li, Wang, Sun, Li, & Chen, 2019; Liu, Wang, Li, Liang, & Yang, 2016; Wang, Liu, Li, & Yang, 2016; Wang, Xu, et al., 2019; Wang, Xue, & Zhang, 2019; Wang, Zhao, et al., 2016). Rice proteins are mostly used in gluten-free baked products and were reported to retain bread moisture during storage (Phongthai, D'Amico, Schoenlechner, & Rawdkuen, 2016). Zein also showed low solubility, but a high ability to stabilize emulsions and foams (Liang & Chen, 2018;Liu, Cao, Ren, Wang, & Zhang, 2019; Wang, Xu, et al., 2019; Wang, Xue, & Zhang, 2019). Zein is one of the plant proteins able to produce fibrous material used in meat analogues formulations (Jones, 2016; Mattice & Marangoni, 2020). Zein has been recognized as a structural protein for various gluten-free baked products due to its ability to form gluten-like viscoelastic network (Schober, Bean, Boyle, & Park, 2008). Zein forms a fibrous protein when mixed with water and is able to hold CO_2 in zein-based doughs during fermentation (Phongthai et al., 2016). Zein can, thus, transmit extensibility and viscoelasticity to the dough, which gives the dough bread-making properties.

The functionality of pulses proteins is closely related to the ratio legumin/vicilin, extraction parameters, and variety (Chao, Jung, & Aluko, 2018; Chihi, Sok, & Saurel, 2018; Ladjal Ettoumi, Chibane, & Romero, 2016; Stone, Avarmenko, Warkentin, & Nickerson, 2015). The solubility of pulses protein is pH-dependent, where the maximum values (about 80%) are reached at alkaline pH and the lowest (2–30%) around the isoelectric point (pH 4–6) (Chao & Aluko, 2018; Eckert et al., 2019; Vogelsang-O'Dwyer et al., 2020; Yin, Zhang, & Yao, 2015). At neutral pH, the solubility of pea (92%) and faba (~82 and 88%) proteins showed values within the same range of soy (85%) protein isolates (Martinez et al., 2016). However, chickpea proteins had low solubility (around 60%) (Boye, Aksay, et al., 2010; Boye, Zare, & Pletch, 2010). These proteins are able to form stable emulsions and foams under acid conditions similar to those of soy protein (Barac, Pesic, Stanojevic, Kostic, & Bivolarevic, 2015; Silva et al., 2019; Stone, Karalash, et al., 2015). The foam stability of pea (63%, v/v), faba bean (64%, v/v), and chickpea (58%, v/v) protein isolates were slightly lower than soy protein (70%, v/v) (Martinez et al., 2016). The emulsion stability of pea, faba bean and chickpea proteins (92%, v/v) was comparable to soy proteins (89%, v/v) (Martinez et al., 2016). However, gelling properties are low but they can be improved using enzymes such as transglutaminase (Klost & Drusch, 2019; Langton et al., 2020; Tontul, Kasimoglu, Asik, Atbakan, & Topuz, 2018). Pulse proteins are not able to form gluten networks, and affect air incorporation and gas retention during leavening, resulting in bread with poor crumb structure and texture (Bresciani & Marti, 2019). Overall, pulses proteins have good functional properties except for gelling properties, but soy protein is still the most used proteins in food applications.

Soy protein has excellent functional features such as gelling, emulsifying ability and water- and oil- holding capacity (Barac et al., 2015; Meinlschmidt, Schweiggert-Weisz, Brode, & Eisner, 2016). Soy protein showed good encapsulation and film-forming capacities enabling a large spectrum of applications including food and beverages as well as nutraceuticals and pharmaceuticals (Lim, Mendes, & Chronakis, 2019; Seung & Rhee, 2004). Soy proteins were reported to improve the rise and quality properties of other gluten-free bakery products formulations acting as structure-forming agent (Espinosa-Ramírez, Garzon, Serna-Saldivar, & Rosell, 2018). On the other hand, the functional properties of sunflower and canola proteins have been found to be poor and further investigations are required to find efficient strategies for improvement (Campbell et al., 2016; González-Pérez & Vereijken, 2007).

3. Impact of protein fortification on bakery products

3.1 Bread

3.1.1 Gluten-containing products

Proteins are added to wheat-based bread for functionality or nutrition motives (enhancing the amino acid profile). Owing to its viscoelastic properties, vital gluten enables the increase the strength of protein network of weak flours (flour strength or W index around 90–160) (Bardini et al., 2018; Giannou & Tzia, 2016). The addition of vital gluten (1%) results in the formation of dough with a stronger and more cohesive network which is able to trap and retain the gases during proofing and baking (Bardini et al., 2018; Boukid et al., 2018; Giannou & Tzia, 2016; Ortolan et al., 2017). As such, the resulting breads made with added vital gluten (1%) have improved volume, texture, crumb grain, and sensory attributes compared to the control (Boukid, Carini, et al., 2018; Giannou & Tzia, 2016; Ortolan et al., 2017). However, when 1% of vital gluten was added to doughs made with strong flours (W index above 300), breads had increased hardness and lowered specific volume (Boukid, Carini, et al., 2018; Giannou & Tzia, 2016; Ortolan et al., 2017). This can be attributed to the formation of compact structure resulting in a firm texture.

On the other hand, adding non-gluten containing proteins improves the nutritional profile of bread, especially the amino acid profile. Nevertheless, depending on the level of addition, gluten-free proteins deriving from pulses or oilseeds might negatively impact dough properties due to the dilution effect of gluten (Hoehnel et al., 2019). Replacing 15% of wheat flour by pea proteins weakened the gluten network and resulted in bread with a low volume. A dough made with 15% of faba bean concentrate (61% protein) had a medium gluten strength but significantly stronger than that made with 15% pea protein isolates (90% protein). Faba protein concentrate enriched bread had a similar specific volume and hardness to that of the control bread (Hoehnel et al., 2019). However, adding 15% of zein (92% protein) had positive impact on the dough properties (consistency and hydration) and resulted in high specific volume and crumb softness (2.63 mL/g and 15 N, respectively) compared to the control (2.55 mL/g and 11 N, respectively) (Hoehnel et al., 2019). This suggests that incorporating up to 15% proteins improves protein without hindering the technological quality of the final product (Hoehnel et al., 2019). Adding over 20% soy protein reduced dough stability and the resulting bread had a smaller volume and increased hardness compared to the control bread made with 100% wheat flour (Zhou, Liu, & Tang, 2018). These changes can be explained by the negative impact of soy protein on dough hydration and gluten network formation (Schmiele, Ferrari Felisberto, Pedrosa Silva Clerici, & Chang, 2017).

3.1.2 Gluten-free products

In gluten-free bread, egg and soy proteins are the most used proteins to improve the nutritional composition and to play a functional role in protein network formation, bread structure, color development (Maillard reaction) and flavor (Han et al., 2019; Suphat Phongthai et al., 2016; Smith, Bean, Selling, Sessa, & Aramouni, 2017). The addition of soy protein (from 2% to 13%) increased water retention capacity resulting in bread with low specific volume and a compact crumb (Masure, Wouters, Fierens, & Delcour, 2019). However, the addition of isolated fraction, ß-conglycinin improved viscoelastic properties and improved gas entrapment during baking (Espinosa-Ramírez et al., 2018). However, in recent years, the use of pulses proteins is increasing in response to consumers request for vegan and no genetically modified crops (Silva et al., 2019; Silva, Balakrishnan, Schmitt, Chassenieux, & Nicolai, 2018; Wouters et al., 2017). Pea protein is emerging as a potential plant-based alternative. The incorporation of 5% pea protein (78.13% protein) improved dough viscoelastic properties due to its ability to absorb water and to entrap air resulting in improved protein network. This result can be attributed to the high-water absorption capacity of pea proteins resulting in lowering the loss of moisture during baking as well as their foaming capacity enabling gases retention resulting in improving bread volume. As such, breads had an increased specific volume, and improved volatile profile compared to the control bread (made from rice and maize flours without protein) (Pico, Reguilón, Bernal, & Gómez, 2019). Dough made with 10% pea protein showed higher elastic and viscous modulus than the control (Ziobro, Juszczak, Witczak, & Korus, 2016). Up to 30% protein addition resulted in bread with increased protein content and decreased starch digestibility (Sahagún & Gómez, 2018a). However, bread crumb was found more compact and harder in texture due to the high water holding capacity of plant proteins (Sahagún & Gómez, 2018a). Regarding cereal proteins, 2% of rice protein increased dough elasticity due to protein water-binding capacity, foam formation and emulsification ability (Phongthai et al., 2016). The resulting bread had increased specific volume and sensory properties (due to the rich volatile profile) (Phongthai et al., 2016). Addition of zein protein at 5% enabled a better gases entrapment which resulted in bread with improved crumb structure and volume (Khuzwayo, Taylor, & Taylor, 2020).

The application of canola and sunflower proteins in gluten-free bread is still limited. Incorporating 6% canola proteins improved the technological and nutritional properties of the dough and the resulted bread had a higher specific volume than the control (Salah, Olkhovatov, & Aïder, 2019). However, incorporating 15% canola protein decreased dough viscosity resulting in dough unable to retain fermentation gases, and consequently the final bread was underdeveloped with a compact structure (Witczak, Chmielewska, Ziobro, Korus, & Juszczak, 2021). Similar results were obtained when canola protein was at 12% and 15% caused a significant increase in bread volume and porosity in comparison to the control sample (Korus, Chmielewska, Witczak, Ziobro, & Juszczak, 2021). Adding 6 or 9% of canola protein increased bread volume, and resulted in an uniform crumb structure due to gas stabilization effects (Korus et al., 2021). Sensory evaluation showed that the canola protein enriched bread was generally acceptable by consumers (Korus et al., 2021). During storage, enriched bread showed a reduced crumb hardening, indicating a retardation of the aging process. The use of sunflower protein (up to 10%) did not hinder the formation of bread structure but imparted bread with dark color which was similarly to the canola enriched bread (Korus et al., 2021; Zorzi, Garske, Flôres, & Thys, 2020). The color change can be attributed to the fact that added amino acids were involved in the creation of Maillard reaction products and were responsible for the browning effect during baking. During storage, bread showed a lower tendency to staling, which might be attributed to the high water holding capacity of added protein (Zorzi et al., 2020).

3.2 Baked goods

3.2.1 Gluten-containing

The application of proteins in baked goods is mainly to enhance protein intake since they are commonly consumed as snacks or breakfasts. Proteins also play a role in the texture owing to their functional properties that vary depending on the source and level of incorporation.

In sponge cake formulation, substituting wheat flour with pea proteins (from 10% to 40%) increased the elastic behavior, water binding capacity and batter stability compared to the control. Pea proteins acted as fillers due to their emulsification, foaming and water holding capacities resulting in the increase of gas and water retention of the batter (Assad Bustillos, Jonchère, Garnier, Réguerre, & Della Valle, 2020; Assad-Bustillos et al., 2020). Furthermore, this fortification improved the nutritional profile of the cake by balancing the amino acid profile and at the same time did not affect protein digestion and bio-accessibility (Assad-Bustillos et al., 2020; Bildstein, Lohmann, Hennigs, Krause, & Hilz, 2008).

In eggless cake, pea protein played a relevant role in mimicking the texture of conventional cake in combination with a mixture of emulsifiers. The formulation having similar properties to control was made from 12.5% pea protein isolates, 0.1% xanthan gum and 1% soy lecithin. This formulation showed similar crumb color and structure to those of the control (Lin, Tay, Yang, Yang, & Li, 2017). As such, pea protein can be used as a valuable ingredient for producing vegan products without compensating for the conventional properties of the traditional product. The addition of soy protein to cookies (up to 10%) resulted in cookies with smaller shape unlike those made with 30% having a wide shape. Cookies made 30% had a higher expansion factor due to high water absorption related to higher water retention of protein. Sensory analysis showed that up to 30% protein addition was acceptable by the panelists (Tang & Liu, 2017).

3.2.2 Gluten-free products

Dairy and soy proteins are the most used proteins in gluten-free products, where they act as fillers and boosters of protein content (Mancebo, Rodriguez, & Gómez, 2016; Sahagún & Gómez, 2018b). However, the increase of plant alternatives is increasingly spreading to satisfy the vegan/flexitarian niches.

The fortification of muffins with chickpea protein isolate (up to 7%) reduced hardness, resulting in softer muffins. However, specific volume and crumb porosity were significantly reduced. The use of transglutaminase and xanthan can help to overcome these defects by strengthening the protein network (Shaabani, Yarmand, Kiani, & Emam-Djomeh, 2018). Incorporating pea protein (up to 17%) in gluten-free muffins batter improved the rheological properties by increasing the elastic and viscous modulus. The resulting viscosity of the batter was comparable to those made with soy protein isolates and casein. As such, muffins made with pea proteins was characterized by high softness, springiness, yellowness and specific volume (Matos, Sanz, & Rosell, 2014). However, the addition of 50% of pea proteins reduced batter density and resulting in cake with a compact structure and hard crumb (Gularte, Gómez, & Rosell, 2012). Protein addition also reduced the starchy ingredients used commonly in formulating gluten-free products and consequently reduced the glycemic index of the final products.

In gluten-free cookies, substituting rice flour with 10% soy protein improved the texture, resulting in soft products compared to the control (Sarabhai et al, 2015). Gluten-free cookies enriched with 20% soy protein had acceptable sensory scores (Adeyeye, Adebayo-Oyetoro, & Omoniyi, 2017). A similar effect was observed in gluten-free cookies made with 20% pea proteins since these proteins increased hydration properties and consistency, and produced soft and crunchy cookies (Mancebo et al., 2016). With 30% pea proteins, the hardness of the cookie increased, and the taste was negatively impacted, resulting in lower sensory scores due to the off-flavors associated with pulses proteins (Mancebo et al., 2016; Sahagún & Gómez, 2018b). Plant proteins produced cookies with a desirable color as compared to those made with whey or casein (Sahagún & Gómez, 2018b). This can be due to the lower browning effect of plant protein as compared to animal proteins. To boost the level of protein addition, combining protein with emulsifiers was suggested to improve the texture of gluten-free cookies and produce protein rich products (Sarabhai et al., 2015). Besides the technological aspect, plant protein fortification in gluten-free baked goods increased protein content and enhanced the amino acid composition.

4. Conclusion and future outlooks

In bakery products, proteins are multifunctional ingredients that can enhance (i) the nutritional value of the product by increasing protein content and improving the amino acid profile and (ii) the technological and sensory properties of the products owing to their functional properties. For a long time, whey, soy, and egg have dominated the baking industry. In recent years, growing trends of plant proteins consumption as healthy and nutritional ingredients is spreading cross food and beverages including the segment of bakery products.

Depending on the level of inclusion, plant protein addition can improve bakery products quality. In the case of bread, adding up to 1% vital gluten on bread improved gluten-containing bread technological quality, while adding non-gluten proteins (up to 15% proteins deriving pulses or oilseeds) resulted in low technological quality due gluten dilution but higher amount of proteins. Gluten- free bread can be fortified with pulses proteins up to 10%, while maintaining an acceptable quality by consumers. In the case of baked goods, adding up to 10% proteins (deriving pulses or oilseeds) improved both gluten-containing and gluten-free products at both nutritional and technological levels. Furthermore, most plant protein sources deriving from pulses can induce off-flavors in bakery products. In this case, several approaches can be adopted such as post-treatments of proteins to attenuate the undesirable flavors, finding the right combinations of proteins for taste regulation or adding suitable masking agents. Considering that plant proteins are versatile, a deeper understanding of the volatile profile of each source can enable a better redirection of their use considering the product requirements. Indeed, each protein has specific compositional and functional attributes to be carefully considered when designing a new product.

The realm of plant proteins keeps getting wider to include emerging sources such as lupine and oat and others non-animal sources (e.g., seaweed, algae, hemp, duckweed, fungi, and insects). The growth of alternative proteins market will be consolidated by innovative technologies to reduce their negative side effects such as allergies and off-flavors. Breeding progress also can boost the selection of protein-rich seeds/grains with desirable attributes.

References

Abrams, E. M., & Gerstner, T. V. (2015). Allergy to cooked, but not raw, peas: A case series and review. *Allergy, Asthma and Clinical Immunology, 11*(1). https://doi.org/10.1186/s13223-015-0077-x.

Adenekan, M. K., Fadimu, G. J., Odunmbaku, L. A., & Oke, E. K. (2018). Effect of isolation techniques on the characteristics of pigeon pea (*Cajanus cajan*) protein isolates. *Food Science & Nutrition, 6*(1), 146–152. https://doi.org/10.1002/fsn3.539.

Adeyeye, S. A. O., Adebayo-Oyetoro, A. O., & Omoniyi, S. A. (2017). Quality and sensory properties of maize flour cookies enriched with soy protein isolate. *Cogent Food & Agriculture, 0*. https://doi.org/10.1080/23311932.2017.1278827.

Aider, M., & Barbana, C. (2011). Canola proteins: Composition, extraction, functional properties, bioactivity, applications as a food ingredient and allergenicity - a practical and critical review. *Trends in Food Science and Technology, 22*(1), 21–39. https://doi.org/10.1016/j.tifs.2010.11.002.

Alavi, F., Chen, L., Wang, Z., & Emam-Djomeh, Z. (2021). Consequences of heating under alkaline pH alone or in the presence of maltodextrin on solubility, emulsifying and foaming properties of faba bean protein. *Food Hydrocolloids, 112*, 106335. https://doi.org/10.1016/j.foodhyd.2020.106335.

Alexandrino, T. D., Ferrari, R. A., de Oliveira, L. M., de Cássia, S. C., Ormenese, R., & Pacheco, M. T. B. (2017). Fractioning of the sunflower flour components: Physical, chemical and nutritional evaluation of the fractions. *LWT - Food Science and Technology, 84*, 426–432. https://doi.org/10.1016/j.lwt.2017.05.062.

Almeida, L. M., Gandolfi, L., Pratesi, R., Uenishi, R. H., de Almeida, F. C., Selleski, N., et al. (2016). Presence of DQ2.2 associated with DQ2.5 increases the risk for celiac disease. *Autoimmune Diseases, 2016*, 5409653. https://doi.org/10.1155/2016/5409653.

Alrahmany, R., Avis, T. J., & Tsopmo, A. (2013). Treatment of oat bran with carbohydrases increases soluble phenolic acid content and influences antioxidant and antimicrobial activities. *Food Research International, 52*(2), 568–574. https://doi.org/10.1016/j.foodres.2013.03.037.

Amagliani, L., O'Regan, J., Kelly, A. L., & O'Mahony, J. A. (2016). Chemistry, structure, functionality and applications of rice starch. *Journal of Cereal Science, 70*, 291–300. Academic Press https://doi.org/10.1016/j.jcs.2016.06.014.

Amagliani, L., O'Regan, J., Kelly, A. L., & O'Mahony, J. A. (2017). The composition, extraction, functionality and applications of rice proteins: A review. *Trends in Food Science and Technology, 64*, 1–12. Elsevier Ltd. https://doi.org/10.1016/j.tifs.2017.01.008.

Anwar, M. M., Ali, S. E., & Nasr, E. H. (2015). Improving the nutritional value of canola seed by gamma irradiation. *Journal of Radiation Research and Applied Sciences, 8*(3), 328–333. https://doi.org/10.1016/j.jrras.2015.05.007.

Anzani, C., Boukid, F., Drummond, L., Mullen, A. M., & Álvarez, C. (2020). Optimising the use of proteins from rich meat co-products and non-meat alternatives: Nutritional, technological and allergenicity challenges. *Food Research International, 137*. Elsevier Ltd. https://doi.org/10.1016/j.foodres.2020.109575.

Aryee, F. N. A., & Nickerson, M. T. (2012). Formation of electrostatic complexes involving mixtures of lentil protein isolates and gum Arabic polysaccharides. *Food Research International, 48*(2), 520–527. https://doi.org/10.1016/j.foodres.2012.05.012.

Assad Bustillos, M., Jonchère, C., Garnier, C., Réguerre, A. L., & Della Valle, G. (2020). Rheological and microstructural characterization of batters and sponge cakes fortified with pea proteins. *Food Hydrocolloids, 101*, 105553. https://doi.org/10.1016/j.foodhyd.2019.105553.

Assad-Bustillos, M., Palier, J., Rabesona, H., Choiset, Y., Della Valle, G., & Feron, G. (2020). Role of the bolus degree of structure on the protein digestibility during in vitro digestion of a pea protein-fortified sponge cake chewed by elderly. *Journal of Texture Studies, 51*(1), 134–143. https://doi.org/10.1111/jtxs.12486.

Assatory, A., Vitelli, M., Rajabzadeh, A. R., & Legge, R. L. (2019). Dry fractionation methods for plant protein, starch and fiber enrichment: A review. *Trends in Food Science and Technology, 86*, 340–351. https://doi.org/10.1016/j.tifs.2019.02.006.

Barac, M. B., Pesic, M. B., Stanojevic, S. P., Kostic, A. Z., & Bivolarevic, V. (2015). Comparative study of the functional properties of three legume seed isolates: Adzuki, pea and soy bean. *Journal of Food Science and Technology, 52*(5), 2779–2787. https://doi.org/10.1007/s13197-014-1298-6.

Bardini, G., Boukid, F., Carini, E., Curti, E., Pizzigalli, E., & Vittadini, E. (2018). Enhancing dough-making rheological performance of wheat flour by transglutaminase and vital gluten supplementation. *LWT, 91*, 467–476. https://doi.org/10.1016/j.lwt.2018.01.077.

Bar-El Dadon, S., Pascual, C. Y., Eshel, D., Teper-Bamnolker, P., Paloma Ibáñez, M. D., & Reifen, R. (2013). Vicilin and the basic subunit of legumin are putative chickpea allergens. *Food Chemistry, 138*(1), 13–18. https://doi.org/10.1016/j.foodchem.2012.10.031.

Bar-El Dadon, S., Pascual, C. Y., & Reifen, R. (2014). Food allergy and cross-reactivity-chickpea as a test case. *Food Chemistry, 165*, 483–488. https://doi.org/10.1016/j.foodchem.2014.05.138.

Beck, M., Jekle, M., Selmair, P. L., Koehler, P., & Becker, T. (2011). Rheological properties and baking performance of rye dough as affected by transglutaminase. *Journal of Cereal Science, 54*(1), 29–36. https://doi.org/10.1016/j.jcs.2011.01.012.

Beloshapka, A., Buff, P., Fahey, G., & Swanson, K. (2016). Compositional analysis of whole grains, processed grains, grain co-products, and other carbohydrate sources with applicability to pet animal nutrition. *Food, 5*(4), 23. https://doi.org/10.3390/foods5020023.

Bernard, J. K. (2016). Oilseed and oilseed meals. *Reference Module in Food Science*. Elsevier. https://doi.org/10.1016/b978-0-08-100596-5.00756-3.

Bildstein, M., Lohmann, M., Hennigs, C., Krause, A., & Hilz, H. (2008). An enzyme-based extraction process for the purification and enrichment of vegetable proteins to be applied in bakery products. *European Food Research and Technology, 228*(2), 177–186. https://doi.org/10.1007/s00217-008-0921-z.

Boukid, F. (2021a). The bright and dark sides of wheat. In *Cereal-based foodstuffs: The backbone of Mediterranean cuisine* (pp. 231–246). Springer International Publishing. https://doi.org/10.1007/978-3-030-69228-5_9.

Boukid, F. (2021b). Oat proteins as emerging ingredients for food formulation: Where we stand? *European Food Research and Technology, 247*(3), 535–544. Springer Science and Business Media Deutschland GmbH. https://doi.org/10.1007/s00217-020-03661-2.

Boukid, F., Carini, E., Curti, E., Bardini, G., Pizzigalli, E., & Vittadini, E. (2018). Effectiveness of vital gluten and transglutaminase in the improvement of physico-chemical properties of fresh bread. *LWT, 92*, 465–470. https://doi.org/10.1016/J.LWT.2018.02.059.

Boukid, F., Folloni, S., Sforza, S., Vittadini, E., & Prandi, B. (2018). Current trends in ancient grains-based foodstuffs: Insights into nutritional aspects and technological applications. *Comprehensive Reviews in Food Science and Food Safety, 17*(1), 123–136. https://doi.org/10.1111/1541-4337.12315.

Boukid, F., Mejri, M., Pellegrini, N., Sforza, S., & Prandi, B. (2017). How looking for celiac-safe wheat can influence its technological properties. *Comprehensive Reviews in Food Science and Food Safety, 16*(5). https://doi.org/10.1111/1541-4337.12288.

Boukid, F., Prandi, B., Sforza, S., Sayar, R., Seo, Y. W., Mejri, M., et al. (2017). Understanding the effects of genotype, growing year, and breeding on tunisian durum wheat allergenicity. 1. The Baker's asthma case. *Journal of Agricultural and Food Chemistry, 65*(28). https://doi.org/10.1021/acs.jafc.7b02040.

Boukid, F., Rosell, C. M., & Castellari, M. (2021). Pea protein ingredients: A mainstream ingredient to (re)formulate innovative foods and beverages. *Trends in Food Science & Technology, 110*, 729–742. https://doi.org/10.1016/j.tifs.2021.02.040.

Boukid, F., Rosell, C. M., Rosene, S., Bover-Cid, S., & Castellari, M. (2021). Non-animal proteins as cutting-edge ingredients to reformulate animal-free foodstuffs: Present status and future perspectives. *Critical Reviews in Food Science and Nutrition*, 1–31. https://doi.org/10.1080/10408398.2021.1901649.

Boukid, F., Vittadini, E., Lusuardi, F., Ganino, T., Carini, E., Morreale, F., et al. (2019). Does cell wall integrity in legumes flours modulate physiochemical quality and in vitro starch hydrolysis of gluten-free bread? *Journal of Functional Foods, 59*, 110–118. https://doi.org/10.1016/j.jff.2019.05.034.

Boukid, F., Zannini, E., Carini, E., & Vittadini, E. (2019). Pulses for bread fortification: A necessity or a choice? *Trends in Food Science and Technology, 88*, 416–428. Elsevier Ltd. https://doi.org/10.1016/j.tifs.2019.04.007.

Boye, J. I., Aksay, S., Roufik, S., Ribéreau, S., Mondor, M., Farnworth, E., et al. (2010). Comparison of the functional properties of pea, chickpea and lentil protein concentrates processed using ultrafiltration and isoelectric precipitation techniques. *Food Research International, 43*(2), 537–546. https://doi.org/10.1016/j.foodres.2009.07.021.

Boye, J., Zare, F., & Pletch, A. (2010). Pulse proteins: Processing, characterization, functional properties and applications in food and feed. *Food Research International, 43*(2), 414–431. https://doi.org/10.1016/j.foodres.2009.09.003.

Bresciani, A., & Marti, A. (2019). Using pulses in baked products: Lights, shadows, and potential solutions. *Food, 8*(10). MDPI Multidisciplinary Digital Publishing Institute https://doi.org/10.3390/foods8100451.

Brückner-Gühmann, M., Benthin, A., & Drusch, S. (2019). Enrichment of yoghurt with oat protein fractions: Structure formation, textural properties and sensory evaluation. *Food Hydrocolloids, 86*, 146–153. https://doi.org/10.1016/j.foodhyd.2018.03.019.

Campbell, L., Rempel, C. B., & Wanasundara, J. P. D. (2016). Canola/rapeseed protein: Future opportunities and directions—Workshop proceedings of IRC 2015. *Plants, 5*(2), 281–284. https://doi.org/10.3390/plants5020017.

Cao, Z., Liu, Z., Zhang, H., Wang, J., & Ren, S. (2020). Protein particles ameliorate the mechanical properties of highly polyunsaturated oil-based whipped cream: A possible mode of action. *Food Hydrocolloids, 99*. https://doi.org/10.1016/j.foodhyd.2019.105350.

Catassi, C., Kryszak, D., Bhatti, B., Sturgeon, C., Helzlsouer, K., Clipp, S. L., et al. (2010). Natural history of celiac disease autoimmunity in a USA cohort followed since 1974. *Annals of Medicine, 42*(7), 530–538. https://doi.org/10.3109/07853890.2010.514285.

Chalamaiah, M., Yu, W., & Wu, J. (2018). Immunomodulatory and anticancer protein hydrolysates (peptides) from food proteins: A review. *Food Chemistry, 245*, 205–222. Elsevier Ltd. https://doi.org/10.1016/j.foodchem.2017.10.087.

Chang, Y. W., Alli, I., Konishi, Y., & Ziomek, E. (2011). Characterization of protein fractions from chickpea (Cicer arietinum L.) and oat (Avena sativa L.) seeds using proteomic techniques. *Food Research International, 44*(9), 3094–3104. https://doi.org/10.1016/j.foodres.2011.08.001.

Chao, D., & Aluko, R. E. (2018). Modification of the structural, emulsifying, and foaming properties of an isolated pea protein by thermal pretreatment. *CYTA - Journal of Food, 16*(1), 357–366. https://doi.org/10.1080/19476337.2017.1406536.

Chao, D., Jung, S., & Aluko, R. E. (2018). Physicochemical and functional properties of high pressure-treated isolated pea protein. *Innovative Food Science and Emerging Technologies, 45*, 179–185. https://doi.org/10.1016/j.ifset.2017.10.014.

Chéreau, D., Videcoq, P., Ruffieux, C., Pichon, L., Motte, J. C., Belaid, S., et al. (2016). Combination of existing and alternative technologies to promote oilseeds and pulses proteins in food applications. *OCL - Oilseeds and Fats, Crops and Lipids, 41*(1). https://doi.org/10.1051/ocl/2016020.

Chihi, M. L., Sok, N., & Saurel, R. (2018). Acid gelation of mixed thermal aggregates of pea globulins and β-lactoglobulin. *Food Hydrocolloids, 85*, 120–128. https://doi.org/10.1016/j.foodhyd.2018.07.006.

Coda, R., Melama, L., Rizzello, C. G., Curiel, J. A., Sibakov, J., Holopainen, U., et al. (2015). Effect of air classification and fermentation by lactobacillus plantarum VTT E-133328 on faba bean (Vicia faba L.) flour nutritional properties. *International Journal of Food Microbiology, 193*, 34–42. https://doi.org/10.1016/j.ijfoodmicro.2014.10.012.

De Angelis, M., Cassone, A., Rizzello, C. G., Gagliardi, F., Minervini, F., Calasso, M., et al. (2010). Mechanism of degradation of immunogenic gluten epitopes from triticum turgidum L. var. durum by sourdough lactobacilli and fungal proteases. *Applied and Environmental Microbiology, 76*(2), 508–518. https://doi.org/10.1128/AEM.01630-09.

de Punder, K., & Pruimboom, L. (2013). The dietary intake of wheat and other cereal grains and their role in inflammation. *Nutrients, 5*(12), 771–787. https://doi.org/10.3390/nu5030771.

Delcour, J. A., Joye, I. J., Pareyt, B., Wilderjans, E., Brijs, K., & Lagrain, B. (2012). Wheat gluten functionality as a quality determinant in cereal-based food products. *Annual Review of Food Science and Technology*, *3*(1), 469–492. https://doi.org/10.1146/annurev-food-022811-101303.

Dhillon, G. S., Kaur, S., Oberoi, H. S., Spier, M. R., & Brar, S. K. (2016). Agricultural-based protein by-products: Characterization and applications. In *Protein byproducts: Transformation from environmental burden into value-added products* (pp. 21–36). Elsevier Inc. https://doi.org/10.1016/B978-0-12-802391-4.00002-1.

Drulyte, D., & Orlien, V. (2019). The effect of processing on digestion of legume proteins. *Food*, *8*(6). https://doi.org/10.3390/foods8060224.

Eckert, E., Han, J., Swallow, K., Tian, Z., Jarpa-Parra, M., & Chen, L. (2019). Effects of enzymatic hydrolysis and ultrafiltration on physicochemical and functional properties of faba bean protein. *Cereal Chemistry*, *96*(4), 725–741. https://doi.org/10.1002/cche.10169.

Espinosa-Ramírez, J., Garzon, R., Serna-Saldivar, S. O., & Rosell, C. M. (2018). Mimicking gluten functionality with β-conglycinin concentrate: Evaluation in gluten free yeast-leavened breads. *Food Research International*, *106*, 64–70. https://doi.org/10.1016/j.foodres.2017.12.055.

Espinosa-Ramírez, J., & Serna-Saldívar, S. O. (2019). Wet-milled chickpea coproduct as an alternative to obtain protein isolates. *LWT*, *115*. https://doi.org/10.1016/j.lwt.2019.108468.

European Parliament and of the Council. (2014). Commission implementing decision of 1 July 2014 authorising the placing on the market of rapeseed protein as a novel food ingredient under regulation (EC) no 258/97 of the European Parliament and of the council. *Off J Eur Union L:196/27*..

FAO. (2019). *Oat production*. http://www.fao.org/home/en/.

Farbo, M. G., Fadda, C., Marceddu, S., Conte, P., Del Caro, A., & Piga, A. (2020). Improving the quality of dough obtained with old durum wheat using hydrocolloids. *Food Hydrocolloids*, *101*. https://doi.org/10.1016/J.FOODHYD.2019.105467.

Fasano, A., Sapone, A., Zevallos, V., & Schuppan, D. (2015). Nonceliac gluten sensitivity. *Gastroenterology*, *148*(6), 1195–1204. https://doi.org/10.1053/j.gastro.2014.12.049.

Felix, M., Lopez-Osorio, A., Romero, A., & Guerrero, A. (2018). Faba bean protein flour obtained by densification: A sustainable method to develop protein concentrates with food applications. *LWT*, *93*, 563–569. https://doi.org/10.1016/j.lwt.2018.03.078.

Fiocchi, A., Dahdah, L., Riccardi, C., Mazzina, O., & Fierro, V. (2016). Preacutionary labelling of cross-reactive foods: The case of rapeseed. *Asthma Research and Practice*, *2*(1). https://doi.org/10.1186/s40733-016-0028-4.

Fišteš, A. (2021). The evolution of milling process. In *Cereal-based foodstuffs: The backbone of Mediterranean cuisine* (pp. 19–45). Springer International Publishing. https://doi.org/10.1007/978-3-030-69228-5_2.

Gao, J., Weng, W., Yan, Y., Wang, Y., & Wang, Q. (2020). Comparison of protein extraction methods from excess activated sludge. *Chemosphere*, *249*, 126107. https://doi.org/10.1016/j.chemosphere.2020.126107.

Garrido, T., Etxabide, A., Leceta, I., Cabezudo, S., De La Caba, K., & Guerrero, P. (2014). Valorization of soya by-products for sustainable packaging. *Journal of Cleaner Production*, *64*, 228–233. https://doi.org/10.1016/j.jclepro.2013.07.027.

Gasim, S., Hamad, S. A. A., Abdelmula, A., & Mohamed Ahmed, I. A. (2015). Yield and quality attributes of faba bean inbred lines grown under marginal environmental conditions of Sudan. *Food Science and Nutrition*, *3*(6), 539–547. https://doi.org/10.1002/fsn3.245.

Geerts, M., Mienis, E., Nikiforidis, C. V., van der Padt, A., & van der Goot, A. J. (2017). Mildly refined fractions of yellow peas show rich behaviour in thickened oil-in-water emulsions. *Innovative Food Science and Emerging Technologies*, *41*, 251–258. https://doi.org/10.1016/j.ifset.2017.03.009.

Giannou, V., & Tzia, C. (2016). Addition of vital wheat gluten to enhance the quality characteristics of frozen dough products. *Food*, *5*(4), 6. https://doi.org/10.3390/foods5010006.

Giuseppin, M. L. F., Laus, M. C., & Schipper, J. (2014). *WO2014011042A1 - potato protein isolates*. https://patents.google.com/patent/WO2014011042A1/en.

GMInsights. (2018). *Plant protein ingredients market for food application* (p. 2025). https://www.gminsights.com/industry-analysis/plant-protein-ingredients-market.

González-Pérez, S., & Vereijken, J. M. (2007). Sunflower proteins: Overview of their physicochemical, structural and functional properties. *Journal of the Science of Food and Agriculture*, *87*(12), 2173–2191. https://doi.org/10.1002/jsfa.2971.

Gorissen, S. H. M., Crombag, J. J. R., Senden, J. M. G., Waterval, W. A. H., Bierau, J., Verdijk, L. B., et al. (2018). Protein content and amino acid composition of commercially available plant-based protein isolates. *Amino Acids*, *50*(12), 1685–1695. https://doi.org/10.1007/s00726-018-2640-5.

Grabowska, K. J., Tekidou, S., Boom, R. M., & van der Goot, A. J. (2014). Shear structuring as a new method to make anisotropic structures from soy-gluten blends. *Food Research International*, *64*, 743–751. https://doi.org/10.1016/j.foodres.2014.08.010.

Gravely, E., & Fraser, E. (2018). Transitions on the shopping floor: Investigating the role of Canadian supermarkets in alternative protein consumption. *Appetite*, *130*, 146–156. https://doi.org/10.1016/j.appet.2018.08.018.

Gularte, M. A., Gómez, M., & Rosell, C. M. (2012). Impact of legume flours on quality and in vitro digestibility of starch and protein from gluten-free cakes. *Food and Bioprocess Technology*, *5*(8), 3142–3150. https://doi.org/10.1007/s11947-011-0642-3.

Han, A., Romero, H. M., Nishijima, N., Ichimura, T., Handa, A., Xu, C., et al. (2019). Effect of egg white solids on the rheological properties and bread making performance of gluten-free batter. *Food Hydrocolloids*, *87*, 287–296. https://doi.org/10.1016/j.foodhyd.2018.08.022.

Harwatt, H., Sabaté, J., Eshel, G., Soret, S., & Ripple, W. (2017). Substituting beans for beef as a contribution toward US climate change targets. *Climatic Change*, *143*(1–2), 261–270. https://doi.org/10.1007/s10584-017-1969-1.

Heusala, H., Sinkko, T., Mogensen, L., & Knudsen, M. T. (2020). Carbon footprint and land use of food products containing oat protein concentrate. *Journal of Cleaner Production*, *276*, 122938. https://doi.org/10.1016/j.jclepro.2020.122938.

Heusala, H., Sinkko, T., Sözer, N., Hytönen, E., Mogensen, L., & Knudsen, M. T. (2020). Carbon footprint and land use of oat and faba bean protein concentrates using a life cycle assessment approach. *Journal of Cleaner Production*, *242*, 118376. https://doi.org/10.1016/j.jclepro.2019.118376.

Hoehnel, A., Axel, C., Bez, J., Arendt, E. K., & Zannini, E. (2019). Comparative analysis of plant-based high-protein ingredients and their impact on quality of high-protein bread. *Journal of Cereal Science*, *89*. https://doi.org/10.1016/j.jcs.2019.102816.

Hsiao, Y. H., Yu, C. J., Li, W. T., & Hsieh, J. F. (2015). Coagulation of β-conglycinin, glycinin and isoflavones induced by calcium chloride in soymilk. *Scientific Reports*, *5*. https://doi.org/10.1038/srep13018.

Huang, L., Jia, S., Zhang, W., Ma, L., & Ding, X. (2020). Aggregation and emulsifying properties of soybean protein isolate pretreated by combination of dual-frequency ultrasound and ionic liquids. *Journal of Molecular Liquids*, *301*, 112394. https://doi.org/10.1016/j.molliq.2019.112394.

Ito, K. (2015). Grain and legume allergy. *Chemical Immunology and Allergy*.. https://doi.org/10.1159/000375468.

Jafari, M., Rajabzadeh, A. R., Tabtabaei, S., Marsolais, F., & Legge, R. L. (2016). Physicochemical characterization of a navy bean (Phaseolus vulgaris) protein fraction produced using a solvent-free method. *Food Chemistry*, *208*, 35–41. https://doi.org/10.1016/j.foodchem.2016.03.102.

Jakobek, L. (2015). Interactions of polyphenols with carbohydrates, lipids and proteins. *Food Chemistry, 175*, 556–567. https://doi.org/10.1016/j.foodchem.2014.12.013.

Jones, O. G. (2016). Recent advances in the functionality of non-animal-sourced proteins contributing to their use in meat analogs. *Current Opinion in Food Science, 7*, 7–13. Elsevier Ltd. https://doi.org/10.1016/j.cofs.2015.08.002.

Karami, Z., Peighambardoust, S. H., Hesari, J., Akbari-Adergani, B., & Andreu, D. (2019). Antioxidant, anticancer and ACE-inhibitory activities of bioactive peptides from wheat germ protein hydrolysates. *Food Bioscience, 32*, 100450. https://doi.org/10.1016/j.fbio.2019.100450.

Katz, Y., Gutierrez-Castrellon, P., González, M. G., Rivas, R., Lee, B. W., & Alarcon, P. (2014). A comprehensive review of sensitization and allergy to soy-based products. *Clinical Reviews in Allergy and Immunology, 46*(3), 272–281. Humana Press Inc https://doi.org/10.1007/s12016-013-8404-9.

Khuzwayo, T. A., Taylor, J. R. N., & Taylor, J. (2020). Influence of dough sheeting, flour pre-gelatinization and zein inclusion on maize bread dough functionality. *LWT, 121*. https://doi.org/10.1016/j.lwt.2019.108993.

Klost, M., & Drusch, S. (2019). Functionalisation of pea protein by tryptic hydrolysis – Characterisation of interfacial and functional properties. *Food Hydrocolloids, 86*, 134–140. https://doi.org/10.1016/j.foodhyd.2018.03.013.

Korus, J., Chmielewska, A., Witczak, M., Ziobro, R., & Juszczak, L. (2021). Rapeseed protein as a novel ingredient of gluten-free bread. *European Food Research and Technology*.. https://doi.org/10.1007/s00217-021-03768-0.

Kotecka-Majchrzak, K., Sumara, A., Fornal, E., & Montowska, M. (2020). Oilseed proteins – Properties and application as a food ingredient. *Trends in Food Science and Technology, 106*, 160–170. Elsevier Ltd. https://doi.org/10.1016/j.tifs.2020.10.004.

Kulkarni, A., Ananthanarayan, L., & Raman, K. (2013). Identification of putative and potential cross-reactive chickpea (Cicer arietinum) allergens through an in silico approach. *Computational Biology and Chemistry, 47*, 149–155. https://doi.org/10.1016/j.compbiolchem.2013.08.003.

Kumar, D., Kumar, S., Verma, A. K., Sharma, A., Tripathi, A., Chaudhari, B. P., et al. (2014). Hypersensitivity linked to exposure of broad bean protein(s) in allergic patients and BALB/c mice. *Nutrition, 30*(7–8), 903–914. https://doi.org/10.1016/j.nut.2013.11.024.

Ladjal Ettoumi, Y., Chibane, M., & Romero, A. (2016). Emulsifying properties of legume proteins at acidic conditions: Effect of protein concentration and ionic strength. *LWT - Food Science and Technology, 66*, 260–266. https://doi.org/10.1016/j.lwt.2015.10.051.

Lan, Y., Ohm, J.-B., Chen, B., & Rao, J. (2019). Phase behavior, thermodynamic and microstructure of concentrated pea protein isolate-pectin mixture: Effect of pH, biopolymer ratio and pectin charge density. *Food Hydrocolloids, 105556*. https://doi.org/10.1016/j.foodhyd.2019.105556.

Langton, M., Ehsanzamir, S., Karkehabadi, S., Feng, X., Johansson, M., & Johansson, D. P. (2020). Gelation of faba bean proteins - effect of extraction method, pH and NaCl. *Food Hydrocolloids, 103*. https://doi.org/10.1016/j.foodhyd.2019.105622.

Lavine, E., & Ben-Shoshan, M. (2015). Allergy to sunflower seed and sunflower butter as proposed vehicle for sensitization. *Allergy, Asthma and Clinical Immunology, 11*(1), 1–3. https://doi.org/10.1186/s13223-014-0065-6.

Le Roux, L., Ménard, O., Chacon, R., Dupont, D., Jeantet, R., Deglaire, A., et al. (2020). Are faba bean and pea proteins potential whey protein substitutes in infant formulas? An in vitro dynamic digestion approach. *Food, 9*(3). https://doi.org/10.3390/foods9030362.

Li, H., & Aluko, R. E. (2010). Identification and inhibitory properties of multifunctional peptides from pea protein hydrolysate. *Journal of Agricultural and Food Chemistry, 58*(21), 11471–11476. https://doi.org/10.1021/jf102538g.

Li, Y., & Damodaran, S. (2017). In vitro digestibility and IgE reactivity of enzymatically cross-linked heterologous protein polymers. *Food Chemistry*, *221*, 1151–1157. https://doi.org/10.1016/j.foodchem.2016.11.044.

Li, T., Wang, L., Sun, D., Li, Y., & Chen, Z. (2019). Effect of enzymolysis-assisted electron beam irradiation on structural characteristics and antioxidant activity of rice protein. *Journal of Cereal Science*, *89*. https://doi.org/10.1016/j.jcs.2019.102789.

Liang, J., & Chen, R. (2018). Impact of cross-linking mode on the physical properties of zein/PVA composite films. *Food Packaging and Shelf Life*, *18*, 101–106. https://doi.org/10.1016/j.fpsl.2018.10.003.

Lim, L. T., Mendes, A. C., & Chronakis, I. S. (2019). Electrospinning and electrospraying technologies for food applications. *Advances in Food and Nutrition Research*, *88*, 167–234. Academic Press Inc. https://doi.org/10.1016/bs.afnr.2019.02.005.

Lin, M., Tay, S. H., Yang, H., Yang, B., & Li, H. (2017). Development of eggless cakes suitable for lacto-vegetarians using isolated pea proteins. *Food Hydrocolloids*, *69*, 440–449. https://doi.org/10.1016/j.foodhyd.2017.03.014.

Liu, Z., Cao, X., Ren, S., Wang, J., & Zhang, H. (2019). Physicochemical characterization of a zein prepared using a novel aqueous extraction technology and tensile properties of the zein film. *Industrial Crops and Products*, *130*, 57–62. https://doi.org/10.1016/j.indcrop.2018.12.071.

Liu, L. H., Hung, T. V., & Bennett, L. (2008). Extraction and characterization of chickpea (Cicer arietinum) albumin and globulin. *Journal of Food Science*, *73*(5), C299–C305. https://doi.org/10.1111/j.1750-3841.2008.00773.x.

Liu, M., Liang, Y., Zhang, H., Wu, G., Wang, L., Qian, H., et al. (2018). Production of a recombinant carrot antifreeze protein by Pichia pastoris GS115 and its cryoprotective effects on frozen dough properties and bread quality. *LWT*, *96*, 543–550. https://doi.org/10.1016/j.lwt.2018.05.074.

Liu, Y., Wang, Z., Li, H., Liang, M., & Yang, L. (2016). In vitro antioxidant activity of rice protein affected by alkaline degree and gastrointestinal protease digestion. *Journal of the Science of Food and Agriculture*, *96*(15), 4940–4950. https://doi.org/10.1002/jsfa.7877.

Majzoobi, M., Raiss Jalali, A., & Farahnaky, A. (2016). Impact of whole oat flour on dough properties and quality of fresh and stored part-baked bread. *Journal of Food Quality*, *39*(6), 620–626. https://doi.org/10.1111/jfq.12237.

Mancebo, C. M., Rodriguez, P., & Gómez, M. (2016). Assessing rice flour-starch-protein mixtures to produce gluten free sugar-snap cookies. *LWT - Food Science and Technology*, *67*, 127–132. https://doi.org/10.1016/j.lwt.2015.11.045.

Mansilla, P. S., Nazar, M. C., & Pérez, G. T. (2020). Flour functional properties of purple maize (Zea mays L.) from Argentina. Influence of environmental growing conditions. *International Journal of Biological Macromolecules*, *146*, 311–319. https://doi.org/10.1016/j.ijbiomac.2019.12.246.

Martínez, M. M., & Gómez, M. (2017). Rheological and microstructural evolution of the most common gluten-free flours and starches during bread fermentation and baking. *Journal of Food Engineering*, *197*, 78–86. https://doi.org/10.1016/j.jfoodeng.2016.11.008.

Martinez, M., Stone, A. K., Yovchev, A. G., Peter, R., Vandenberg, A., & Nickerson, M. T. (2016). Effect of genotype and environment on the surface characteristics and functionality of air-classified faba bean protein concentrates. *European Food Research and Technology*, *242*(11), 1903–1911. https://doi.org/10.1007/s00217-016-2690-4.

Masure, H. G., Wouters, A. G. B., Fierens, E., & Delcour, J. A. (2019). Impact of egg white and soy proteins on structure formation and crumb firming in gluten-free breads. *Food Hydrocolloids*, *95*, 406–417. https://doi.org/10.1016/j.foodhyd.2019.04.062.

Mathai, J. K., Liu, Y., & Stein, H. H. (2017). Values for digestible indispensable amino acid scores (DIAAS) for some dairy and plant proteins may better describe protein quality than values calculated using the concept for protein digestibility-corrected amino acid scores (PDCAAS). *British Journal of Nutrition, 117*(4), 490–499. https://doi.org/10.1017/S0007114517000125.

Matos, M. E., Sanz, T., & Rosell, C. M. (2014). Establishing the function of proteins on the rheological and quality properties of rice based gluten free muffins. *Food Hydrocolloids, 35*, 150–158. https://doi.org/10.1016/j.foodhyd.2013.05.007.

Mattice, K. D., & Marangoni, A. G. (2020). Comparing methods to produce fibrous material from zein. *Food Research International, 128*. https://doi.org/10.1016/j.foodres.2019.108804.

McGraw, N. J., Krul, E. S., Grunz-Borgmann, E., & Parrish, A. R. (2016). Soy-based renoprotection. *World Journal of Nephrology, 5*(3), 233. https://doi.org/10.5527/wjn.v5.i3.233.

Meinlschmidt, P., Schweiggert-Weisz, U., Brode, V., & Eisner, P. (2016). Enzyme assisted degradation of potential soy protein allergens with special emphasis on the technofunctionality and the avoidance of a bitter taste formation. *LWT - Food Science and Technology, 68*, 707–716. https://doi.org/10.1016/j.lwt.2016.01.023.

Meinlschmidt, P., Sussmann, D., Schweiggert-Weisz, U., & Eisner, P. (2016). Enzymatic treatment of soy protein isolates: Effects on the potential allergenicity, technofunctionality, and sensory properties. *Food Science & Nutrition, 4*(1), 11–23. https://doi.org/10.1002/fsn3.253.

Mennini, M., Dahdah, L., Mazzina, O., & Fiocchi, A. (2016). Lupin and other potentially cross-reactive allergens in Peanut allergy. *Current Allergy and Asthma Reports, 16*(12). Current Medicine Group LLC 1. https://doi.org/10.1007/s11882-016-0668-8.

Millar, K. A., Barry-Ryan, C., Burke, R., McCarthy, S., & Gallagher, E. (2019). Dough properties and baking characteristics of white bread, as affected by addition of raw, germinated and toasted pea flour. *Innovative Food Science and Emerging Technologies, 56*, 102189. https://doi.org/10.1016/j.ifset.2019.102189.

Mondor, M., Ippersiel, D., Lamarche, F., & Boye, J. I. (2004). Production of soy protein concentrates using a combination of electroacidification and ultrafiltration. *Journal of Agricultural and Food Chemistry, 52*(23), 6991–6996. https://doi.org/10.1021/jf0400922.

Mune Mune, M. A., Minka, S. R., & Henle, T. (2018). Investigation on antioxidant, angiotensin converting enzyme and dipeptidyl peptidase IV inhibitory activity of Bambara bean protein hydrolysates. *Food Chemistry, 250*, 162–169. https://doi.org/10.1016/j.foodchem.2018.01.001.

Murtagh, G. J., Archer, D. B., Dumoulin, M., Ridout, S., Matthews, S., Arshad, S. H., et al. (2003). In vitro stability and immunoreactivity of the native and recombinant plant food 2S albumins Ber e 1 and SFA-8. *Clinical and Experimental Allergy, 33*(8), 1147–1152. https://doi.org/10.1046/j.1365-2222.2003.01736.x.

Nordlund, E., Katina, K., Mykkänen, H., & Poutanen, K. (2016). Distinct characteristics of rye and wheat breads impact on their in vitro gastric disintegration and in vivo glucose and insulin responses. *Food, 5*(2), 1–13. https://doi.org/10.3390/foods5020024.

Nosworthy, M. G., Neufeld, J., Frohlich, P., Young, G., Malcolmson, L., & House, J. D. (2017). Determination of the protein quality of cooked Canadian pulses. *Food Science and Nutrition, 5*(4), 896–903. https://doi.org/10.1002/fsn3.473.

Oest, M., Bindrich, U., Voß, A., Kaiser, H., & Rohn, S. (2020). Rye bread defects: Analysis of composition and further influence factors as determinants of dry-baking. *Food, 9*(12), 1900. https://doi.org/10.3390/foods9121900.

Öhgren, C., Fabregat, N., & Langton, M. (2016). Quality of bread baked from frozen dough – Effects of rye, and sugar content, kneading time and proofing profile. *LWT - Food Science and Technology, 68*, 626–633. https://doi.org/10.1016/j.lwt.2015.12.069.

Ortolan, F., Corrêa, G. P., da Cunha, R. L., & Steel, C. J. (2017). Rheological properties of vital wheat glutens with water or sodium chloride. *LWT - Food Science and Technology*, *79*, 647–654. https://doi.org/10.1016/J.LWT.2017.01.059.

Oseyko, M., Romanovska, T., & Shevchyk, V. (2020). Justification of the amino acid composition of sunflower proteins for dietary and functional products. *Ukrainian Food Journal*, *9*(2), 394–404. https://nuft.edu.ua/doi/doc/ufj/2020/2/11.pdf.

Ozdal, T., Capanoglu, E., & Altay, F. (2013). A review on protein–phenolic interactions and associated changes. *Food Research International*, *51*(2), 954–970. https://doi.org/10.1016/j.foodres.2013.02.009.

Padhi, E. M. T., & Ramdath, D. D. (2017). A review of the relationship between pulse consumption and reduction of cardiovascular disease risk factors. *Journal of Functional Foods*, *38*, 635–643. Elsevier Ltd. https://doi.org/10.1016/j.jff.2017.03.043.

Paredes-López, O., Ordorica-Falomir, C., & Olivares-Vázquez, M. R. (1991). Chickpea protein isolates: Physicochemical, functional and nutritional characterization. *Journal of Food Science*, *56*(3), 726–729. https://doi.org/10.1111/j.1365-2621.1991.tb05367.x.

Patil, S. P., Niphadkar, P. V., & Bapat, M. M. (2001). Chickpea: A major food allergen in the Indian subcontinent and its clinical and immunochemical correlation. *Annals of Allergy, Asthma and Immunology*, *87*(2), 140–145. https://doi.org/10.1016/S1081-1206(10)62209-0.

Pedro, A.C., Granato, D., & Rosso, N.D. (2016). Extraction of anthocyanins and polyphenols from black rice (Oryza sativa L.) by modeling and assessing their reversibility and stability. *Food Chemistry*, *191*, 12–20. https://doi.org/10.1016/j.foodchem.2015.02.045.

Pelgrom, P. J. M., Boom, R. M., & Schutyser, M. A. I. (2015). Method development to increase protein enrichment during dry fractionation of starch-rich legumes. *Food and Bioprocess Technology*, *8*(7), 1495–1502. https://doi.org/10.1007/s11947-015-1513-0.

Pelgrom, P. J. M., Wang, J., Boom, R. M., & Schutyser, M. A. I. (2015). Pre- and post-treatment enhance the protein enrichment from milling and air classification of legumes. *Journal of Food Engineering*, *155*, 53–61. https://doi.org/10.1016/j.jfoodeng.2015.01.005.

Phongthai, S., D'Amico, S., Schoenlechner, R., & Rawdkuen, S. (2016). Comparative study of rice bran protein concentrate and egg albumin on gluten-free bread properties. *Journal of Cereal Science*, *72*, 38–45. https://doi.org/10.1016/j.jcs.2016.09.015.

Pico, J., Reguilón, M. P., Bernal, J., & Gómez, M. (2019). Effect of rice, pea, egg white and whey proteins on crust quality of rice flour-corn starch based gluten-free breads. *Journal of Cereal Science*, *86*, 92–101. https://doi.org/10.1016/j.jcs.2019.01.014.

Pisapia, L., Camarca, A., Picascia, S., Bassi, V., Barba, P., Del Pozzo, G., et al. (2016). HLA-DQ2.5 genes associated with celiac disease risk are preferentially expressed with respect to non-predisposing HLA genes: Implication for anti-gluten T cell response. *Journal of Autoimmunity*, *70*, 63–72. https://doi.org/10.1016/j.jaut.2016.03.016.

Prosekov, A., Babich, O., Kriger, O., Ivanova, S., Pavsky, V., Sukhikh, S., et al. (2018). Functional properties of the enzyme-modified protein from oat bran. *Food Bioscience*, *24*, 46–49. https://doi.org/10.1016/j.fbio.2018.05.003.

Puumalainen, T. J., Puustinen, A., Poikonen, S., Turjanmaa, K., Palosuo, T., & Vaali, K. (2015). Proteomic identification of allergenic seed proteins, napin and cruciferin, from cold-pressed rapeseed oils. *Food Chemistry*, *175*, 381–385. https://doi.org/10.1016/j.foodchem.2014.11.084.

Rani, S., Pooja, K., & Pal, G. K. (2018). Exploration of rice protein hydrolysates and peptides with special reference to antioxidant potential: Computational derived approaches for bio-activity determination. *Trends in Food Science and Technology*, *80*, 61–70. Elsevier Ltd. https://doi.org/10.1016/j.tifs.2018.07.013.

ReportsandData. (2021). *Canola protein market | size, share | Analysis & Demand*. https://www.reportsanddata.com/report-detail/canola-protein-market.

Riaz, M. N. (2011). Texturized vegetable proteins. In *Handbook of food proteins* (pp. 395–418). Elsevier. https://doi.org/10.1533/9780857093639.395.
Rodehutscord, M., Rückert, C., Maurer, H. P., Schenkel, H., Schipprack, W., Bach Knudsen, K. E., et al. (2016). Variation in chemical composition and physical characteristics of cereal grains from different genotypes. *Archives of Animal Nutrition, 70*(2), 87–107. https://doi.org/10.1080/1745039X.2015.1133111.
Roy, F., Boye, J. I., & Simpson, B. K. (2010). Bioactive proteins and peptides in pulse crops: Pea, chickpea and lentil. *In Food Research International, 43*(2), 432–442. https://doi.org/10.1016/j.foodres.2009.09.002.
Ruiz, G. A., Xiao, W., Van Boekel, M., Minor, M., & Stieger, M. (2016). Effect of extraction pH on heat-induced aggregation, gelation and microstructure of protein isolate from quinoa (Chenopodium quinoa Willd). *Food Chemistry, 209*, 203–210. https://doi.org/10.1016/j.foodchem.2016.04.052.
Rutherfurd, S. M., Fanning, A. C., Miller, B. J., & Moughan, P. J. (2015). Protein digestibility-corrected amino acid scores and digestible indispensable amino acid scores differentially describe protein quality in growing male rats. *The Journal of Nutrition, 145*(2), 372–379. https://doi.org/10.3945/jn.114.195438.
Sá, A. G. A., Moreno, Y. M. F., & Carciofi, B. A. M. (2020). Plant proteins as high-quality nutritional source for human diet. *Trends in Food Science and Technology, 97*, 170–184. Elsevier Ltd. https://doi.org/10.1016/j.tifs.2020.01.011.
Sahagún, M., & Gómez, M. (2018a). Assessing influence of protein source on characteristics of gluten-free breads optimising their hydration level. *Food and Bioprocess Technology, 11*(9), 1686–1694. https://doi.org/10.1007/s11947-018-2135-0.
Sahagún, M., & Gómez, M. (2018b). Influence of protein source on characteristics and quality of gluten-free cookies. *Journal of Food Science and Technology, 55*(10), 4131–4138. https://doi.org/10.1007/s13197-018-3339-z.
Salah, K., Olkhovatov, E. A., & Aïder, M. (2019). Effect of canola proteins on rice flour bread and mathematical modelling of the baking process. *Journal of Food Science and Technology, 56*(8), 3744–3753. https://doi.org/10.1007/s13197-019-03842-2.
Saldanha do Carmo, C., Silventoinen, P., Nordgård, C. T., Poudroux, C., Dessev, T., Zobel, H., et al. (2020). Is dehulling of peas and faba beans necessary prior to dry fractionation for the production of protein- and starch-rich fractions? Impact on physical properties, chemical composition and techno-functional properties. *Journal of Food Engineering, 278*. https://doi.org/10.1016/j.jfoodeng.2020.109937.
Sapone, A., Bai, J. C., Ciacci, C., Dolinsek, J., Green, P. H. R., Hadjivassiliou, M., et al. (2012). Spectrum of gluten-related disorders: Consensus on new nomenclature and classification. *BMC Medicine, 10*. https://doi.org/10.1186/1741-7015-10-13.
Sarabhai, S., Indrani, D., Vijaykrishnaraj, M., Milind, Arun Kumar, V., & Prabhasankar, P. (2015). Effect of protein concentrates, emulsifiers on textural and sensory characteristics of gluten free cookies and its immunochemical validation. *Journal of Food Science and Technology, 52*(6), 3763–3772. https://doi.org/10.1007/s13197-014-1432-5.
Schmiele, M., Ferrari Felisberto, M. H., Pedrosa Silva Clerici, M. T., & Chang, Y. K. (2017). Mixolab™ for rheological evaluation of wheat flour partially replaced by soy protein hydrolysate and fructooligosaccharides for bread production. *LWT - Food Science and Technology, 76*, 259–269. https://doi.org/10.1016/j.lwt.2016.07.014.
Schober, T. J., Bean, S. R., Boyle, D. L., & Park, S. H. (2008). Improved viscoelastic zein-starch doughs for leavened gluten-free breads: Their rheology and microstructure. *Journal of Cereal Science, 48*(3), 755–767. https://doi.org/10.1016/j.jcs.2008.04.004.
Schutyser, M. A. I., Pelgrom, P. J. M., van der Goot, A. J., & Boom, R. M. (2015). Dry fractionation for sustainable production of functional legume protein concentrates. *Trends in Food Science and Technology, 45*(2), 327–335. https://doi.org/10.1016/j.tifs.2015.04.013.

Schutyser, M. A. I., & van der Goot, A. J. (2011). The potential of dry fractionation processes for sustainable plant protein production. In *Vol. 22. Trends in Food Science and Technology* (pp. 154–164), Issue 4. https://doi.org/10.1016/j.tifs.2010.11.006.

Seung, Y. C., & Rhee, C. (2004). Mechanical properties and water vapor permeability of edible films made from fractionated soy proteins with ultrafiltration. *LWT - Food Science and Technology*, *37*(8), 833–839. https://doi.org/10.1016/j.lwt.2004.03.009.

Shaabani, S., Yarmand, M. S., Kiani, H., & Emam-Djomeh, Z. (2018). The effect of chickpea protein isolate in combination with transglutaminase and xanthan on the physical and rheological characteristics of gluten free muffins and batter based on millet flour. *LWT - Food Science and Technology*, *90*, 362–372. https://doi.org/10.1016/j.lwt.2017.12.023.

Shewry, P. R., Halford, N. G., Belton, P. S., & Tatham, A. S. (2002). The structure and properties of gluten: An elastic protein from wheat grain. *Philosophical Transactions of the Royal Society B: Biological Sciences*, *357*(1418), 133–142. The Royal Society https://doi.org/10.1098/rstb.2001.1024.

Shewry, P. R., Tatham, A. S., Forde, J., Kreis, M., & Miflin, B. J. (1986). The classification and nomenclature of wheat gluten proteins: A reassessment. *Journal of Cereal Science*, *4*(2), 97–106. https://doi.org/10.1016/S0733-5210(86)80012-1.

Silva, J. V. C., Balakrishnan, G., Schmitt, C., Chassenieux, C., & Nicolai, T. (2018). Heat-induced gelation of aqueous micellar casein suspensions as affected by globular protein addition. *Food Hydrocolloids*, *82*, 258–267. https://doi.org/10.1016/j.foodhyd.2018.04.002.

Silva, J. V. C., Jacquette, B., Amagliani, L., Schmitt, C., Nicolai, T., & Chassenieux, C. (2019). Heat-induced gelation of micellar casein/plant protein oil-in-water emulsions. *Colloids and Surfaces A: Physicochemical and Engineering Aspects*, *569*, 85–92. https://doi.org/10.1016/j.colsurfa.2019.01.065.

Silva, F.d. O., Miranda, T. G., Justo, T., Frasão, B.d. S., Conte-Junior, C. A., Monteiro, M., et al. (2018). Soybean meal and fermented soybean meal as functional ingredients for the production of low-carb, high-protein, high-fiber and high isoflavones biscuits. *LWT*, *90*, 224–231. https://doi.org/10.1016/j.lwt.2017.12.035.

Singh, G. D., Wani, A. A., Kaur, D., & Sogi, D. S. (2008). Characterisation and functional properties of proteins of some Indian chickpea (Cicer arietinum) cultivars. *Journal of the Science of Food and Agriculture*, *88*(5), 778–786. https://doi.org/10.1002/jsfa.3144.

Singhal, A., Stone, A. K., Vandenberg, A., Tyler, R., & Nickerson, M. T. (2016). Effect of genotype on the physicochemical and functional attributes of faba bean (Vicia faba L.) protein isolates. *Food Science and Biotechnology*, *25*(6), 1513–1522. https://doi.org/10.1007/s10068-016-0235-z.

Smith, B. M., Bean, S. R., Selling, G., Sessa, D., & Aramouni, F. M. (2017). Effect of salt and ethanol addition on zein–starch dough and bread quality. *Journal of Food Science*, *82*(3), 613–621. https://doi.org/10.1111/1750-3841.13637.

Stępniewska, S., Hassoon, W. H., Szafrańska, A., Cacak-Pietrzak, G., & Dziki, D. (2019). Procedures for breadmaking quality assessment of rye wholemeal flour. *Food*, *8*(8). https://doi.org/10.3390/foods8080331.

Stone, A. K., Avarmenko, N. A., Warkentin, T. D., & Nickerson, M. T. (2015). Functional properties of protein isolates from different pea cultivars. *Food Science and Biotechnology*, *24*(3), 827–833. https://doi.org/10.1007/s10068-015-0107-y.

Stone, A. K., Karalash, A., Tyler, R. T., Warkentin, T. D., & Nickerson, M. T. (2015). Functional attributes of pea protein isolates prepared using different extraction methods and cultivars. *Food Research International*, *76*(P1), 31–38. https://doi.org/10.1016/j.foodres.2014.11.017.

Tabtabaei, S., Jafari, M., Rajabzadeh, A. R., & Legge, R. L. (2016). Solvent-free production of protein-enriched fractions from navy bean flour using a triboelectrification-based approach. *Journal of Food Engineering*, *174*, 21–28. https://doi.org/10.1016/j.jfoodeng.2015.11.010.

Tan, S. H., Mailer, R. J., Blanchard, C. L., & Agboola, S. O. (2011). Canola proteins for human consumption: Extraction, profile, and functional properties. *Journal of Food Science*, *76*(1), R16. https://doi.org/10.1111/j.1750-3841.2010.01930.x.

Tang, X., & Liu, J. (2017). A comparative study of partial replacement of wheat flour with whey and soy protein on rheological properties of dough and cookie quality. *Journal of Food Quality*, *2017*. https://doi.org/10.1155/2017/2618020.

Taylor, S. L., Remington, B. C., Panda, R., Goodman, R. E., & Baumert, J. L. (2015). Detection and control of soybeans as a food allergen. In *Handbook of food allergen detection and control* (pp. 341–366). Elsevier Ltd. https://doi.org/10.1533/9781782420217.3.341.

Tontul, İ., Kasimoglu, Z., Asik, S., Atbakan, T., & Topuz, A. (2018). Functional properties of chickpea protein isolates dried by refractance window drying. *International Journal of Biological Macromolecules*, *109*, 1253–1259. https://doi.org/10.1016/j.ijbiomac.2017.11.135.

Tuso, P. J., Ismail, M. H., Ha, B. P., & Bartolotto, C. (2013). Nutritional update for physicians: Plant-based diets. *The Permanente Journal*, *17*(2), 61–66. https://doi.org/10.7812/TPP/12-085.

Ukleja-Sokołowska, N., Gawrońska-Ukleja, E., Zbikowska-Gotz, M., Bartuzi, Z., & Sokołowski, Ł. (2016). Sunflower seed allergy. *International Journal of Immunopathology and Pharmacology*, *29*(3), 498–503. https://doi.org/10.1177/0394632016651648.

Uvackova, L., Skultety, L., Bekesova, S., McClain, S., & Hajduch, M. (2013). MSE based multiplex protein analysis quantified important allergenic proteins and detected relevant peptides carrying known epitopes in wheat grain extracts. *Journal of Proteome Research*, *12*(11), 4862–4869. https://doi.org/10.1021/pr400336f.

Van Der Goot, A. J., Pelgrom, P. J. M., Berghout, J. A. M., Geerts, M. E. J., Jankowiak, L., Hardt, N. A., et al. (2016). Concepts for further sustainable production of foods. *Journal of Food Engineering*, *168*, 42–51. Elsevier Ltd. https://doi.org/10.1016/j.jfoodeng.2015.07.010.

Vatansever, S., & Hall, C. (2020). Flavor modification of yellow pea flour using supercritical carbon dioxide + ethanol extraction and response surface methodology. *Journal of Supercritical Fluids*, *156*, 104659. https://doi.org/10.1016/j.supflu.2019.104659.

Vatansever, S., Rao, J., & Hall, C. (2020). Effects of ethanol modified supercritical carbon dioxide extraction and particle size on the physical, chemical, and functional properties of yellow pea flour. *Cereal Chemistry*, *97*(6), 1133–1147. https://doi.org/10.1002/cche.10334.

Verma, A. K., Kumar, S., Das, M., & Dwivedi, P. D. (2013). A comprehensive review of legume allergy. *Clinical Reviews in Allergy and Immunology*, *45*(1), 30–46. https://doi.org/10.1007/s12016-012-8310-6.

Verma, A. K., Kumar, S., Tripathi, A., Chaudhari, B. P., Das, M., & Dwivedi, P. D. (2012). Chickpea (Cicer arietinum) proteins induce allergic responses in nasobronchial allergic patients and BALB/c mice. *Toxicology Letters*, *210*(1), 24–33. https://doi.org/10.1016/j.toxlet.2012.01.011.

Verma, A. K., Sharma, A., Kumar, S., Gupta, R. K., Kumar, D., Gupta, K., et al. (2016). Purification, characterization and allergenicity assessment of 26 kDa protein, a major allergen from Cicer arietinum. *Molecular Immunology*, *74*, 113–124. https://doi.org/10.1016/j.molimm.2016.05.002.

Verma, D. K., & Srivastav, P. P. (2017). Proximate composition, mineral content and fatty acids analyses of aromatic and non-aromatic indian rice. *Rice Science*, *24*(1), 21–31. https://doi.org/10.1016/j.rsci.2016.05.005.

Vogelsang-O'Dwyer, M., Petersen, I. L., Joehnke, M. S., Sørensen, J. C., Bez, J., Detzel, A., et al. (2020). Comparison of Faba bean protein ingredients produced using dry fractionation and isoelectric precipitation: Techno-functional, nutritional and environmental performance. *Food*, *9*(3). https://doi.org/10.3390/foods9030322.

Wanasundara, J. P. D., Tan, S., Alashi, A. M., Pudel, F., & Blanchard, C. (2017). Proteins from canola/rapeseed: Current status. In *Sustainable protein sources* (pp. 285–304). Elsevier Inc. https://doi.org/10.1016/B978-0-12-802778-3.00018-4.

Wang, Z., Liu, Y., Li, H., & Yang, L. (2016). Rice proteins, extracted by alkali and α-amylase, differently affect in vitro antioxidant activity. *Food Chemistry*, *206*, 137–145. https://doi.org/10.1016/j.foodchem.2016.03.042.

Wang, R., Xu, P., Chen, Z., Zhou, X., & Wang, T. (2019). Complexation of rice proteins and whey protein isolates by structural interactions to prepare soluble protein composites. *LWT*, *101*, 207–213. https://doi.org/10.1016/j.lwt.2018.11.006.

Wang, L., Xue, J., & Zhang, Y. (2019). Preparation and characterization of curcumin loaded caseinate/zein nanocomposite film using pH-driven method. *Industrial Crops and Products*, *130*, 71–80. https://doi.org/10.1016/j.indcrop.2018.12.072.

Wang, J., Zhao, J., De Wit, M., Boom, R. M., & Schutyser, M. A. I. (2016). Lupine protein enrichment by milling and electrostatic separation. *Innovative Food Science and Emerging Technologies*, *33*, 596–602. https://doi.org/10.1016/j.ifset.2015.12.020.

Wangorsch, A., Kulkarni, A., Jamin, A., Spiric, J., Bräcker, J., Brockmeyer, J., et al. (2020). Identification and characterization of IgE-reactive proteins and a new allergen (Cic a 1.01) from chickpea (*Cicer arietinum*). *Molecular Nutrition & Food Research*, *64*(19), 2000560. https://doi.org/10.1002/mnfr.202000560.

Warnakulasuriya, S., Pillai, P. K. S., Stone, A. K., & Nickerson, M. T. (2018). Effect of the degree of esterification and blockiness on the complex coacervation of pea protein isolate and commercial pectic polysaccharides. *Food Chemistry*, *264*, 180–188. https://doi.org/10.1016/j.foodchem.2018.05.036.

Warsame, A. O., Michael, N., O'Sullivan, D. M., & Tosi, P. (2020). Identification and quantification of major faba bean seed proteins. *Journal of Agricultural and Food Chemistry*, *68*(32), 8535–8544. https://doi.org/10.1021/acs.jafc.0c02927.

WHO/FAO/UNU. (2007). Protein and amino acid requirements in human nutrition - PubMed. *World Health Organization Technical Report Series*, *935*, 1–265. https://pubmed.ncbi.nlm.nih.gov/18330140/.

Wijewardana, C., Reddy, K. R., & Bellaloui, N. (2019). Soybean seed physiology, quality, and chemical composition under soil moisture stress. *Food Chemistry*, *278*, 92–100. https://doi.org/10.1016/j.foodchem.2018.11.035.

Witczak, M., Chmielewska, A., Ziobro, R., Korus, J., & Juszczak, L. (2021). Rapeseed protein as a novel ingredient of gluten-free dough: Rheological and thermal properties. *Food Hydrocolloids*, *118*. https://doi.org/10.1016/j.foodhyd.2021.106813.

Wouters, A. G. B., Rombouts, I., Fierens, E., Brijs, K., Blecker, C., & Delcour, J. A. (2017). Impact of ethanol on the air-water interfacial properties of enzymatically hydrolyzed wheat gluten. *Colloids and Surfaces A: Physicochemical and Engineering Aspects*, *529*, 659–667. https://doi.org/10.1016/j.colsurfa.2017.06.013.

Wu, J., & Muir, A. D. (2008). Comparative structural, emulsifying, and biological properties of 2 major canola proteins, cruciferin and napin. *Journal of Food Science*, *73*(3). https://doi.org/10.1111/j.1750-3841.2008.00675.x.

Wu, S., Myers, D. J., & Johnson, L. A. (1997). Factors affecting yield and composition of zein extracted from commercial corn gluten meal. *Cereal Chemistry*, *74*(3), 258–263. https://doi.org/10.1094/CCHEM.1997.74.3.258.

Xing, Q., Dekker, S., Kyriakopoulou, K., Boom, R. M., Smid, E. J., & Schutyser, M. A. I. (2020). Enhanced nutritional value of chickpea protein concentrate by dry separation and solid state fermentation. *Innovative Food Science and Emerging Technologies*, *59*. https://doi.org/10.1016/j.ifset.2019.102269.

Yin, B., Zhang, R., & Yao, P. (2015). Influence of pea protein aggregates on the structure and stability of pea protein/soybean polysaccharide complex emulsions. *Molecules*, *20*(3), 5165–5183. https://doi.org/10.3390/molecules20035165.

Zhao, X., Zhang, X., Liu, H., Zhang, G., & Ao, Q. (2018). Functional, nutritional and flavor characteristic of soybean proteins obtained through reverse micelles. *Food Hydrocolloids*, *74*, 358–366. https://doi.org/10.1016/j.foodhyd.2017.08.024.

Zhou, J., Liu, J., & Tang, X. (2018). Effects of whey and soy protein addition on bread rheological property of wheat flour. *Journal of Texture Studies*, *49*(1), 38–46. https://doi.org/10.1111/jtxs.12275.

Zia-Ul-Haq, M., Iqbal, S., Ahmad, S., Imran, M., Niaz, A., & Bhanger, M. I. (2007). Nutritional and compositional study of desi chickpea (Cicer arietinum L.) cultivars grown in Punjab, Pakistan. *Food Chemistry*, *105*(4), 1357–1363. https://doi.org/10.1016/j.foodchem.2007.05.004.

Ziobro, R., Juszczak, L., Witczak, M., & Korus, J. (2016). Non-gluten proteins as structure forming agents in gluten free bread. *Journal of Food Science and Technology*, *53*(1), 571–580. https://doi.org/10.1007/s13197-015-2043-5.

Zorzi, C. Z., Garske, R. P., Flôres, S. H., & Thys, R. C. S. (2020). Sunflower protein concentrate: A possible and beneficial ingredient for gluten-free bread. *Innovative Food Science and Emerging Technologies*, *66*, 102539. https://doi.org/10.1016/j.ifset.2020.102539.

Xu J., Wang W., Zhao Y., (2021). Phenolic compounds in whole grain sorghum and their health benefits, Foods, 10, 1921, 10.3390/foods10081921.

CHAPTER FOUR

Guiding the formulation of soft cereal foods for the elderly population through food oral processing: Challenges and opportunities

Melissa Assad-Bustillos[a,*], Gilles Feron[b], and Guy Della Valle[c]
[a]Food Process Engineering Group, Institute of Food, Nutrition and Health, ETH Zürich, Zürich, Switzerland
[b]Centre des Sciences du Goût et de l'Alimentation, AgroSup Dijon, CNRS, INRAE, Université Bourgogne Franche-Comté, Dijon, France
[c]INRAE Biopolymères Interactions et Assemblages, Nantes, France
*Corresponding author: e-mail address: melissa.assad-bustillos@hest.ethz.ch

Contents

Advances in Food and Nutrition Research, Volume 99
ISSN 1043-4526
https://doi.org/10.1016/bs.afnr.2021.11.003

Abstract

As the elderly population is growing steadily, more age-friendly food products that allow them to cover their nutritional needs and are enjoyable need to be designed. Since their oral physiology is considerably altered, the study of Food Oral Processing has become an essential discipline in food development, as it takes into consideration the complex interactions between food structure, oral processing, physiology and perception. Cereals are staple foods in many countries, and their consumption as bakery products is popular among the elderly population. In addition, when fortified with pulse proteins, they can help meet the protein needs of seniors and help fight against sarcopenia. For these reasons, this chapter presents an overview of the various aspects involved in the oral processing and formulation of soft cereal foods, translating them into challenges and opportunities that are of relevance to the design of realistic soft cereal foods targeted for the elderly that are nutritious and sensory appealing. This review focuses on the healthy elderly population and does not intend to cover the needs of the dependent elderly suffering from chronical diseases.

1. Introduction and context: Why develop food for the elderly?

The world's population keeps growing at a fast pace, and so does the proportion of elderly people: in 2050, one out of five persons is expected to be 60 years old or over. In fact, the evolution of demographic factors such as the reduction in fertility and the improvement in life expectancy are responsible for this trend, which is now recognized as global.

As this transition takes place, the demand for goods and services adapted to the elderly such as food, housing, transportation and social protection has increased. According to the United Nations, preparing for the economic and social shifts associated with an aging population has become essential to fulfil the 2030 Agenda for Sustainable Development.[a] Also, the improvement of the quality of life of older citizens has been listed as one of the seven priorities of the European Union's strategy for research and innovation.[b]

Among other consequences, aging is often accompanied by deterioration of the oral physiology and impairment of perception, which can lead to loss of appetite and decrease of food intake. As a result, the nutritional and health

[a] The United Nations 2030 Sustainable Development Agenda was adopted by world leaders in September 2015. It features goals and recommendations over the next fifteen years that universally apply to all countries. It aims at ending all forms of poverty, fight inequalities and tackle climate change.

[b] See Horizon 2020 societal challenges: "Health, demographic change and wellbeing."

status of the elderly is at stake. Therefore, it is important to develop age-friendly food products with an improved nutritional value and pleasant sensory properties.

In order to meet this challenge, the study of the in-mouth mechanisms during chewing, known as "Food Oral Processing" (FOP), has revealed to be promising. Indeed, recent advances in this field have allowed a better understanding of the perception of foods during eating.

In this context, the present chapter focuses on reviewing the oral breakdown mechanisms during chewing for the purpose of developing cereal foods adapted to the oral capacities of the elderly.

Cereal foods are of particular interest, since they are staple foods in many countries, are affordable and can be consumed regardless of culture and beliefs. Plus, they are widely consumed among the elderly population, who tends to orient toward a more "traditional" dietary pattern.

In addition, these foods have a high nutritional value: when fortified with pulse proteins, they can help meet the protein needs of seniors and help fight against sarcopenia, which is the loss of muscle mass. Indeed, pulses possess a quality profile of essential amino acids, complementary to that of cereals. For that purpose, this chapter also focuses on the fortification of cereal bakery products with pulse proteins.

In sum, the main goals of this chapter are (i) to provide insight into the various aspects involved in the oral processing and formulation of soft[c] cereal foods, and (ii) to translate them into challenges and opportunities that are of relevance to the design of realistic soft cereal foods targeted for the elderly that are nutritious and sensory appealing.

2. The role of food oral processing in perception and digestion

As early as 1993, Peleg (1993) had already pointed out the need for understanding the relationship between the "mechanical and geometrical properties of food and its perceived texture" in order to provide guidelines to develop special products targeted at the elderly. Since then, advances in the understanding of food oral processing (FOP) have highlighted the importance of food structure and its mechanical properties in the bolus formation mechanisms, as well as in the perception of flavor, and texture

[c] Soft cereal foods (bread, cakes…) are opposed to brittle ones (biscuits, crackers, flakes…), since they do not break and fragment in their elastic domain (Callister, 2007).

(Chen, 2009). Today, thanks to this new understanding of texture from the study of FOP and bolus formation mechanisms on one side, and the structural and mechanical properties of foods on the other, new opportunities to develop food for the elderly have emerged by combining both approaches. Indeed, thanks to the growing interest in the study of FOP, the influence of oral processing on sensory perception has been the subject of numerous studies, with an emphasis on texture and flavor in solid and semi-solid foods, as they are the main sensory criteria that determine consumer liking and acceptance. In this section, an introduction to the main concepts of FOP will be given, and the main methods used to assess FOP will be reviewed. Next, the relationships between FOP, food structure and sensory perception will be discussed, with a particular focus on soft cereal foods.

2.1 Food oral processing: General considerations

Of all vertebrates, mammals are distinguished by the mechanical breakdown that takes place in their mouth prior to the chemical digestion in the gut. It is generally agreed that this need for mastication can be explained by their high energy expenditure and metabolic rate; food comminution adds new surface area on which digestive enzymes can increase the rate of chemical breakdown of nutrients and energy (Lucas, 2004).

In human nutrition, besides from being fundamental for the assimilation of nutrients, the phase of oral processing (FOP) is essential for the perception of texture and flavor of food. It is therefore considered a key stage for its acceptance and appreciation (Chen, 2009; Salles et al., 2011).

That way, during FOP, the combined actions of the teeth, the tongue and saliva lead to the transformation of foods into a bolus that is ready to be swallowed (Prinz & Lucas, 1997). In addition, the incorporation of enzymes present in saliva, such as α-amylase, allows further degradation of the food to facilitate gastro-intestinal digestion (Freitas, Le Feunteun, Panouillé, & Souchon, 2017; Pentikäinen et al., 2014).

Thus, bolus formation is governed mainly by two processes: fragmentation and hydration. As a result, FOP is highly dependent on the oral physiology of the individual, but also on the properties and structure of the food (Chen, 2009; Salles et al., 2011).

A qualitative model integrating both the structure of the food, the degree of hydration and the chewing time, was proposed by Hutchings and Lillford (1988) (Fig. 1). This model is considered as the paradigm in the field of oral processing. However, the degree of structure and lubrication were not

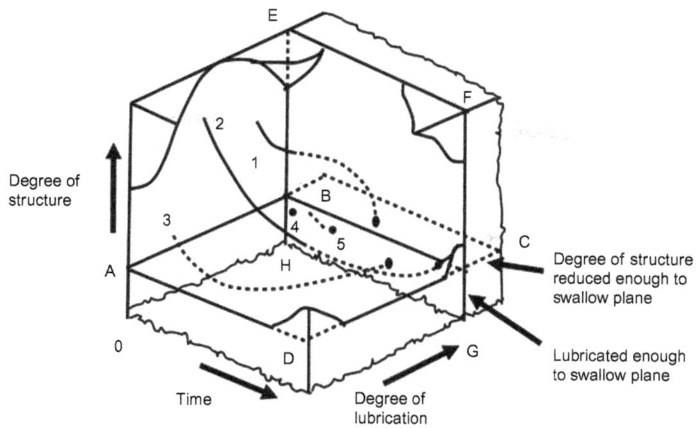

Fig. 1 Each food requires reaching a degree of structure and lubrication to achieve swallowing. The "mouth process model": (1) tender juicy steak; (2) tough dry meat; (3) dry sponge-cake; (4) oyster; and (5) liquids (Chen, 2009). *Model proposed by Hutchings, J. B., & Lillford, P. J. (1988). The perception of food texture—The philosophy of the breakdown path.* Journal of Texture Studies, *19(2), 103–115. https://doi.org/10.1111/j.1745-4603.1988.tb00928.x.*

clearly defined. Prinz and Lucas (1997) improved this model, by considering the balance of cohesion and adhesion forces for the mechanisms of agglomeration and formation of the bolus.

Given the complexity of FOP mechanisms, and its implications in the sensory perception of foods, the interest in the field has grown over the past three decades, as reflected by the abundant literature available on the subject. Much of this information, including the main organs and structures involved in FOP in terms of anatomy and functionality, has been gathered and reviewed (Chen, 2009, 2015; Wang & Chen, 2017). Today, FOP has an enormous potential in the design of foods with optimal texture and perception for the elderly, and the research efforts must be put toward this objective (Chen, 2016).

2.2 FOP assessment: The importance of bolus properties

The assessment of FOP encompasses aspects related to the chewing behavior (chewing duration, number of cycles, muscular effort), and the characterization of bolus properties (hydration, rheology, and particle size) along with their evolution during oral processing.

Chewing behavior is commonly studied via electro-myography (EMG), a non-invasive method where surface electrodes are placed on the *masseter* and *temporalis* muscles to measure the bioelectrical signal propagation during muscle contraction (Vinyard & Fiszman, 2016). This method provides an indicative assessment of muscular effort, along with information of the chewing time, number of chewing cycles, chewing frequency, burst duration, and muscular activities. However, as pointed out by Gao and Zhou (2021), the use of EMG has some disadvantages, such as not being able to measure absolute force due to muscle length and electrode placement variations, which makes it difficult to compare different subjects and products. Also, electrodes can draw the attention of subjects and modify the natural chewing duration consequently.

The characterization of bolus properties is of capital importance, since they play an important role in the triggering of swallowing (Peyron et al., 2011) and in driving the sensory perception toward the middle and the end of oral processing (Devezeaux de Lavergne, van de Velde, & Stieger, 2017; Jourdren, Saint-Eve, et al., 2016). Since chewing is a dynamic process, it is common to assess the bolus properties at various stages of the total chewing duration, commonly two or three, including the swallowing point, in order to capture their evolution (Fig. 2).

Bolus hydration is an indication of saliva incorporation and is the most basic property that can be measured almost universally for a wide variety of foods. The instrumental method to measure it simply quantifies bolus water content, usually obtained after dehydration (Panouillé, Saint-Eve, & Souchon, 2016).

Bolus rheological properties provide information on the bulk behavior of the bolus during chewing, and are relevant to the perception of texture (Witt & Stokes, 2015). They have been assessed in small (Panouillé, Saint-Eve, Déléris, Le Bleis, & Souchon, 2014) and large deformations (Le Bleis, Chaunier, Della Valle, Panouillé, & Réguerre, 2013), by using classic dynamic rotational rheometry in the former, and capillary rheometry in the latter. In large deformations, it is convenient to assess the flow behavior at shear rates close to the real values involved during chewing and/or the beginning of swallowing, which typically range between 10 and 10^2 s^{-1} (Chen, 2009; Zhu, Mizunuma, & Michiwaki, 2014). Nevertheless, as mastication progresses, with the continuous particle size reduction, the study of the tribological properties of the food-saliva mixture could become more relevant than bulk properties to understand texture perception (Chen & Stokes, 2012). Also, the mechanical properties of the bolus have been determined by uniaxial compression tests usually performed with texture analyzers (simple or double compression, penetration test, spreading test, and

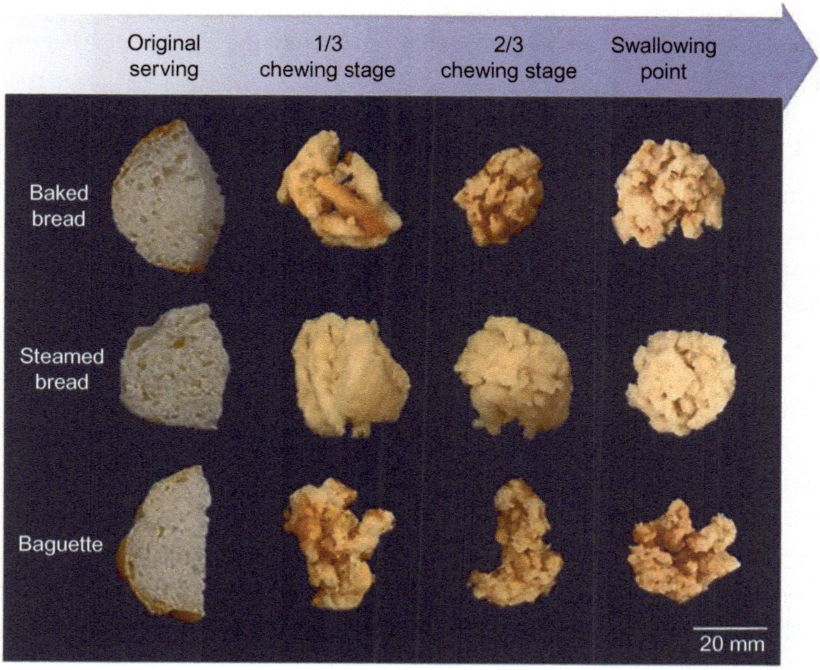

Fig. 2 Picture of three types of bread and their bolus to illustrate the transformation at early (1/3 of total chewing duration), middle (3/2 of total chewing duration) and late (swallowing point) stages of oral processing (Gao & Zhou, 2021).

back extrusion) (Panouillé et al., 2016); however, these methods are often judged as inappropriate or too simple to be able to ascertain the mechanisms of an observed behavior or to find a correlation with sensory properties (Witt & Stokes, 2015).

Finally, the bolus particle size has been studied by using different methods, such as drying, sieving and weighing the recovered fractions. Others have used laser diffraction when particles are smaller than 1 mm (Gao, Wong, Lim, Henry, & Zhou, 2015; Hoebler et al., 1998; Hoebler, Devaux, Karinthi, Belleville, & Barry, 2000). More recently, methods based on image acquisition by optical scanning or camera, followed by their quantitative analysis, have proved to be effective and less time consuming; for the same reason they are becoming more popular (Jourdren, Panouillé, et al., 2016; Le Bleis et al., 2013; Pentikäinen et al., 2014; Tournier, Grass, Zope, Salles, & Bertrand, 2012).

Overall, the diversity of methods used in rheology and particle size analysis has made it difficult to compare results between studies. Moreover, it is important to point out that inter-individual variability has shown to have a large influence on bolus properties and chewing behavior (Panouillé et al., 2016).

2.3 Food structure, oral processing and texture perception

Whether it is solid, semi-solid or liquid, every food has a defined structure with different levels of matter organization that depend on the nature of the matrix (i.e., natural, like meat or plants; or reassembled after processing, like bread and biscuits) (Lillford, 2017). In soft cereal foods, which enter in the second category, the composition and processing dictate their structure and mechanical behavior; these in turn are key determinants of their texture.

According to Szczesniak (2002), texture is a sensory property that is so complex it can presumably only be perceived by humans. However, it is generally recognized that texture encompasses all structural characteristics and mechanical properties of food, detected by instrumentation and/or by the senses (vision, hearing and touch). More importantly, texture is the main attribute that determines the liking of a food and the consumer's decision to purchase it (Guinard & Mazzucchelli, 1996). So, in order to design food with desirable sensory properties, it has become necessary to properly establish the relationships between structure and texture perception.

In this regard, the study of FOP is thought to be the missing piece of the puzzle and has already provided some insight on how the structural and mechanical properties of the foods influence FOP, and how the latter impacts texture perception.

2.3.1 Impact of food structure and composition on FOP

It has been shown that the composition and structure of cereal foods has a considerable impact on FOP (Agrawal, Lucas, Prinz, & Bruce, 1997; Wilson & Brown, 1997). For example, Gao et al. (2015) found that a greater masticatory effort was needed to breakdown a porous structure, such as bread baguette, compared to a dense structure, like the one in traditional Chinese bread. Moreover, many studies have shown that bolus hydration ratio depends on the structure of bread (Jourdren, Panouillé, et al., 2016; Le Bleis, Chaunier, Montigaud, & Della Valle, 2016), as well as the hydration kinetics: the more porous is the structure, the higher is the hydration rate (Mathieu et al., 2016) (Fig. 3).

Other factors, such as the presence of crust or fat also have an influence on masticatory behavior and bolus formation mechanisms. For instance, the

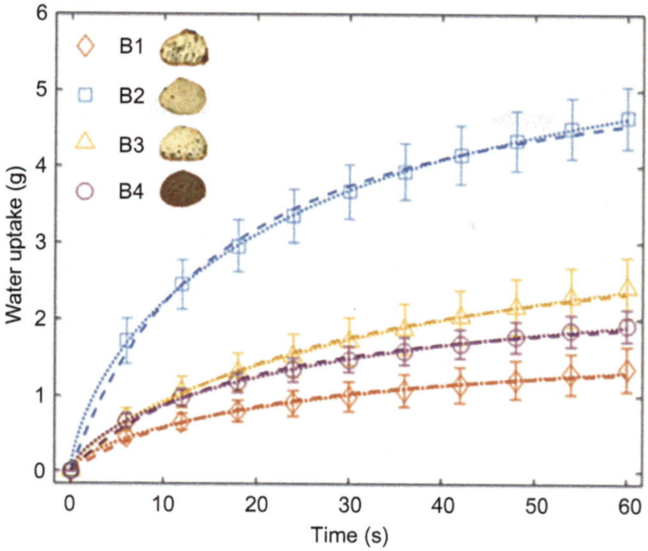

Fig. 3 Kinetics of hydration for "baguette" type breads with different structure and density. The decreasing order of densities is: B4 > B1 > B3 > B2 (Mathieu et al., 2016).

spreading of 0.8 g butter on Melba toast significantly reduced the number of masticatory cycles required for swallowing (Engelen, Fontijn-Tekamp, & van der Bilt, 2005). Also, it was seen that the presence of crust in several types of breads generated more heterogeneous and less hydrated boli (Jourdren, Panouillé, et al., 2016).

2.3.2 Impact of FOP on texture perception

Oral processing is known to have an influence in texture perception; it is important to understand it since it has a major influence in consumer preferences. For example, in a study by Brown and Braxton (2000), it was found that individuals with similar masticatory abilities assessed some texture descriptors in the same way, with a tendency to prefer foods that were "easier to eat" according to individual profiles. Jeltema, Beckley, and Vahalik (2015) have proposed to categorize consumers according to their preferred oral manipulation strategies into four groups: *Crunchers, Chewers, Suckers and Smooshers*. They hypothesized that these strategies are related to the liking of the foods and the degree of satisfaction of the consumer.

It is generally recognized that the perception of texture is a dynamic process that depends on the initial structural properties of the food during the early stages of mastication (Kim et al., 2012; Young, Cheong, Hedderley, Morgenstern, & James, 2013), and on the bolus properties toward the middle and the end of oral processing (Devezeaux de Lavergne et al., 2017; Jourdren, Saint-Eve, et al., 2016).

When comparing breads of different structure and composition, Panouillé et al. (2014) found that texture perception evolved during mastication. Thus, at the beginning, the "soft," "hard" and "chewy" attributes made it possible to discriminate the samples, but when approaching the swallowing point, they were all dominated by the "sticky," "crumbly" and "hydrated" descriptors. These results were coherent with the hydration ratio of the boli, which were not significantly different at the swallowing point. Similar results were obtained by Jourdren, Saint-Eve, et al. (2016) who highlighted the importance of bolus properties, and particularly the hydration ratio, on the perception of bread texture.

Given its dynamic nature, many authors have used temporal sensory techniques to characterize texture perception, such as Temporary Check-All-That Apply (TCATA), Time Intensity (TI), Progressive Profiling (PP) or Temporal Dominance of Sensations (TDS) (Fiszman & Tarrega, 2018). However, choosing the appropriate method to measure texture perception is always a challenge (Panouillé et al., 2016). In the case of specific populations, such as the elderly, the use of temporal techniques may be too sophisticated and has been judged as inappropriate (Methven, Jiménez-Pranteda, & Lawlor, 2016). Plus, besides from getting descriptive data about texture and flavor perception, it is important to focus on other sensory criteria such as the ease of oral manipulation and bolus formation. For that purpose, the concept of oral comfort (Fig. 4) was proposed by Vandenberghe-Descamps, Labouré, et al. (2017), and successfully used by the same authors to evaluate the impact of physiology upon the perception of oral comfort in cereal and meat based products (Vandenberghe-Descamps, Sulmont-Rossé, Septier, Feron, & Labouré, 2017).

2.4 FOP, flavor release and perception

Flavor (taste and aroma) perception during food consumption is determined principally by the amount of volatile (aroma) and non-volatile (taste) compounds that are released in the mouth, which in turn depends on the bolus properties, particularly the particle size, and on the masticatory behavior (Salles et al., 2011).

General question

This food is…

Very uncomfortable	Uncomfortable	Moderately comfortable	Comfortable	Very comfortable
□	□	□	□	□

Bolus formation

To cut this food with your incisor is…

Impossible	Very difficult	Difficult	Moderately easy	Easy	Very easy
□	□	□	□	□	□

Same scale for cutting with premolars, masticating, humidification with saliva and swallowing the food bolus

The time needed to form the food bolus is…

Impossible to swallow	Very long	Long	Moderately brief	Brief	Very brief
□	□	□	□	□	□

Mouth pain

Does eating the food bring a burning or spicy sensation?

Not at all	Little	A lot	Extremely
□	□	□	□

Same scale for muscular pain, articular pain, dental pain and gum pain

Texture

Is this food sticky?

Not at all	Little	A lot	Extremely
□	□	□	□

Same scale for stringy, greasy, dry, doughy, melting, firm and hard

Taste

Is this food intense in taste?

Not at all	Little	A lot	Extremely
□	□	□	□

Same scale for salty, sweet, acidic and bitter

Fig. 4 Food comfortability questionnaire developed based on the results of focus groups organized with elderly people (Vandenberghe-Descamps, Labouré, Septier, Feron, & Sulmont-Rossé, 2017).

The majority of the studies available on the subject have been conducted on model cheeses (Feron et al., 2014; Repoux et al., 2012; Tarrega, Yven, Sémon, & Salles, 2008), polysaccharide and protein gels (Hansson, Giannouli, & van Ruth, 2003; Mestres, Moran, Jordan, & Buettner, 2005; Mosca, van de Velde, Bult, van Boekel, & Stieger, 2015), and only a few of them on soft cereal foods (Jourdren et al., 2017; Panouillé et al., 2014; Tournier, Grass, Septier, Bertrand, & Salles, 2014).

For example, with regards to the release of taste compounds, a study conducted by Mosca et al. (2015) investigated the sweetness intensity perception of protein/polysaccharide gels related to the number of fragments in the bolus. It was seen that gels with a low fracture stress led to boli with higher number of fragments, which had a larger contact surface and thus were perceived as sweeter.

Regarding sodium release, which is responsible for the perception of saltiness, very similar results were observed by de Loubens et al. (2011), who found the number and size of fragments from artificial boli were directly related to their sodium release kinetics and helped to explain differences in salty perception *in vivo*.

In bread, sodium release is modulated by masticatory parameters such as the number of chewing cycles and the muscle activity (Tournier et al., 2014). Also, Panouillé et al. (2014) showed that the perception of saltiness was influenced by both oral processing characteristics and the initial structure properties of the samples. Indeed, the densest bread was perceived as less salty and displayed a less complex texture perception sequence (Fig. 5).

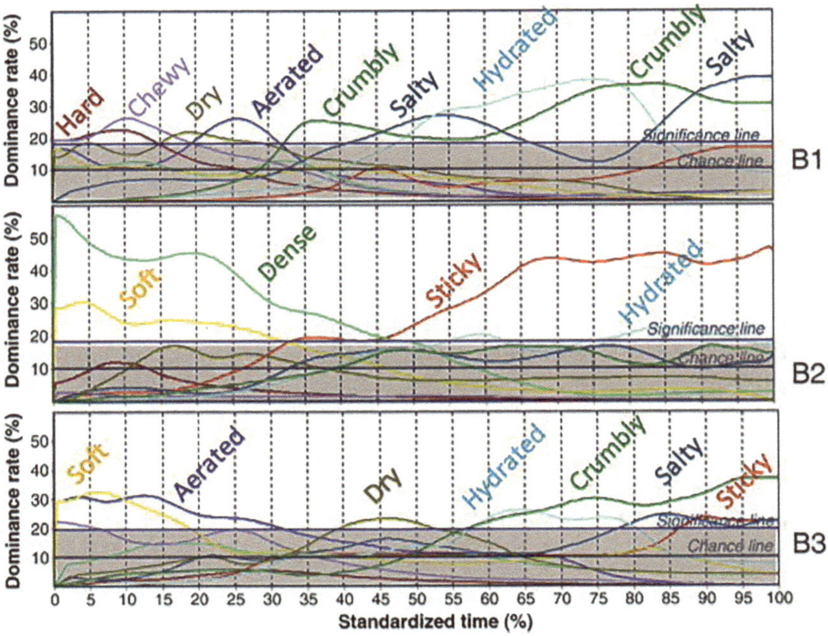

Fig. 5 Variation of dominance rates as a function of standardized time for the three breads (B1 = reference, B2 = dense, B3 = fat), obtained with the Temporal Dominance of Sensation (TDS) method. The lower and upper horizontal lines indicate the chance and significance levels ($\alpha = 5\%$), respectively. Only TDS curves above the significance level are considered to be significantly dominant (Panouillé et al., 2014).

Moreover, the rheological properties of the boli obtained from the densest bread were correlated with a lower tan δ (i.e., a higher level of structure).

Concerning the aroma release profile, differences can be mainly explained by the chewing behavior of the subjects, where: the higher the chewing activity of the subject, the higher the aroma release (Tarrega et al., 2008). These conclusions were supported by Feron et al. (2014), who showed that most of the variability observed in the aroma release profiles of model cheeses was explained by the chewing behavior of the subjects, rather than bolus properties or salivary flow. However, in some cases, and mainly for soft cheeses, stimulated saliva flow was negatively correlated with aroma release. A similar observation was made by Jourdren et al. (2017), who found that saliva addition had a dilution effect on the volatile compounds of bread, reducing aroma release during oral processing. Thus, bread structures with high hydration capacity were characterized by a low aroma release profile. Actually, the volume of saliva in the mouth can have simultaneously an extraction and a dilution effect depending on the solubility of the molecule and its hydrophobic character (Salles et al., 2011).

2.5 FOP and the digestibility of nutrients

The whole digestive process is oriented toward the mechanical reduction of the size of ingested foods. Briefly, food digestion consists of bolus formation, followed by swallowing and transportation of the bolus through the esophagus until the stomach. Once there, the peristaltic movements assure the continuity of the size reduction process, all while being chemically broken down. This is allowed by the acidic and enzymatic conditions of the gastric secretions (Kong & Singh, 2008). This process, known as the gastric digestion, occurs upon the ingestion of every meal, and persists until the stomach is emptied. Thereafter, the resulting fluid (chyme) reaches the small intestine, where the action of pancreatic enzymes allows the nutrient release for its subsequent absorption (Klein, Cohn, & Alpers, 2006).

Today, the majority of digestive phenomena have been extensively studied, and the role of particle size reduction in increasing the surface area for enzymatic attack and chemical digestion has been well established. Yet, the role of particle breakdown during oral processing and its relationship with the release of nutrients during their subsequent digestion is still to be determined. In this regard, some studies have investigated the influence of bolus properties, in particular the degree of breakdown and the saliva uptake in the digestibility of nutrients, such as lipids (Grassby et al., 2014; Grundy et al., 2015), carbohydrates (Gao et al., 2019; Joubert et al., 2017; Pentikäinen et al., 2014), and to a lesser extent, that of proteins (Assad-Bustillos, Tournier, Feron, et al., 2019; Rémond et al., 2007).

In cereal foods, given their high starch and protein contents, the focus will be made on the digestibility of these two biomolecules.

2.5.1 Carbohydrates

The Glycemic Index (GI) is the ability of a food to raise blood glucose rapidly, and depends on the kinetics of starch enzymatic hydrolysis, which occurs by the action of α-amylase, both salivary and pancreatic.[d] Amylase breaks down amylose and amylopectin by cleaving their α-1-4 glycosidic bonds. Its optimum activity is reached at pH 7 and 37 °C (Fried, Abramson, & Meyer, 1987).

Starch hydrolysis depends on factors such as the crystallinity, the degree of swelling and gelatinization of the starch granule, and largely on the initial structure of the food matrix which rules the accessibility of the enzyme to the substrate (Fardet, Leenhardt, Lioger, Scalbert, & Rémésy, 2006). In fact, denser foods are associated with lower values of GI, since a more compact structure hinders the diffusion of digestive fluids and thus the accessibility of enzymes. Heat processes like drying or precooking can induce the presence of a highly cross-linked protein network that "protects" starch from being digested and thus lowers the GI, like it was seen in the case of pasta (Petitot & Micard, 2010). The increase of bread density partially explains that the addition of fibers lowers the GI of wheat-based foods (Saulnier, Micard, & Della Valle, 2014).

The role of oral processing on starch digestion of different kinds of bread has been investigated by Pentikäinen et al. (2014), who found that rye breads, which were denser than wheat bread, displayed a trend toward lower starch hydrolysis, despite the fact they required a longer chewing time and were fragmented into smaller (an likely higher number) particles. Additionally, Joubert et al. (2017) found the starch hydrolysis degree was higher for industrial bread than whole-meal, since its density was lower, allowing a higher absorption from saliva. These findings suggest that bread initial structure and saliva absorption play a preponderant role in the hydrolysis of starch almost independently of the bolus particle size. However, recently, Gao et al. (2019) found, using the INFOGEST protocol (Brodkorb et al.,

[d] For many years the enzymatic hydrolysis of starch was considered to occur essentially by pancreatic α-amylase; however, this has recently been contested and the salivary amylase seems to play an important role on starch digestion, not only during the oral phase, but also during gastric digestion (Freitas, Le Feunteun, Panouillé, & Souchon, 2017; Pentikäinen et al., 2014).

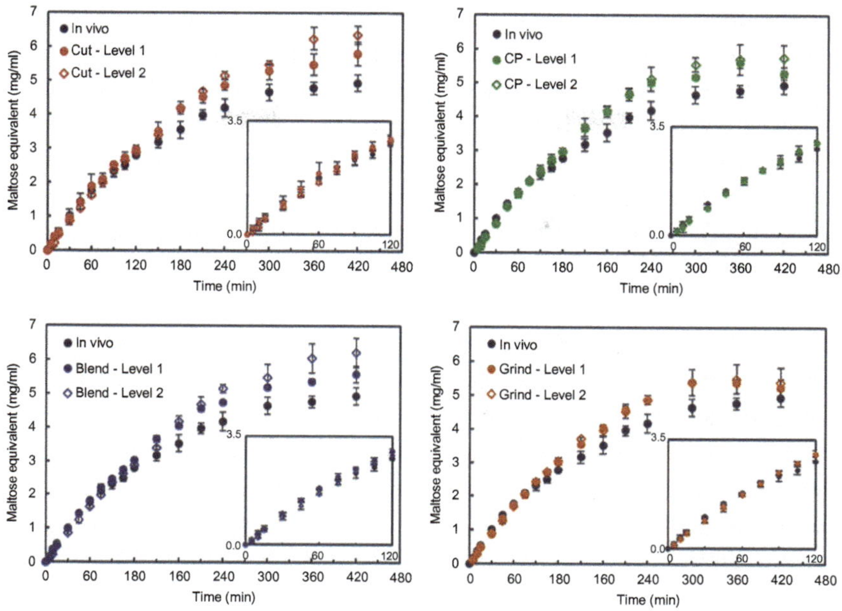

Fig. 6 The digestion kinetics of bread prepared from in vivo mastication (less disintegrated) and in vitro oral processing (2 levels of disintegration). The inserted figures provide a zoom-in view of the digestion happened during the first 120 min of digestion (Gao et al., 2019).

2019; Minekus et al., 2014), that more disintegrated boli had a higher starch hydrolysis rate (Fig. 6). These results show that the oral processing phase, which has been commonly overlooked in digestibility studies, has an influence on the hydrolysis kinetics, not only of starch, but also of other nutrients, such as proteins.

2.5.2 Proteins

Despite the proteolytic activity of saliva (Neyraud et al., 2012), it is generally accepted that protein hydrolysis happens essentially in the stomach, thanks to the combined action of hydrochloric acid (HCl) and pepsin, the gastric protease, which exhibits maximum activity at pH 2 (Malagelada, Go, & Summerskill, 1979).

The products of gastric proteolysis consist of a mixture of large polypeptides, oligopeptides and free amino acids (Kong & Singh, 2008). Protein digestion is completed in the small intestine by the action of pancreatic proteases and peptidases before the absorption of small peptides and free amino acids (Klein et al., 2006).

Therefore, the proteolysis kinetics (bioaccessibility) is a key determining factor for the subsequent postprandial absorption of amino acids (bioavailability), which induces protein synthesis. Both, bioaccessibility and bioavailability constitute ways to assess protein digestibility, which is influenced by factors like the processing conditions and interaction with other food components.

Overall, cooking and heat treatments are known to improve the protein digestibility (Fennema, 2017), but very high temperatures or prolonged exposure to heat can result in a reduction of bioaccessibility due to the formation of highly cross-linked aggregates (De Zorzi, Curioni, Simonato, Giannattasio, & Pasini, 2007; Pasini, Simonato, Giannattasio, Peruffo, & Curioni, 2001) and losses of essential amino acids due to Maillard reactions (Fennema, 2017). This is of particular relevance for soft cereal foods, since the highly cross-linked structure that is formed during baking, involving different protein sources like wheat and egg (Wilderjans, Luyts, Goesaert, Brijs, & Delcour, 2010), generates concern about the bioaccessibility of the added proteins for fortification purposes. In the case of pulse proteins, one of the few studies that has been reported in cereal foods concerns gluten-free cakes (Gularte, Gómez, & Rosell, 2012). Even if the principal focus of this study was starch hydrolysis, the authors also compared the digestibility of cakes prepared with different pulse flours, assessed as the pH decline during *in vitro* enzymatic hydrolysis. All of the studied pulse flours, including pea, showed high *in vitro* digestibility, as reflected by their rapid decline in pH, with the exception of bean. Similar results were found by Assad-Bustillos, Tournier, Septier, Della Valle, and Feron (2019), who reported that pea proteins in fortified sponge-cakes are bioaccessible under the standardized conditions of the INFOGEST protocol (Brodkorb et al., 2019; Minekus et al., 2014).

Surprisingly the effect of chewing and bolus formation on protein digestibility has been little studied. In meat proteins, Rémond et al. (2007) showed an influence of the masticatory efficiency and the bolus degree of breakdown on the postprandial metabolism and on the assimilation of amino acids; interestingly, in the study by Assad-Bustillos, Tournier, Feron, et al. (2019) found that the degree of structure of the food bolus assessed by shear viscosity had no significant effect on the concentration of NH_2 released during *in vitro* gastric digestion (Fig. 7). Similar results were obtained by Peyron et al. (2021), who demonstrated that the release of peptides *in vitro* from pork sausages was lower for the elderly digestive condition, but irrespective of masticatory efficiency.

In all, more studies that investigate the impact of oral processing and bolus formation on the digestibility of proteins are needed, especially in protein fortified cereal foods. Also, studying the impact of oral processing in

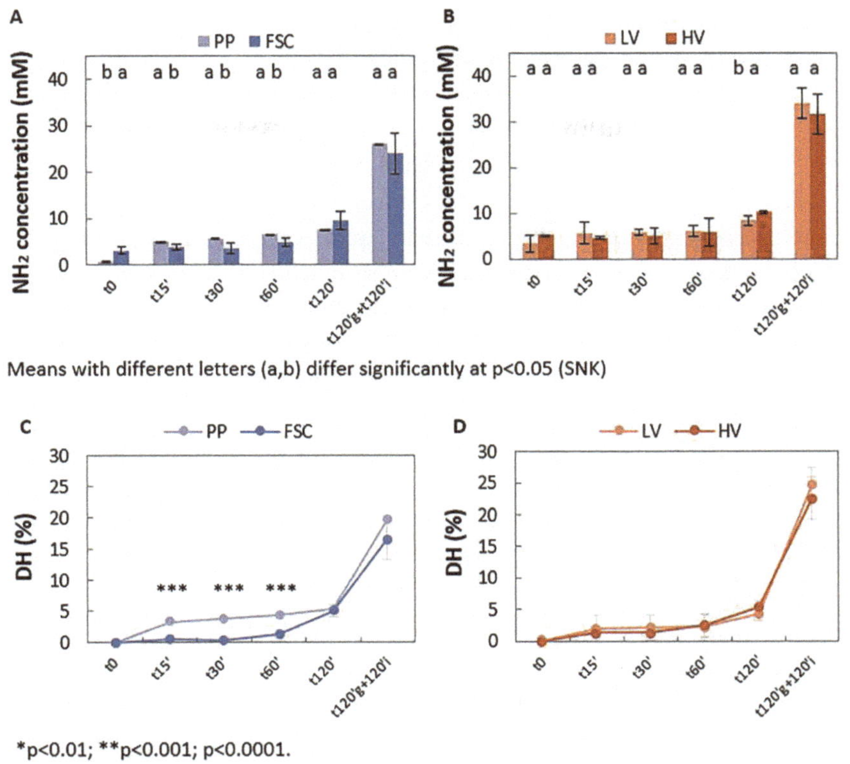

Fig. 7 Evolution of the (A, B) total NH_2 concentration of the digest a before digestion (t0), during the gastric (t15′, t30′, t60′, t120′) and after the intestinal (t120′g+120′i) phases; and the (C, D) degree of hydrolysis (DH) expressed as the percentage of NH_2 cleaved during digestion. (A, C) show protein isolates (PP) and fortified sponge cake (FSC); (B, D) show low (LV) and high viscosity (HV) bolus pools (Assad-Bustillos, Palier, Rabesona, et al., 2019).

protein digestibility is of capital importance, especially when addressing to the elderly population, whose oral physiology is commonly impaired (Peyron, Woda, Bourdiol, & Hennequin, 2017; Vandenberghe-Descamps et al., 2016) and have higher protein needs than younger adults (Bauer et al., 2013; Deutz et al., 2014), as discussed in the following section.

3. The elderly: Physiology and nutritional needs

3.1 Toward an aging population

Aging, or senescence, is a process common to all living organisms, defined as the decline of the physiological functions at all levels: cellular, tissue, organic and systemic (Rose, 1991). It is a complex mechanism influenced by many factors (genetic, environmental, etc.); yet the exact reasons as to why aging occurs remain to be understood (Hayflick, 2007).

The age of senescence has several definitions. Even though the current life expectancy in many European countries is close to 80 years (UN, 2017b), most developed countries have accepted the social criterion of 65 years to define an elderly person. This corresponds to the age of retirement from professional activity in many parts of the world. However, it is impossible to define a universal age criterion, since the experience of aging varies from person to person and the chronological age does not necessarily determine the physiological condition of the aging organism. Moreover, the individual perception of this process differs depending on the attitude, living conditions, diseases and environmental influences (Rémond et al., 2015).

From a demographic point of view, a population becomes old when the life expectancy increases and fertility decreases (Lutz, Sanderson, & Scherbov, 2008). Within the past decade, many countries in the world have started to display this trend, mainly in Europe, where currently the healthy life year expectancy at birth is around 62 years old (Eurostat, 2016). This has generated concern about the health status and quality of life of the forthcoming aging population.

Ever since the second half of the past century, the world's population has increased by more than twofold, going from 3 to over 7 billion, and is expected to reach 10 billion by 2050 (UN, 2017b). This unprecedented growth rate can be explained by the remarkable progress of medical sciences and the improvement of hygiene conditions that allowed higher survival rates for children and a better preservation of health for the population in general (Lutz et al., 2008).

Also, during the 19th century the industrial revolution brought the mass production and distribution of many services and goods such as food, clothing, housing and education. This meant people's access to nourishment increased considerably, and so did their quality of life and wealth. For instance, in the UK, historians estimate that by the end of the 20th century, the average citizen's supply of such goods was multiplied by fifteen as compared to their ancestors in the early 1700s (McCloskey, 2008).

Globally, all of these factors have led to the decline of mortality, an increase of life expectancy and a reduction of the fertility rate to historical levels. Although these trends represent significant achievements for the human race, they are responsible for a major demographic transition: the aging of the population. This means the number of older persons is growing faster than the number of people in all younger age groups (Dugarova, 2017; UN, 2017a).

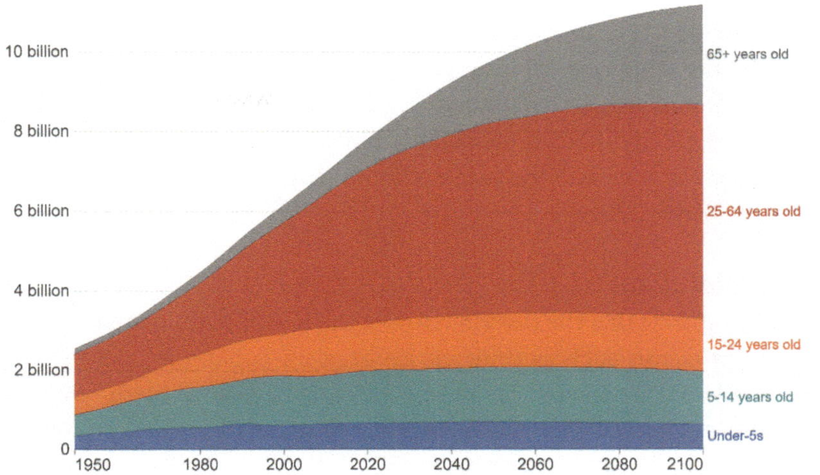

Fig. 8 World population by broad age group, with historical estimates from 1950 to 2015 and projections to 2100 based on the United Nations medium population scenario (Roser & Ortiz-Ospina, 2013).

As a matter of fact, in 2017 the global population aged 60 years or over already numbered 962 million, which is more than twice as large as in 1980. Likewise, the number of older persons is expected to double again by 2100, when it is projected to reach nearly 2.5 billion, which represents more than 20% of the total projected population (Fig. 8). In some countries, especially in developed regions like Europe and North America, where 25% of the population is already aged 60 years or over, this proportion will be even higher; plus, the population aged 80 years and over will reach 11% (Eurostat, 2016; UN, 2017a).

3.2 Consequences of aging on physiology and food oral processing

Among the functional impairments of masticatory ability that occur with age are the decrease in muscle strength and the deterioration of the oral health status. The latter is characterized by tooth loss, dental and periodontal diseases and reduced salivary flow, among others (Henshaw & Calabrese, 2001).

3.2.1 Muscle strength

Laguna, Sarkar, Artigas, and Chen (2015) have shown that the averages of jaw strength (bite force) and tongue-palate compression were negatively correlated with age. Conversely, regarding the compressive force of the lips, no correlation with age was established, but the values obtained for the elderly were lower than those of young people. Plus, in another study conducted by the same research group, negative correlations between age and hand, finger and tongue strength were observed. Also, it was seen that the bite force of elderly subjects depended on their dental status and influenced the number of chews and the liking of foods (Laguna, Hetherington, Chen, Artigas, & Sarkar, 2016).

3.2.2 Dental status

It has been shown that the deterioration of the dental status depends on age, but also on factors such as drug intake or socio-economic grade (Peltzer et al., 2014). In the United States, the average number of teeth remaining is 19.3 among people aged 65 to 74, while the rate of complete tooth loss (edentulism) in the same age group is 24% (NIDCR, 2004). A similar trend is observed in other American and European countries such as Canada or Sweden (Emami, de Souza, Kabawat, & Feine, 2013). In France, a rate of 30% was estimated in persons aged over 70 years old (Peyron, Veyrune, & Woda, 2013).

Today, thanks to the better availability and awareness of dental care, edentulism rates in the elderly are decreasing; however, there is still a large gap between high and low income households as well as for developed *vs.* developing countries (Peltzer et al., 2014; Slade, Akinkugbe, & Sanders, 2014).

3.2.3 Salivary flow

For many years the effects of aging on salivation were controversial, as it was unclear whether hyposalivation in the elderly was caused by systemic diseases and medical treatments, or by the sole effect of aging (Nagler, 2004). However, it is now accepted that salivary flow decreases naturally with age (Nagler & Hershkovich, 2005). In a recent study, it was shown that a group of healthy elderly aged 65 and over had a reduced salivary flow as compared to healthy younger adults independently of drug intake and dental status (Vandenberghe-Descamps et al., 2016) (Fig. 9). The box-plot distributions reveal a median difference between young vs. old adult participants for both the resting (0.5 vs. 0.19 mL/min) and stimulated (2.47 vs. 1.52 mL/min) salivary flow, but also a larger variability for the young adult participants.

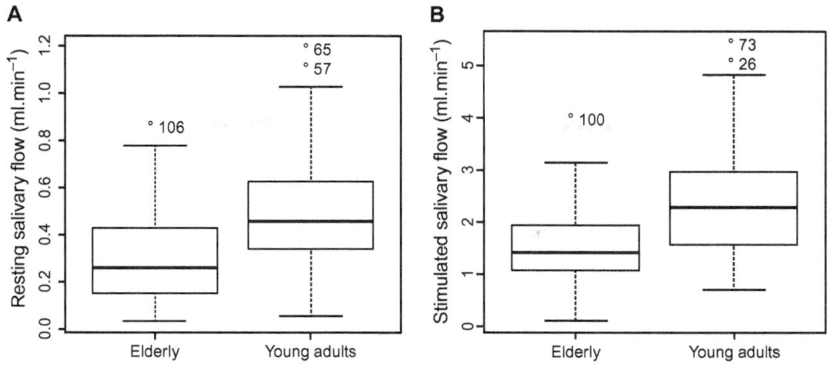

Fig. 9 Resting (A) and stimulated (B) salivary flow of elderly (>65 yo) vs. young subjects (Vandenberghe-Descamps et al., 2016).

Even if authors like Mese and Matsuo (2007) have stated that hyposalivation can be associated with altered taste, difficulty in eating, chewing and swallowing, there have been very few studies to support these claims, and surprisingly the effect of saliva is one of the most overlooked in research about mastication and bolus formation (Muñoz-González et al., 2018).

3.2.4 Masticatory performance and chewing behavior

Despite the cited physiological changes, particularly in muscle strength, total muscular work during chewing does not differ between the young and the elderly for a majority of foods, such as apple, cheese, hard bread, peanuts and rice (Kohyama, Mioche, & Martin, 2002; Mioche, Bourdiol, Monier, Martin, & Cormier, 2004; Peyron, Blanc, Lund, & Woda, 2004). On the other hand, the chewing duration and the number of cycles are higher for seniors, meaning the elderly adapt their chewing behavior to compensate for a lower masticatory efficiency.

However, it was pointed out by Peyron et al. (2017) that most of the masticatory impairments observed in the elderly are rather caused by a deterioration of the dental status than by the effect of aging *per se*. This is in accordance with the results from Fontijn-Tekamp, van der Bilt, Abbink, and Bosman (2004), which indicate that, on average, people with a poor dental status have a bad masticatory performance and swallow larger food particles independently of their age.

Furthermore, it has been shown that dental status effects on the loss of masticatory efficacy are all the more important when the number of functional units[e] is low (Kohyama, Mioche, & Bourdiol, 2003), which confirms the importance of the dental status over the effect of aging itself.

Moreover, since tooth loss is accompanied by the disappearance of the periodontal ligament, the sensibility and tactile perception are lost as well. In fact, mechanoreceptors in this tissue are in charge of giving detailed information on the spatial relationship and load modulation in the process of food fragmentation. As a consequence, the pressure and exerted forces upon chewing are perturbed, resulting in masticatory impairments (Johansson, Svensson, & Trulsson, 2006). This is the same reason why denture wearers have difficulties to adapt to the texture of foods (Veyrune & Mioche, 2000) and display more severe masticatory deficiencies than their counterparts with a natural dentition (Fontijn-Tekamp et al., 2000).

3.2.5 Bolus properties and perception of oral comfort

The decline in oral physiology that comes with aging influences not only the chewing behavior, but also the bolus properties and by consequence, the sensory perception and acceptability of foods. Surprisingly, research focusing in understanding the impact of the impaired physiology of the elderly on these aspects of oral processing is still scarce, especially for soft cereal foods.

In healthy subjects, it is generally accepted that salivation ability plays a major role on the chewing behavior and bolus properties of bread, while masticatory efficiency shows minimum influence (Gao & Zhou, 2021). In the few studies available to date, similar results have been observed for the elderly.

For example, the bolus hydration is largely influenced by the salivary flow rate of individuals. Le Bleis et al. (2016) proposed that saliva uptake could be predicted based on the theoretical volume of saliva present in the mouth and the median particle size of bread bolus. For the elderly, Assad-Bustillos, Tournier, Septier, et al. (2019) reported a similar relationship for sponge-cake and brioche, although the prediction power was not increased by including bolus particle size. In the same study, it was found that the elderly with a higher salivary flow rate, achieved the bolus viscosity needed to trigger swallowing in a shorter duration. Moreover, the elderly

[e] A functional unit is defined as a pair of occluding teeth. A linear relationship has been found between masticatory performance and the number of post-canine occluding teeth (van der Bilt, Olthoff, Bosman, & Oosterhaven, 1993).

with a satisfactory dental status produced a bolus with a higher apparent viscosity than their counterparts with a poor dental status. In contrast, those with a satisfactory dental status, but a low salivary flow rate, still needed a longer chewing duration. This confirms the preponderant role of salivary flow over dental status in determining the bolus viscosity, hydration and chewing duration in soft-cereal products.

For healthy individuals, the particle size of bread bolus required to achieve swallowing is mainly due to differences in the chewing time rather than physiology (Gao & Zhou, 2021). For the elderly, a significant influence of dental status was found for sponge-cake, where the individuals with a satisfactory dental status produced boli with smaller particles, but neither dental status nor salivary flow rate affected the bolus particle size of brioche (Assad-Bustillos, Tournier, Feron, et al., 2019). However, when the same authors used a Gompertz model fitting that was applied directly to the particle size distribution curves, rather than only focusing on one value of the curve (D_{50}, commonly), a better discrimination between chewing sequences was achieved and an effect of the dental status was finally revealed, showing that the individuals with a poor dental status produced boli with a larger particle size.

Finally, it was seen that the perception of oral comfort of sponge-cakes (Fig. 10A) by the elderly was primarily driven by the salivary flow rate, irrespective of the dental status (Assad-Bustillos, Tournier, Septier, et al., 2019). However, this relationship could not be confirmed for brioche (Fig. 10B), which has a higher fat content and was perceived as highly comfortable, suggesting the lubrication provided by the fat could compensate for oral comfort in individuals with a lower salivary flow. While this hypothesis needs to be confirmed, other studies suggest the inclusion of fat or toppings to dry foods (i.e., bread and toast) helps to facilitate oral processing, and is therefore a guideline to privilege when designing cereal foods for elderly people (Engelen et al., 2005; van Eck et al., 2019).

3.3 Consequences of aging on food intake

Many cross-sectional studies have shown an association between dental status and nutrient intake (Krall, Hayes, & Garcia, 1998; Papas et al., 1989), but in order to be able to establish causality, longitudinal studies that measure dietary status before and after tooth loss and salivary hypofunction are needed (Kiesswetter et al., 2018; Muñoz-González et al., 2018). However, this type of study is difficult to perform because of the large number of data required and the need for a consistent follow-up over time.

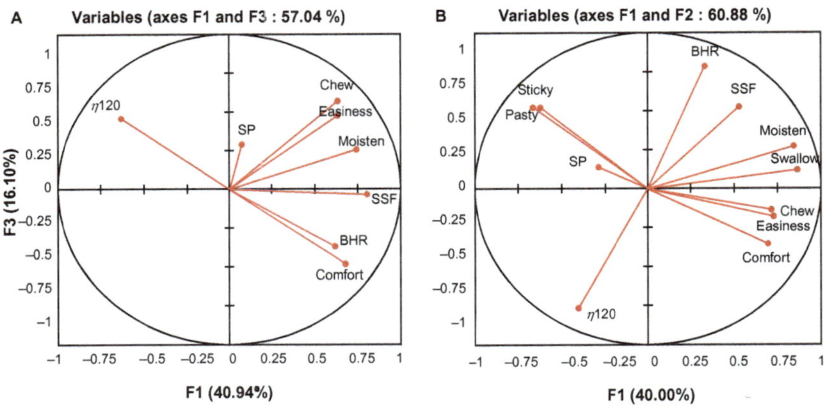

Fig. 10 PCA correlation circle for oral health status, bolus properties, chewing duration and oral comfort variables. (A) Sponge-cake, (B) Brioche. Comfort = Overall oral comfort perception score; Chew = Perceived as easy to chew; Moisten = Perceived as easy to moisten; Swallow = Easy to swallow; Easiness = Perceived as easy to eat; Pasty = Perceived pastiness; Sticky = Perceived stickiness; BHR = Bolus hydration ratio or added saliva; η120 = Bolus apparent viscosity at 120 s^{-1}; SSF = Stimulated salivary flow rate; SP = Swallowing point or total chewing duration (Assad-Bustillos, Tournier, Septier, et al., 2019).

Currently, it is possible to hypothesize that poor oral health and other age related deficiencies have a negative impact on food intake and the nutritional status of the elderly; yet, it is important to keep in mind there are many other factors that influence food choice directly or indirectly, such as the psychological (i.e., anxiety and depression), social (i.e., knowledge of cooking and loneliness) and economic status; as well as the use of medications, digestive difficulties, and others (Peyron et al., 2017; Rémond et al., 2015; Schwartz, Vandenberghe-Descamps, Sulmont-Rossé, Tournier, & Feron, 2017).

Overall, the link between age and physiological degradation is undeniable, although research should be oriented toward understanding the impact of these changes on food intake in order to find strategies that help to preserve the nutritional and health status of elderly people.

3.4 The nutritional needs of the elderly: A special focus on protein

Besides from all the typical macro and micro-nutrients that are needed by adults to maintain a good health, in the elderly a special focus is made on the need for supplementation for two particular nutrients: protein and vitamin D. In this short review, only the specific needs of protein in the elderly will be addressed.

The constant synthesis and degradation of proteins that occur in the human body assure protein maintenance in case of damage caused by oxidative stress or protein misfolding. This protein turnover is affected by a number of variables, including age, physical activity level, sex, hormones, disease, and diet (Lonnie et al., 2018). Imbalances in protein turnover can lead to excessive gains or losses in body protein mass, which translates in a gain or reduction of lean muscular mass, respectively (Phillips, 2017).

When protein is ingested, the subsequent enzymatic hydrolysis generates an excess of postprandial amino acids (hyperaminoacidemia) that results in the stimulation of protein synthesis (anabolism). During fasting periods, the lack of amino-acids (hypoaminoacidemia) results in a reduced stimulation of protein synthesis and protein breakdown is, comparatively, elevated (catabolism) (Fig. 11A) (Fardet, Souchon, & Dupont, 2013).

In healthy older persons, there is a lowered anabolic response that leads to a progressive loss of muscle mass and function, named sarcopenia (Fig. 11B). The main reason for this is an increase in the "anabolic threshold," meaning the usual nutrient intake is not sufficient to promote protein synthesis (Dardevet et al., 2012). Other factors such as insulin resistance, impaired digestion, inflammation, are known to contribute to the apparition of sarcopenia (Lonnie et al., 2018; Phillips, 2017).

Sarcopenic muscle loss can be observed as early as the fifth decade of life, and it proceeds at a rate of approximately 0.8% per year (Lonnie et al., 2018). Nevertheless, this process can be slowed down and even be avoided by increasing protein intake and physical activity level (Dickerson, 2016; Moore et al., 2015).

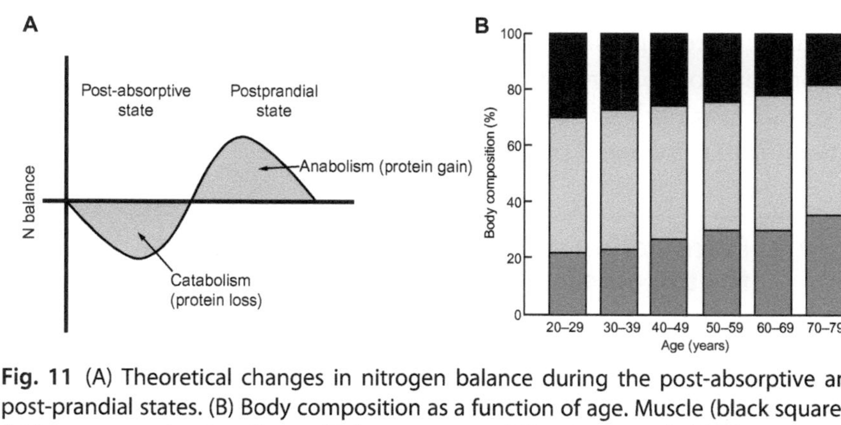

Fig. 11 (A) Theoretical changes in nitrogen balance during the post-absorptive and post-prandial states. (B) Body composition as a function of age. Muscle (black squares); fat (gray squares); other tissues (light gray squares) (Dardevet et al., 2003).

According to the World Health Organization (WHO), the current international recommended dietary allowance (RDA) for protein is 0.8 g/kg of body weight/day, regardless of age (WHO, FAO, & UNU, 2007). Yet, there is evidence that the safe minimal protein intake would be at least 1.2 g/kg/day for older persons (Bauer et al., 2013; Deutz et al., 2014). A further increase is recommended for individuals with acute or chronic illnesses (1.2–1.5 g protein/kg/day) and severe illnesses, injuries, or malnutrition (2.0 protein g/kg/day).

Among dietary proteins, leucine-rich proteins have been shown to be the most effective to stimulate muscle protein synthesis, presenting positive effects during aging or disuse when compared to proteins with lower leucine content and whose digestion rate is slower (Dardevet et al., 2002; Magne, Savary-Auzeloux, Rémond, & Dardevet, 2013). So, leucine, besides being an indispensable amino acid, is considered also as a trigger for protein synthesis.

Since the elderly have a greater protein requirement to stimulate protein synthesis than their younger counterparts, protein supplementation, and especially rich in leucine, could be useful to help the elderly stimulate and retain muscle mass. When it comes to using plant proteins, they need to be combined (i.e., pulses with cereals) in order to re-equilibrate the amino acid composition when ingested. This strategy, if followed correctly, has been shown to be safe enough to cover the requirement of the essential amino acids and is therefore optimal for anabolism in the healthy young adult population; however, since they are not particularly rich in leucine, it is still recommended to maintain a portion of high-quality animal proteins in the diet (Dardevet et al., 2021). Currently, most of the protein supplements available on the market are often found as thickened drinks that produce taste fatigue in the long-term (Gosney, 2003). Thus, it is clear that the design of more appealing food products targeted for the elderly is a great opportunity for industry, even though the inclusion of protein, especially in cereal foods, without negatively impacting negatively their sensory properties is still considered a challenge.

4. Formulation of soft cereal foods targeted for the elderly

Besides constituting the dietary basis in many countries, cereal foods are affordable, nutritious and can be consumed regardless of culture and beliefs. They are widely consumed among the elderly population, who tends to orient toward a more "traditional" dietary pattern (Andreeva et al., 2016).

Over the past 30 years, the relationships between structure and properties of cereal foods have been widely investigated. Thanks to this research, it is now well acknowledged that the structure of cereal foods can be controlled through processing using simple models (Della Valle et al., 2014). Consequently, it is possible to tailor products with specific mechanical and, thus, textural features (Gao, Wang, Dong, & Zhou, 2017). In this regard, a relevant application of this knowledge could be the design of cereal foods fortified with protein in order to fit the needs of the elderly population. However, due to their wide range of compositions and processes, there is still a lack of knowledge on the structure-properties relations for many of them, which can be partially addressed through the application of mechanical modeling (Guessasma, Chaunier, Della Valle, & Lourdin, 2011).

To this extent, soft cereal products are good candidates for the development since they have pleasant sensory properties and a relative flexibility regarding formulation, opening the possibility for modifications including the fortification with proteins to increase their nutritional value. This section focuses on the principal aspects involved in the processing, characterization and mechanical modeling of soft cereal foods, including their application to the simulation of chewing and FOP. At last, the impact of protein fortification from pulses on product properties and FOP of soft cereal foods will be discussed.

4.1 Soft cereal foods: Processing, structural and mechanical properties

Many solid food systems present a cellular structure (Fig. 12). In cereal foods, this structure is originated by the air expansion that occurs during their processing (i.e., fermentation, extrusion, and baking). Thus, since they consist of a gas dispersed in a continuous solid phase, they are considered as solid foams (Della Valle, Chiron, & Saulnier, 2019; Guessasma et al., 2011).

As it is common in solid foams, most cereal foods have relatively low densities ($<0.5\,g\,cm^{-3}$), because the air occupies an important part of their volume. Also, since their processing often involves high temperatures ($>100\,°C$), their water content is also rather low ($<50\%$) (Della Valle et al., 2019). Both their low density and water content highlight the importance of studying their hydration and fragmentation during chewing.

Moreover, like all solid foams, the cellular structure of cereal foods is related to their mechanical properties by power laws (Gibson & Ashby, 1997), whether they show brittle (i.e., biscuits, crackers, flakes) or ductile (i.e., bread, brioche and sponge-cake) mechanical behavior. Cereal foams within the second category, also referred as *soft* cereal products, are

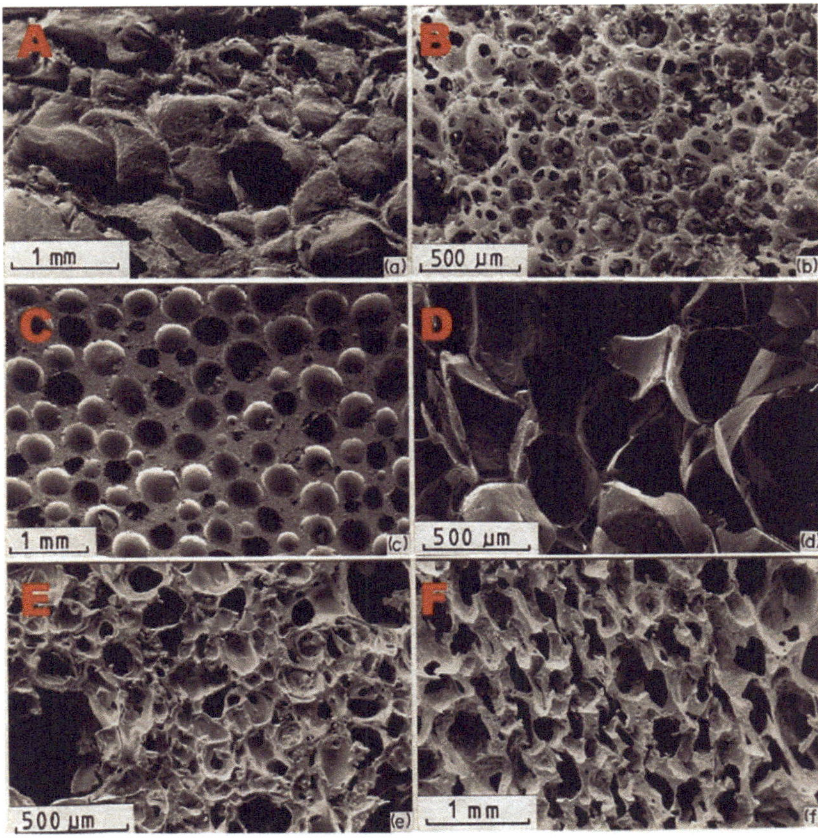

Fig. 12 Cellular structure of common solid foods observed via scanning electron microscopy (SEM). (A) Bread; (B) Meringue; (C) Aero-chocolate bar; (D) Potato chip; (E) Malteser®; (F) Sponge-cake. *Adapted from Gibson, L. J., & Ashby, M. F. (1997).* Cellular solids: Structure and properties *(2nd ed.). Cambridge University Press.*

characterized by their ability to endure large deformations before fracture (Callister, 2007). Such behavior in turn, largely depends on their composition[f] and processing conditions, which will be reviewed in the following section.

4.2 Composition, processing and classification

Soft cereal foods cover a wide range of compositions, which include different ingredients such as flour, water, salt, and a leavening agent in different

[f] Like all polymer based materials, the transition from ductile to fragile is governed by the glass transition temperature (Tg) of their components, which in turn depends largely on their water content (Guessasma et al., 2011; Nicholls, Appelqvist, Davies, Ingman, & Lillford, 1995).

proportions. Additionally, depending on the product type, they can include fat, sugar, milk and eggs. Briefly, the elaboration process of a soft solid foam consists in three main steps: (i) the transition from a divided solid medium to a continuous one (flour to dough or batter), where air is incorporated by mechanical action; followed by ii) the air expansion, which occurs by the action of leavening agents (either chemical or biological); and finally (iii) the stabilization, which corresponds to the thermal setting of the visco-elastic medium formed, to become a soft solid foam (Della Valle et al., 2014; Sahu & Niranjan, 2009). Indeed, the composition and processing of soft cereal foods define their final properties and allow their classification into bakery or pastry products (Table 1).

In spite of these differences, their principal constituent is cereal flour. Of all the available cereal sources in the world, wheat (*Triticum spp.*) is the most widely cultivated commercial crop and constitutes the most important carbohydrate source for humans, way above rice, maize and potatoes (Curtis, 2002). The constituents of wheat flour and their functionality have been recently reviewed by Della Valle et al. (2019).

Table 1 Ingredients used in the processing of soft cereal foods.

Basic ingredients (Flour, water, salt)	**Yes**	**Yes**
Air incorporation method	Kneading	Beating, whipping, rolling, sheeting
Leavening agent	Biological (yeast or sourdough)	Chemical (phosphates, carbonates)
Additional ingredients	• Lipids (0–45%) • Sugar (0–15%) • Eggs (10–45%) • Milk (2–5%) • Emulsifiers and other additives (>1%)	• Sugar (16–25%) • Eggs: whole or separated whites or yolks (30–50%) • Lipids (10–18%) • Milk (2–5%) • Emulsifiers and other additives (>1%)
Examples	Brioche, milk roll, Panettone…	Cakes, cookies, croissants…

The percentages are expressed by 100 g of flour.

Adapted from Lassoued, N. (2005). *Structure alvéolaire des produits céréaliers de cuisson en lien avec les propriétés rhéologiques et thermiques de la pâte: Effet de la composition*. Université de Nantes. Completed with data from Conforti, F. (2014). Cake manufacture. In W. Zhou (Ed.), *Bakery products: Science and technology* (2nd ed., pp. 566–583). Wiley-Blackwell.

4.3 Structure development at different scales and characterization

The structure of soft cereal foams is developed during their processing at various levels of matter organization. At the macroscopic scale, the structures of the batter or dough, and of the baked products, can be monitored through the measurement of their density, often carried out by volume displacement methods.

At the mesoscopic level, the hydrothermal transitions of starch and proteins can be represented through state diagrams that can be used to retrace the paths followed by the product during processing (Della Valle et al., 2019) (Fig. 13). This scale is also known as the cellular intermediate scale, since it is mainly represented by the cellular structure of the solid foam. The latter is described by the distribution of the air cells size and cell wall thickness and can be accurately determined by 3D imaging (and analysis) acquired by X-ray micro-tomography (XR-μT), a non-destructive technique that has been widely applied to characterize the internal structure of plants (Schoeman,

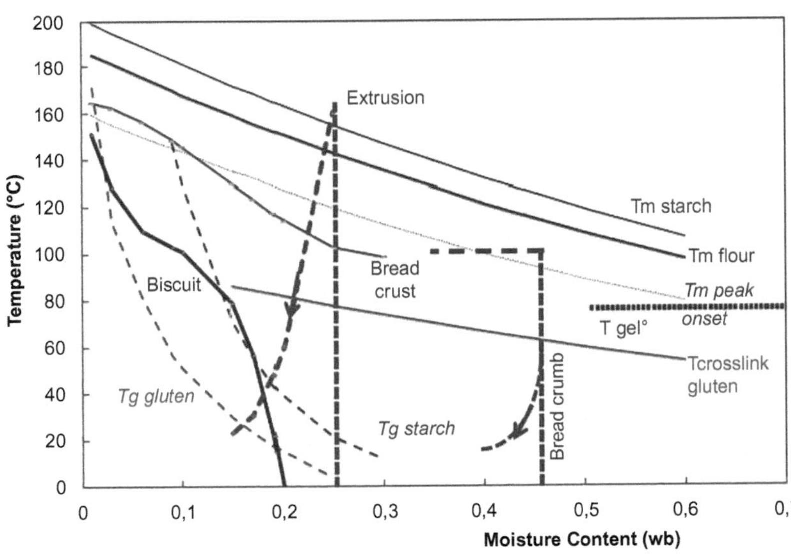

Fig. 13 Example of state diagram for starch and gluten changes during biscuit, bread (crumb and crust) baking and extrusion. Red and green continuous lines reflect transition temperature variations for starch, and gluten, respectively (continuous: starch melting and gluten crosslinking; dotted: · · · —long-glass transition, – – – —short starch gelatinization). Gray lines refer to the hydrothermal pathways, including cooling (arrows), for each product mentioned in the graph: bread crumb and crust, extruded snack (dotted lines) and biscuit (continuous line) (Della Valle et al., 2019).

Williams, du Plessis, & Manley, 2016) and bread at different stages of their processing (Babin et al., 2006; Turbin-Orger et al., 2012, 2015).

At microscopic levels, the cell wall material is considered as composite material, made of more or less amorphous starch and aggregated proteins, containing more or less polysaccharides from dietary fibers coming from the outer part of the kernel (whole grain products) or from the starchy endosperm (white flour) blended with other ingredients (water, fat, sugar and other dietary fibers) (Della Valle et al., 2019). A useful method to characterize the cell wall material is confocal laser scanning microscopy (CLSM), which thanks to the fluorescence achieved mainly by specific dying (Monici, 2005) allows the localization of starch, proteins and lipids in many cereal foods such as bread and dough (Brooker, 1996; Dubreil, Biswas, & Marion, 2002; Turbin-Orger et al., 2015), cakes and batters (Díaz-Ramírez et al., 2013; Eslava-Zomeño, Quiles, & Hernando, 2016; Hesso et al., 2015; Rodríguez-García, Puig, Salvador, & Hernando, 2012) and laminated dough (Bousquieres, Deligny, Challois, & Lucas, 2014; Bousquieres, Deligny, Riaublanc, & Lucas, 2014).

4.4 Mechanical properties of breads and cakes and their characterization

The mechanical properties of soft cereal foods, like all solid foams, are dictated by their density, their cellular structure and the intrinsic material properties (Gibson & Ashby, 1997). Together with their composition, including moisture content, they determine the texture properties of the product.

Texture can be measured instrumentally *via* their stress-strain response, which provides the relation between the amount of deformation of the material when subjected to an external stress and *vice versa*. Stress is the force acting per unit area, while strain (deformation) is the change of the dimension of the sample relative to its original dimension (Callister, 2007).

The stress-strain behavior can be measured using a universal loading machine in all types of deformation conditions (compression, tension, shearing, etc.). For food materials, it is common to use compression mode because it is more closely related to the action of the teeth. Furthermore, tensile tests are more difficult to perform and gripping problems can occur (Goh, 2002; van Vliet, 2014).

Although recently overlooked, notice that stress-strain curves are independent of the sample geometry and size, as opposed to other instrumental popular methods, such as the Texture Profile Analysis (TPA) (Friedman, Whitney, & Szczesniak, 1963), which only give force *vs.* displacement data under uncontrolled strain conditions. From stress-strain results, significant information of the food material behavior during chewing can be attained, while the method is well suited to comparative studies between different products. In addition, the property data can be used for theoretical or numerical simulations (Skamniotis, 2017).

As shown in Fig. 14, stress–strain curve of solid foams can be divided into three regions: (i) the elastic region is the beginning of the curve and, if a sample is loaded only within this region, no permanent deformation will occur; this means it will retain its original shape when unloaded. If the sample is loaded past the elastic region, the cell walls collapse and experience permanent deformation by a combination of several mechanisms: elastic buckling, plastic yielding and bending fracture (Gibson & Ashby, 1997). Since the most common deformation mechanism occurring in this region is plasticity, it can be known as (ii) the "plastic" region. Finally, the (iii) densification stage occurs when the cell walls are crushed together, and as a consequence, the stress increases steeply in this region. The Young's modulus (E), also known as the elastic modulus, is calculated as the slope of the elastic region. The transition point between the two first stages is the yield stress (σ_y). If a material is loaded past the yield stress and then unloaded, this causes a localized increase in the yield stress (Berthaume, 2016).

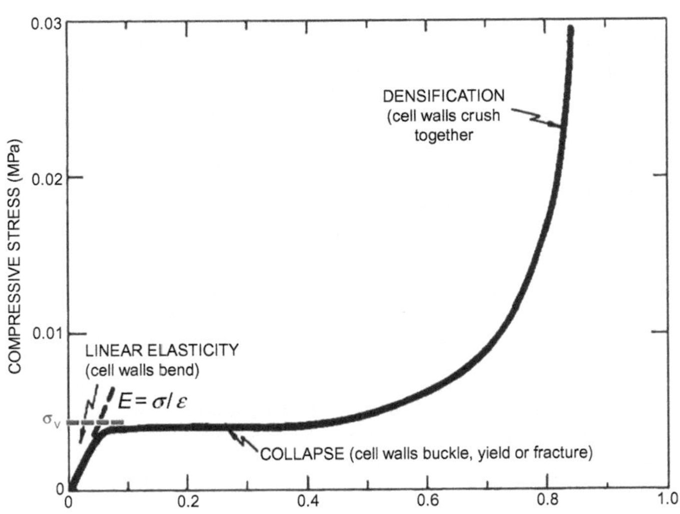

Fig. 14 Typical stress-strain curve of solid foam obtained by compression. *Adapted from Ashby, M. F., & Medalist, R. F. M. (1983). The mechanical properties of cellular solids.* Metallurgical Transactions A, 14*(9), 1755–1769. doi:10.1007/BF02645546.*

4.5 Modeling the mechanical response of cereal foods in conditions close to FOP

Given the complexity of structural changes, the wide variety of formulations and processing conditions involved in cereal food processing, there is a need to use models in order to favor the design of products with target properties and/or to optimize their process conditions (Della Valle et al., 2014).

In this regard, since the beginning of the century, various modeling approaches taken from material science have been used to predict a large number of mechanical responses in cereal based and other food materials, such as extruded corn, starches and biscuits (Guessasma et al., 2011). Overall, such models have contributed to improving the understanding of their structure-properties relationships and, although they are not widely used in practice, they have opened prospect to the prediction of target structures (Della Valle et al., 2019). Today, with the concomitant advances of FOP understanding, they could help to establish the operating parameters and/or material composition to attain a desired perception of texture or flavor (Cleary & Harrison, 2016).

To this extent, models that aim to predict the food behavior during chewing have gained attention; although they are still considered a challenge (Cleary & Harrison, 2016; Harrison, Cleary, Eyres, Sinnott, & Lundin, 2014). This is due to the high complexity of the chewing mechanisms, which require the use of more advanced and sophisticated modeling techniques.

4.5.1 Phenomenological models and others

The first attempts to predict food fragmentation were based on statistical models that consider chewing as a process of selection and probability of breakage (Lucas & Luke, 1983a, 1983b; van der Bilt, Olthoff, van der Glas, van der Weelen, & Bosman, 1987). These models used the fundamental breakdown theory of Epstein (1947) to simulate the breakdown of solids during human mastication. They separate the mastication process into two distinct processes. The first process is the arrangement of the particles in the mouth so that they can be broken apart (selection). The second process concerns the actual fracture of the particles and their subsequent size distribution (breakage). However, these models fail to take into account any of the food mechanical properties. For that reason, Agrawal et al. (1997) proposed the use of two key mechanical properties to predict the response of a solid food to fracture: the toughness (R) and the Young's modulus (E) (Fig. 15A). Although these studies showed there is an important relationship between food initial properties and fragmentation, these predictions are limited to

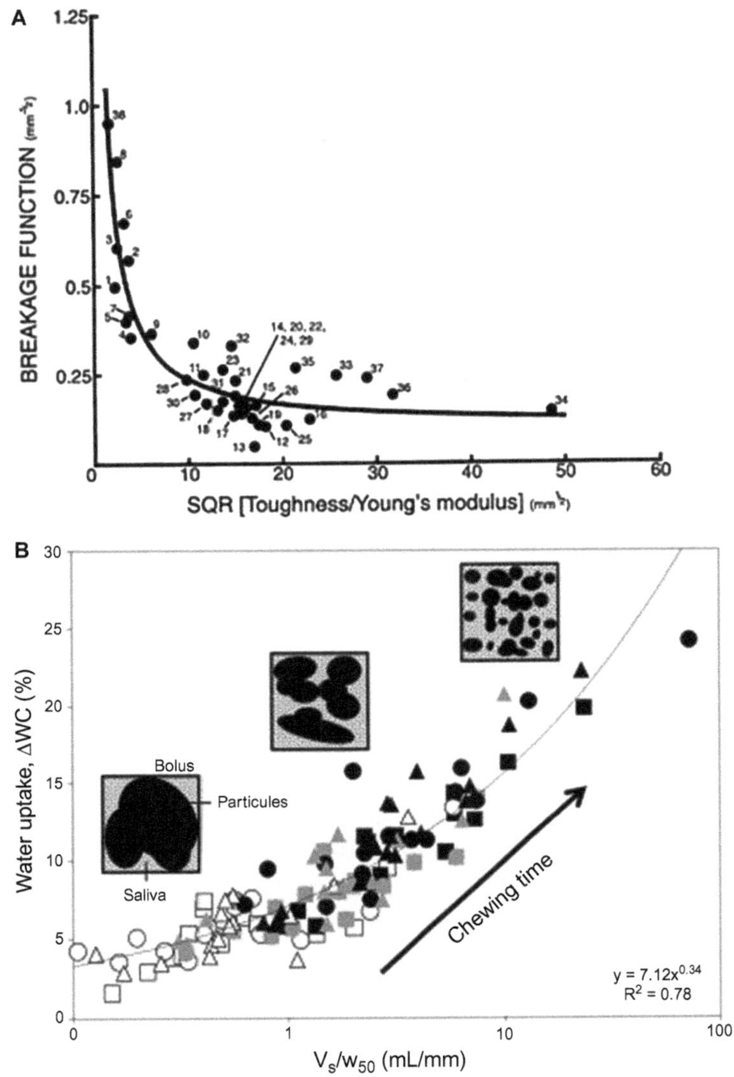

Fig. 15 (A) The "breakage function." Each number represents a different food: 1–9 nuts; 10–27 cheeses; 28–32 raw fruits and vegetables; 33–35 breads; 37–38 others (Lucas, Prinz, Agrawal, & Bruce, 2002). (B) Variations of bolus water uptake during chewing with the ratio of the theoretical saliva volume (V_s) to the particle median width W_{50} for the three mastication stages (open symbols = 1st stage; gray symbols = 2nd stage; full symbols = 3rd stage) of three different fiber fortified breads (square, circle, and triangle) (Le Bleis et al., 2016).

the first bite and mostly to products exhibiting a brittle behavior such as crispy cereals and snacks. Therefore, a model able to predict particle breakdown accurately in soft solid foods has yet to be established.

Fragmentation can also be modeled from empirical laws that are fitted to experimental data obtained from the particle size of *ex vivo* collected boli. For example, de Loubens et al. (2011) and Wright, Sprunt, Smith, and Hills (2003) modeled aroma release from fragmentation during mastication; also Le Bleis et al. (2016) did so for the breakdown of breads enriched with fiber (Fig. 15B), and more recently Assad-Bustillos, Tournier, et al. (2020) explained bolus breakdown in sponge-cake and brioche fortified with pulse proteins with a similar approach. In this mode, the bolus viscosity reduction was also related to the salivary flow and chewing duration and a coefficient of interaction of saliva with food was defined. It was seen that the interaction with saliva was reduced when the fat content was higher and when pulse proteins were added. This highlighted the importance of being able to ascertain the different interactions between food and saliva in cereal products and likely also in other type of foods.

4.5.2 Numerical models

Based on the ability of numerical methods to capture the large strain fracture behavior of foods, models able to reproduce food breakdown during chewing have been developed within the past decade (Harrison et al., 2014; Harrison & Cleary, 2014; Harrison, Cleary, et al., 2014; Skamniotis, Patel, Charalambides, & Elliott, 2016). They raise great potential for efficient product development and optimization, eliminating the need for physical simulations or *in vivo* experiments (Skamniotis et al., 2016).

For example, Skamniotis et al. (2016) have proposed a Finite Element Method (FEM) based model able to reproduce the cereal based pet-foods fracture after the "first bite" using realistic loading conditions and teeth morphologies. Assad-Bustillos, Guessasma, Réguerre, and Della Valle (2020) implemented a FEM model based on the X-ray micro-tomography 3D images of sponge-cake and brioche, and although it was limited to deformation within the elastic domain, the model was able to explain the differences in the fragmentation mechanisms of the two products, showing brioche, which displayed a less efficient stress transfer, was less prone to fragmentation than sponge-cake.

More sophisticated computational models of food breakdown and taste release based on smoothed particle hydrodynamics (SPH) have been developed (Cleary & Harrison, 2016; Harrison & Cleary, 2014; Harrison, Cleary, et al., 2014; Harrison, Eyres, et al., 2014). They consisted of realistic and moving representations of the oral cavity, including teeth, jaw and soft tissues applied to the chewing of a simple gel type homogeneous food. They provided a first attempt at validating predictions of the fragment size distribution after two chewing cycles (Harrison, Eyres, et al., 2014). Moreover, the authors used the model to study the mastication of a simple inhomogeneous food consisting of a cubic chocolate shell enclosing a liquid caramel core, and the release of aroma compounds into saliva and thermal-based softening were also included. This opens up the possibility for various applications such as the prediction of aroma release and the modeling of perception (Cleary & Harrison, 2016).

4.6 Protein fortification of soft cereal foods: Impact on product properties and FOP mechanisms

Due to their nutritional value and low environmental impact, there is a growing interest in including pulses and their derived ingredients in the development of new food products, particularly cereal foods targeted for vulnerable populations (FAO, 2016). Indeed, pulses and cereals have a complimentary amino acid profile that makes them an excellent source of high-quality protein. In this context, soft cereal foods fortified with protein from pulses have a huge potential to satisfy the nutritional needs of the elderly population while preserving palatability.

However, as it was seen from the complexity of the composition and processing of soft cereal foods, the inclusion of atypical components is expected to disrupt their structure and in consequence, affect the texture properties of the final products. Therefore, a good understanding of the composition, structure, and functionality of pulse proteins is needed; all of them are reviewed elsewhere in this book. In this chapter, a general overview of the consequences and impact of the addition of pulse proteins in food properties and oral processing mechanisms of soft cereal foods is given.

4.6.1 Impact of fortification on food properties

Given its nutritional interest, in the past years, many studies have been carried out to explore the consequences of the fortification with pulse flours or isolates on the properties of different soft cereal foods (Table 2). Most of them usually investigated the maximum amount of flour/isolate that could be added without impacting the texture and sensory acceptability. However, there is still little information about their mechanisms of action within the food matrix that could allow correcting or preventing those negative changes.

To summarize, it appears the addition of pulse proteins increases the need for added water in bread dough, which increases the final moisture content

Table 2 Summary of the effects of fortification with pulses (mainly flours) in different types of soft cereal matrices.

Soft cereal food	Pulse flour/isolate	Levels of fortification	Consequences on dough/batter/bread properties	References
Pita bread	Lentil flour Navy bean flour Pinto bean flour	25%, 50%, 75%, 100% (wheat flour substitution)	↗Dough water absorption ↗Hardness (TPA) ↘Break force to rupture (at high levels of substitution) Darker crust color Coarser particle size of flours produced softer bread 25% of pinto or navy bean were as acceptable as control	Borsuk, Arntfield, Lukow, Swallow, and Malcolmson (2012)
Sponge-cake	Soy protein isolate (SPI)	0%, 5%, 10%, 15%, 20%, 30% (total formulation)	↗Batter viscosity and density ↘Cake density ↗Crumb heterogeneity ↗Hardness (TPA) Darker crust and crumb (more reddish) 20% of SPI preserved texture and acceptability	Majzoobi, Ghiasi, Habibi, Hedayati, and Farahnaky (2014)
Sponge-cake	Pin-milled pea flour	25%, 50% and 100% (wheat flour substitution)	No changes observed up to 50% substitution ↗Batter viscosity ↘Bubble size and ↗homogeneous distribution (batter) ↗Cake density ↗Hardness (TPA) ↘Staling 25% of substitution allows the preservation of texture and acceptability	Gómez, Doyagüe, and de la Hera (2012)

Continued

Table 2 Summary of the effects of fortification with pulses (mainly flours) in different types of soft cereal matrices.—cont'd

Soft cereal food	Pulse flour/isolate	Levels of fortification	Consequences on dough/batter/bread properties	References
Sponge-cake	Pea protein isolate	0%, 10%, 20%, 30% and 40% (wheat flour substitution)	↗Batter rheological properties (viscosity and G') ↗Batter density ↗Batter stability ↗Cake density Pea protein formed a network of interconnected "bridged" particles in the continuous phase that swelled and connected to each other and were the main driver for the increase of rheological properties.	Assad Bustillos, Jonchère, Garnier, Réguerre, and Della Valle (2020)
Bread	Chickpea flour	10%, 20%, 30% (wheat flour substitution)	↗Dough water absorption and development time ↘Dough extensibility Dough resistance to deformation: At 10%: ↗ At 20% ↘ ↗Bread density ↗Hardness (TPA) Darker crumb and crust 10% was the maximum level to avoid negative impacts in in texture and acceptability.	Mohammed, Ahmed, and Senge (2012)
Bread	Pea flour	5%, 10% (total formulation)	↗Dough water absorption ↗ Dough stickiness (TPA) ↗Bread density 5% preserved acceptability	Kamaljit, Baljeet, and Amarjeet (2010)

of the baked product. In cakes, the batter viscosity and density are both increased. In both cakes and breads, the heterogeneity of the crumb, as well as the density are increased. The mechanical properties (hardness) always increase, but in some cases the force at rupture decreases (the product is more "brittle"). In most cases, pulse flours have been commonly used to substitute wheat flour at 15–20%, yet it appears that the optimum fortification levels are rather between 5% and 10% of flour. The use of high protein isolates is still little explored. To conclude, since the addition of protein appears to have a non-negligible impact on the texture of products, the fortification of soft cereal foods with pulse proteins is still a challenge.

4.6.2 Impact of fortification on FOP and oral comfort perception

To date, few studies have investigated the impact of protein fortification on the chewing behavior, bolus properties and sensory perception of soft cereal foods. In the elderly, the same authors that investigated the impact of oral physiology on the FOP mechanisms of sponge-cake and brioche (Assad-Bustillos, Tournier, Feron, et al., 2019; Assad-Bustillos, Tournier, Septier, et al., 2019), have recently published a follow-up study featuring the protein fortified version of the same products (Assad-Bustillos, Tournier, et al., 2020).

Overall, both fortified sponge-cake and brioche were perceived as highly comfortable, although only the fortified brioche was significantly less comfortable than its non-fortified counterpart. Unlike their standard versions, the perception of oral comfort of the fortified foods depended more on the dental status, than on the salivary flow. Also, the bolus apparent viscosity was related to the perception of oral comfort in the fortified sponge-cake, while the chewing duration and the bolus particle size at the beginning of chewing contributed to explaining oral comfort in the fortified brioche (Fig. 16). These differences, and in the general the higher difficulty to fragment the fortified brioche, can be explained by the coefficient of interaction of food with saliva defined by the authors, which decreased when pulse proteins were added.

Moreover, in a separate study by the same research group, it was seen that structure of fortified brioche had smaller gas cells and slightly larger cell walls than the standard one, which made the food sample more difficult to fragment. Additionally, the fortified brioche in particular, displayed a less efficient stress transfer, being less prone to fragmentation (Assad-Bustillos, Guessasma, et al., 2020).

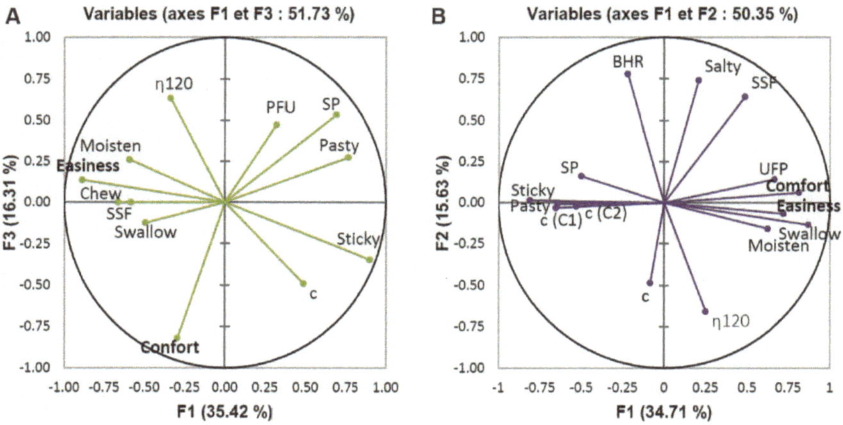

Fig. 16 PCA correlation circle for oral health status, bolus properties, chewing duration and oral comfort variables for (A) fortified sponge cake and (B) fortified brioche. Comfort = Overall comfort perception; Chew = Perceived as easy to chew; Moisten = Perceived as easy to moisten; Swallow = Perceived as easy to swallow; Easiness = Perceived as easy to eat; Pasty = Perceived pastiness; Sticky = Perceived stickiness; BHR = Bolus hydration ratio or added saliva; η120 = Bolus apparent viscosity at 120 s^{-1}; SSF = Stimulated salivary flow rate; SP = Swallowing point or total chewing duration; c = c Gompertz parameter (Assad-Bustillos, Tournier, et al., 2020).

5. Conclusions and perspectives

Beyond the characterization of food properties alone, the study Food Oral Processing (FOP) has become crucial to the development of sensory appealing food products and shows great potential to be used in the development of food for elderly, in particular soft cereal foods. Indeed, except from bread, the chewing behavior and bolus formation mechanisms of these type of foods have been little characterized, and their relationships between mechanical properties and fragmentation are less clearly established than in other cereal foods showing brittle behavior (wafers, biscuits, flakes, etc.).

The oral physiology of the elderly is commonly degraded, especially the muscular strength, dental status and salivary flow. In addition, they require higher protein intakes than young adults. The latter should be of 25–30 g of high-quality protein on a daily basis, stressing the need for a supplementation. Currently, most of the existing supplements are found in a medical context, and are not sensory appealing, opening an opportunity for the food industry to fill this gap.

Today, thanks to the few existing studies that have been carried out to understand the influence of physiology on oral processing in elderly subjects, the remarkable influence of salivary flow on the chewing behavior, bolus properties and the perception of oral comfort in soft cereal foods has been highlighted. However, more studies are needed that include other physiological aspects such as the tongue (which induces important shearing and compressive deformations of food against the palate) or saliva composition, which may also be accountable for differences in perception. Moreover, since swallowing may also play an important role in the perception of oral comfort, the elongational component of the food bolus must be determined. Surface properties, such as tribology, should also be studied in order to take into account for their role in lubrication and swallowing.

It is clear that pulse proteins show a particular nutritional interest due to the amino-acid equilibrium and complementarity with cereal foods, although their incorporation in soft cereal foods without altering sensory properties is still considered a challenge. It was seen that the interaction with saliva was reduced when the fat content was higher and when pulse proteins were added. This suggests that the need for added saliva in this type of foods can be reduced by modifying their composition, i.e., by adding less hydrophilic components such as fats to enhance lubrication. More in-depth and comprehensive studies that address the structure-texture relationships in these matrices are needed, with an emphasis in enhancing lubrication in order to improve oral comfort perception.

In conclusion, the design of protein-fortified soft cereal foods for the elderly should be done by considering the interactions of food structure, oral processing, physiology and perception. Moreover, to guarantee the nutritional value of the developed foods, another important issue to be addressed is the digestibility of the added proteins, since FOP is thought to have an impact in their assimilation.

Finally, there are many factors that can affect the elderly's food intake, and in this regard, the conclusions of this review may not be able to extend to the elderly who are not fully autonomous or suffering from a chronical disease (i.e., dysphagia). Research in the field of dependent elderly population exists but exceeds the limits and the scope of this chapter.

Overall, the contents of this chapter have provided guidelines for developing realistic cereal foods fortified with protein with acceptable sensory properties and have the potential to meet the nutritional needs of healthy elderly.

References

Agrawal, K. R., Lucas, P. W., Prinz, J. F., & Bruce, I. C. (1997). Mechanical properties of foods responsible for resisting food breakdown in the human mouth. *Archives of Oral Biology*, *42*(1), 1–9. https://doi.org/10.1016/S0003-9969(96)00102-1.

Andreeva, V., Allès, B., Feron, G., Gonzalez, R., Sulmont-Rossé, C., Galan, P., et al. (2016). Sex-specific sociodemographic correlates of dietary patterns in a large sample of french elderly individuals. *Nutrients*, *8*(8), 484. https://doi.org/10.3390/nu8080484.

Assad Bustillos, M., Jonchère, C., Garnier, C., Réguerre, A. L., & Della Valle, G. (2020). Rheological and microstructural characterization of batters and sponge cakes fortified with pea proteins. *Food Hydrocolloids*, *101*(June 2019), 105553. https://doi.org/10.1016/j.foodhyd.2019.105553.

Assad-Bustillos, M., Guessasma, S., Réguerre, A. L., & Della Valle, G. (2020). Impact of protein reinforcement on the deformation of soft cereal foods under chewing conditions studied by X-ray tomography and finite element modelling. *Journal of Food Engineering*, *286*(January). https://doi.org/10.1016/j.jfoodeng.2020.110108, 110108.

Assad-Bustillos, M., Palier, J., Rabesona, H., Choiset, Y., Della Valle, G., & Feron, G. (2019). Role of the bolus degree of structure on the protein digestibility during in vitro digestion of a pea protein-fortified sponge cake chewed by elderly. *Journal of Texture Studies*, *51*(1), 134–143. https://doi.org/10.1111/jtxs.12486. jtxs.12486.

Assad-Bustillos, M., Tournier, C., Feron, G., Guessasma, S., Reguerre, A. L., & Della Valle, G. (2019). Fragmentation of two soft cereal products during oral processing in the elderly: Impact of product properties and oral health status. *Food Hydrocolloids*, *91*(January), 153–165. https://doi.org/10.1016/j.foodhyd.2019.01.009.

Assad-Bustillos, M., Tournier, C., Palier, J., Septier, C., Feron, G., & Della Valle, G. (2020). Oral processing and comfort perception of soft cereal foods fortified with pulse proteins in the elderly with different oral health status. *Food & Function*, *11*(5), 4535–4547. https://doi.org/10.1039/C9FO02993A.

Assad-Bustillos, M., Tournier, C., Septier, C., Della Valle, G., & Feron, G. (2019). Relationships of oral comfort perception and bolus properties in the elderly with salivary flow rate and oral health status for two soft cereal foods. *Food Research International*, *118*(September), 13–21. https://doi.org/10.1016/j.foodres.2017.11.057.

Babin, P., Della Valle, G., Chiron, H., Cloetens, P., Hoszowska, J., Pernot, P., et al. (2006). Fast X-ray tomography analysis of bubble growth and foam setting during breadmaking. *Journal of Cereal Science*, *43*(3), 393–397. https://doi.org/10.1016/J.JCS.2005.12.002.

Bauer, J., Biolo, G., Cederholm, T., Cesari, M., Cruz-Jentoft, A. J., Morley, J. E., et al. (2013). Evidence-based recommendations for optimal dietary protein intake in older people: A position paper from the PROT-AGE Study Group. *Journal of the American Medical Directors Association*, *14*(8), 542–559. https://doi.org/10.1016/j.jamda.2013.05.021.

Berthaume, M. A. (2016). Food mechanical properties and dietary ecology. *American Journal of Physical Anthropology*, *159*, 79–104. https://doi.org/10.1002/ajpa.22903.

Borsuk, Y., Arntfield, S., Lukow, O. M., Swallow, K., & Malcolmson, L. (2012). Incorporation of pulse flours of different particle size in relation to pita bread quality. *Journal of the Science of Food and Agriculture*, *92*(10), 2055–2061. https://doi.org/10.1002/jsfa.5581.

Bousquieres, J., Deligny, C., Challois, S., & Lucas, T. (2014). Using confocal laser scanning microscopy to examine the breakdown of fat layers in laminated dough. *Food Research International*, *62*, 359–365. https://doi.org/10.1016/j.foodres.2014.02.019.

Bousquieres, J., Deligny, C., Riaublanc, A., & Lucas, T. (2014). CLSM study of layers in laminated dough: Roll out of layers and elastic recoil. *Journal of Cereal Science*, *60*(1), 82–91. https://doi.org/10.1016/j.jcs.2014.01.018.

Brodkorb, A., Egger, L., Alminger, M., Alvito, P., Assunção, R., Ballance, S., et al. (2019). INFOGEST static in vitro simulation of gastrointestinal food digestion. *Nature Protocols*, *14*(4), 991–1014. https://doi.org/10.1038/s41596-018-0119-1.

Brooker, B. E. (1996). The role of fat in the stabilisation of gas cells in bread dough. *Journal of Cereal Science*, *24*(3), 187–198. https://doi.org/10.1006/jcrs.1996.0052.

Brown, W. E., & Braxton, D. (2000). Dynamics of food breakdown during eating in relation to perceptions of texture and preference: A study on biscuits. *Food Quality and Preference*, *11*(4), 259–267. https://doi.org/10.1016/S0950-3293(99)00014-2.

Callister, W. D. (2007). *Materials science and engineering: An introduction* (7th ed.). John Wiley & Sons, Ltd.

Chen, J. (2009). Food oral processing—A review. *Food Hydrocolloids*, *23*(1), 1–25. https://doi.org/10.1016/j.foodhyd.2007.11.013.

Chen, J. (2015). Food oral processing: Mechanisms and implications of food oral destruction. *Trends in Food Science & Technology*, *45*(2), 222–228. https://doi.org/10.1016/j.tifs.2015.06.012.

Chen, J. (2016). Food for elderly: Challenges and opportunities. *Journal of Texture Studies*, *47*(4), 255–256. https://doi.org/10.1111/jtxs.12203.

Chen, J., & Stokes, J. R. (2012). Rheology and tribology: Two distinctive regimes of food texture sensation. *Trends in Food Science & Technology*, *25*(1), 4–12. https://doi.org/10.1016/J.TIFS.2011.11.006.

Cleary, P. W., & Harrison, S. (2016). Modeling food digestion in the oral cavity. In *Reference module in food science* (pp. 1–10). https://doi.org/10.1016/B978-0-08-100596-5.03238-8.

Curtis, B. C. (2002). *Wheat in the world*. Retrieved March 18, 2019, from FAO website http://www.fao.org/3/y4011e/y4011e04.htm.

Dardevet, D., Mosoni, L., Savary-Auzeloux, I., Peyron, M.-A., Polakof, S., & Rémond, D. (2021). Important determinants to take into account to optimize protein nutrition in the elderly: Solutions to a complex equation. *Proceedings of the Nutrition Society*, *80*(2), 207–220. https://doi.org/10.1017/S0029665120007934.

Dardevet, D., Rémond, D., Peyron, M.-A., Papet, I., Savary-Auzeloux, I., & Mosoni, L. (2012). Muscle wasting and resistance of muscle anabolism: The "anabolic threshold concept" for adapted nutritional strategies during sarcopenia. *The Scientific World Journal*, *2012*, 1–6. https://doi.org/10.1100/2012/269531.

Dardevet, D., Rieu, I., Fafournoux, P., Sornet, C., Combaret, L., Bruhat, A., et al. (2003). Leucine: A key amino acid in ageing-associated sarcopenia? *Nutrition Research Reviews*, *16*(01), 61. https://doi.org/10.1079/NRR200252.

Dardevet, D., Sornet, C., Bayle, G., Prugnaud, J., Pouyet, C., & Grizard, J. (2002). Postprandial stimulation of muscle protein synthesis in old rats can be restored by a leucine-supplemented meal. *The Journal of Nutrition*, *132*(1), 95–100. https://doi.org/10.1093/jn/132.1.95.

de Loubens, C., Panouillé, M., Saint-Eve, A., Déléris, I., Tréléa, I. C., & Souchon, I. (2011). Mechanistic model of in vitro salt release from model dairy gels based on standardized breakdown test simulating mastication. *Journal of Food Engineering*, *105*(1), 161–168. https://doi.org/10.1016/j.jfoodeng.2011.02.020.

De Zorzi, M., Curioni, A., Simonato, B., Giannattasio, M., & Pasini, G. (2007). Effect of pasta drying temperature on gastrointestinal digestibility and allergenicity of durum wheat proteins. *Food Chemistry*, *104*(1), 353–363. https://doi.org/10.1016/J.FOODCHEM.2006.11.057.

Della Valle, G., Chiron, H., Cicerelli, L., Kansou, K., Katina, K., Ndiaye, A., et al. (2014). Basic knowledge models for the design of bread texture. *Trends in Food Science & Technology*, *36*(1), 5–14. https://doi.org/10.1016/J.TIFS.2014.01.003.

Della Valle, G., Chiron, H., & Saulnier, L. (2019). Food structure development in cereal and snack products. In *Handbook on food structure development*. Cambridge: Royal Society of Chemistry (In Press).

Deutz, N. E. P., Bauer, J. M., Barazzoni, R., Biolo, G., Boirie, Y., Bosy-Westphal, A., et al. (2014). Protein intake and exercise for optimal muscle function with aging: Recommendations from the ESPEN Expert Group. *Clinical Nutrition*, *33*(6), 929–936. https://doi.org/10.1016/J.CLNU.2014.04.007.

Devezeaux de Lavergne, M., van de Velde, F., & Stieger, M. (2017). Bolus matters: The influence of food oral breakdown on dynamic texture perception. *Food & Function*, *8*(2), 464–480. https://doi.org/10.1039/C6FO01005A.

Díaz-Ramírez, M., Calderón-Domínguez, G., Chanona-Pérez, J. J., Janovitz-Klapp, A., López-Santiago, R., Farrera-Rebollo, R. R., et al. (2013). Modelling sorption kinetic of sponge cake crumb added with milk syrup. *International Journal of Food Science & Technology*, *48*(8), 1649–1660. https://doi.org/10.1111/ijfs.12135.

Dickerson, R. N. (2016). Nitrogen balance and protein requirements for critically ill older patients. *Nutrients*, *8*(4), 226. https://doi.org/10.3390/nu8040226.

Dubreil, L., Biswas, S. C., & Marion, D. (2002). Localization of puroindoline-a and lipids in bread dough using confocal scanning laser microscopy. *Journal of Agricultural and Food Chemistry*, *50*(21), 6078–6085. https://doi.org/10.1021/jf020277d.

Dugarova, E. (2017). *Ageing, older persons and the 2030 agenda for sustainable development*. United Nations Publications. https://doi.org/10.1016/j.tcs.2014.12.001.

Emami, E., de Souza, R. F., Kabawat, M., & Feine, J. S. (2013). The impact of edentulism on oral and general health. *International Journal of Dentistry*, *2013*, 1–7. https://doi.org/10.1155/2013/498305.

Engelen, L., Fontijn-Tekamp, A., & van der Bilt, A. (2005). The influence of product and oral characteristics on swallowing. *Archives of Oral Biology*, *50*(8), 739–746. https://doi.org/10.1016/j.archoralbio.2005.01.004.

Epstein, B. (1947). The mathematical description of certain breakage mechanisms leading to the logarithmico-normal distribution. *Journal of the Franklin Institute*, *244*(6), 471–477. https://doi.org/10.1016/0016-0032(47)90465-1.

Eslava-Zomeño, C., Quiles, A., & Hernando, I. (2016). Designing a clean label sponge cake with reduced fat content. *Journal of Food Science*, *81*(10), C2352–C2359. https://doi.org/10.1111/1750-3841.13446.

Eurostat. (2016). *Population structure and ageing—Statistics explained*. Retrieved May 15, 2017, from http://ec.europa.eu/eurostat/statistics-explained/index.php/Population_structure_and_ageing.

FAO. (2016). International year of pulses 2016. Retrieved April 27, 2017, from http://www.fao.org/pulses-2016/en/.

Fardet, A., Leenhardt, F., Lioger, D., Scalbert, A., & Rémésy, C. (2006). Parameters controlling the glycaemic response to breads. *Nutrition Research Reviews*, *19*(01), 18. https://doi.org/10.1079/NRR2006118.

Fardet, A., Souchon, I., & Dupont, D. (2013). *Structure des aliments et effets nutritionnels*. Editions Quae.

Fennema, O. (2017). In S. Damodaran, & K. L. Parkin (Eds.), *Fennema's food chemistry* (5th ed.). CRC Press.

Feron, G., Ayed, C., Qannari, E. M., Courcoux, P., Laboure, H., & Guichard, E. (2014). Understanding aroma release from model cheeses by a statistical multiblock approach on oral processing. *PLoS One*, *9*(4), e93113. https://doi.org/10.1371/journal.pone.0093113.

Fiszman, S., & Tarrega, A. (2018). The dynamics of texture perception of hard solid food: A review of the contribution of the temporal dominance of sensations technique. *Journal of Texture Studies*, *49*(2), 202–212. https://doi.org/10.1111/jtxs.12273.

Fontijn-Tekamp, F. A., Slagter, A. P., van Der Bilt, A., van'T Hof, M. A., Witter, D. J., Kalk, W., et al. (2000). Biting and chewing in overdentures, full dentures, and natural dentitions. *Journal of Dental Research, 79*(7), 1519–1524. https://doi.org/10.1177/00220345000790071501.

Fontijn-Tekamp, F. A., van der Bilt, A., Abbink, J. H., & Bosman, F. (2004). Swallowing threshold and masticatory performance in dentate adults. *Physiology & Behavior, 83*(3), 431–436. https://doi.org/10.1016/j.physbeh.2004.08.026.

Fried, M., Abramson, S., & Meyer, J. H. (1987). Passage of salivary amylase through the stomach in humans. Digestive Diseases and Sciences, 32(10), 1097–1103. Retrieved from http://www.ncbi.nlm. nih.gov/pubmed/3652896.

Freitas, D., Le Feunteun, S., Panouillé, M., & Souchon, I. (2017). The important role of salivary α-amylase in gastric digestion of wheat bread starch. *Food & Function, 9*(1), 200–208. https://doi.org/10.1039/C7FO01484H.

Friedman, H. H., Whitney, J. E., & Szczesniak, A. S. (1963). The texturometer: A new instrument for objective texture measurement. *Journal of Food Science, 28*(4), 390–396. https://doi.org/10.1111/j.1365-2621.1963.tb00216.x.

Gao, J., Lin, S., Jin, X., Wang, Y., Yin, J., Dong, Z., et al. (2019). In vitro digestion of bread: How is it influenced by the bolus characteristics? *Journal of Texture Studies, 50*(3), 257–268. https://doi.org/10.1111/jtxs.12391.

Gao, J., Wang, Y., Dong, Z., & Zhou, W. (2017). Structural and mechanical characteristics of bread and their impact on oral processing: A review. *International Journal of Food Science & Technology, 1–15.* https://doi.org/10.1111/ijfs.13671.

Gao, J., Wong, J. X., Lim, J. C.-S., Henry, J., & Zhou, W. (2015). Influence of bread structure on human oral processing. *Journal of Food Engineering, 167*, 147–155. https://doi.org/10.1016/j.jfoodeng.2015.07.022.

Gao, J., & Zhou, W. (2021). Oral processing of bread: Implications of designing healthier bread products. *Trends in Food Science & Technology, 112*, 720–734. https://doi.org/10.1016/j.tifs.2021.04.030.

Gibson, L. J., & Ashby, M. F. (1997). *Cellular solids: Structure and properties* (2nd ed.). Cambridge University Press.

Goh, S. M. (2002). *An engineering approach to food texture studies.* London: Imperial College.

Gómez, M., Doyagüe, M. J., & de la Hera, E. (2012). Addition of pin-milled pea flour and air-classified fractions in layer and sponge cakes. *LWT - Food Science and Technology, 46*(1), 142–147. https://doi.org/10.1016/j.lwt.2011.10.014.

Gosney, M. (2003). Are we wasting our money on food supplements in elder care wards? *Journal of Advanced Nursing, 43*(3), 275–280. https://doi.org/10.1046/j.1365-2648.2003.02710.x.

Grassby, T., Picout, D. R., Mandalari, G., Faulks, R. M., Kendall, C. W. C., Rich, G. T., et al. (2014). Function modelling of nutrient bioaccessibility in almond. *Food & Function.* https://doi.org/10.1039/c4fo00659c.

Grundy, M. M. L., Grassby, T., Mandalari, G., Waldron, K. W., Butterworth, P. J., Berry, S. E. E., et al. (2015). Effect of mastication on lipid bioaccessibility of almonds in a randomized human study and its implications for digestion kinetics, metabolizable energy, and postprandial lipemia. *American Journal of Clinical Nutrition, 101*(1), 25–33. https://doi.org/10.3945/ajcn.114.088328.

Guessasma, S., Chaunier, L., Della Valle, G., & Lourdin, D. (2011). Mechanical modelling of cereal solid foods. *Trends in Food Science & Technology, 22*(4), 142–153. https://doi.org/10.1016/j.tifs.2011.01.005.

Guinard, J. X., & Mazzucchelli, R. (1996). The sensory perception of texture and mouthfeel. Trends in Food Science & Technology, 7(July), 213–219. Retrieved from http://www.scopus.com/inward/record.url?eid=2-s2.0-0030186762&partnerID=40&md5=7a83776aaa97458db82ed5c0861a2971.

Gularte, M. A., Gómez, M., & Rosell, C. M. (2012). Impact of legume flours on quality and in vitro digestibility of starch and protein from gluten-free cakes. *Food and Bioprocess Technology*, *5*(8), 3142–3150. https://doi.org/10.1007/s11947-011-0642-3.

Hansson, A., Giannouli, P., & van Ruth, S. (2003). The influence of gel strength on aroma release from pectin gels in a model mouth and in vivo, monitored with proton-transfer-reaction mass spectrometry. *Journal of Agricultural and Food Chemistry*, *51*(16), 4732–4740. https://doi.org/10.1021/jf030050y.

Harrison, S. M., & Cleary, P. W. (2014). Towards modelling of fluid flow and food breakage by the teeth in the oral cavity using smoothed particle hydrodynamics (SPH). *European Food Research and Technology*, *238*(2), 185–215. https://doi.org/10.1007/s00217-013-2077-8.

Harrison, S. M., Cleary, P. W., Eyres, G. D., Sinnott, M., & Lundin, L. (2014). Challenges in computational modelling of food breakdown and flavour release. *Food & Function*, *5*(11), 2792–2805. https://doi.org/10.1039/C4FO00786G.

Harrison, S. M., Eyres, G., Cleary, P. W., Sinnott, M. D., Delahunty, C., & Lundin, L. (2014). Computational modeling of food oral breakdown using smoothed particle hydrodynamics. *Journal of Texture Studies*, *45*(2), 97–109. https://doi.org/10.1111/jtxs.12062.

Hayflick, L. (2007). Biological aging is no longer an unsolved problem. *Annals of the New York Academy of Sciences*, *1100*, 1–13. https://doi.org/10.1196/annals.1395.001.

Henshaw, M. M., & Calabrese, M. P. H. J. M. (2001). Oral health and nutrition in the elderly. *Nutririon in Clinical Care*, *4*(1), 34–42.

Hesso, N., Garnier, C., Loisel, C., Chevallier, S., Bouchet, B., & Le-Bail, A. (2015). Formulation effect study on batter and cake microstructure: Correlation with rheology and texture. *Food Structure*, *5*, 31–41. https://doi.org/10.1016/j.foostr.2015.03.002.

Hoebler, C., Devaux, M. F., Karinthi, A., Belleville, C., & Barry, J. L. (2000). Particle size of solid food after human mastication and in vitro simulation of oral breakdown. *International Journal of Food Sciences and Nutrition*, *51*(5), 353–366. https://doi.org/10.1080/096374800426948.

Hoebler, C., Karinthi, A., Devaux, M.-F. F., Guillon, F., Gallant, D. J. G., Bouchet, B., et al. (1998). Physical and chemical transformations of cereal food during oral digestion in human subjects. *The British Journal of Nutrition*, *80*(5), 429–436. https://doi.org/10.1017/S0007114598001494.

Hutchings, J. B., & Lillford, P. J. (1988). The perception of food texture—The philosophy of the breakdown path. *Journal of Texture Studies*, *19*(2), 103–115. https://doi.org/10.1111/j.1745-4603.1988.tb00928.x.

Jeltema, M., Beckley, J., & Vahalik, J. (2015). Model for understanding consumer textural food choice. *Food Science & Nutrition*, *3*(3), 202–212. https://doi.org/10.1002/fsn3.205.

Johansson, A. S., Svensson, K. G., & Trulsson, M. (2006). Impaired masticatory behavior in subjects with reduced periodontal tissue support. *Journal of Periodontology*, *77*(9), 1491–1497. https://doi.org/10.1902/jop.2006.050355.

Joubert, M., Septier, C., Brignot, H., Salles, C., Panouillé, M., Feron, G., et al. (2017). Chewing bread: Impact on alpha-amylase secretion and oral digestion. *Food & Function*, *8*(2), 607–614. https://doi.org/10.1039/C6FO00963H.

Jourdren, S., Panouillé, M., Saint-Eve, A., Déléris, I., Forest, D., Lejeune, P., et al. (2016). Breakdown pathways during oral processing of different breads: Impact of crumb and crust structures. *Food & Function*, 7(3), 1446–1457. https://doi.org/10.1039/c5fo01286d.

Jourdren, S., Saint-Eve, A., Panouillé, M., Lejeune, P., Déléris, I., & Souchon, I. (2016). Respective impact of bread structure and oral processing on dynamic texture perceptions through statistical multiblock analysis. *Food Research International*, *87*, 142–151. https://doi.org/10.1016/j.foodres.2016.06.021.

Jourdren, S., Saint-Eve, A., Pollet, B., Panouillé, M., Lejeune, P., Guichard, E., et al. (2017). Gaining deeper insight into aroma perception: An integrative study of the oral processing of breads with different structures. *Food Research International, 92*, 119–127. https://doi.org/10.1016/j.foodres.2017.01.001.

Kamaljit, K., Baljeet, S., & Amarjeet, K. (2010). Preparation of bakery products by incorporating pea flour as a functional ingredient. *American Journal of Food Technology, 5*(2), 130–135. https://doi.org/10.3923/ajft.2010.130.135.

Kiesswetter, E., Poggiogalle, E., Migliaccio, S., Donini, L. M., Sulmont-Rossé, C., Feart, C., et al. (2018). Functional determinants of dietary intake in community-dwelling older adults: A DEDIPAC (DEterminants of DIet and Physical ACtivity) systematic literature review. *Public Health Nutrition, 21*(10), 1886–1903. https://doi.org/10.1017/S1368980017004244.

Kim, E. H. J., Corrigan, V. K., Wilson, A. J., Waters, I. R., Hedderley, D. I., & Morgenstern, M. P. (2012). Fundamental fracture properties associated with sensory hardness of brittle solid foods. *Journal of Texture Studies, 43*(1), 49–62. https://doi.org/10.1111/j.1745-4603.2011.00316.x.

Klein, S., Cohn, S. M., & Alpers, D. H. (2006). Alimentary tract in nutrition. In M. E. Shils, & M. Shike (Eds.), *Modern nutrition in health and disease* (10th ed., pp. 1115–1142). Lippincott Williams & Wilkins.

Kohyama, K., Mioche, L., & Bourdiol, P. (2003). Influence of age and dental status on chewing behaviour studied by EMG recordings during consumption of various food samples. *Gerodontology, 20*(1), 15–23. https://doi.org/10.1111/j.1741-2358.2003.00015.x.

Kohyama, K., Mioche, L., & Martin, J.-F. (2002). Chewing patterns of various texture foods studied by electromyography in young and elderly populations. *Journal of Texture Studies, 33*, 269–283. https://doi.org/10.1111/j.1745-4603.2002.tb01349.x.

Kong, F., & Singh, R. P. (2008). Disintegration of solid foods in human stomach. *Journal of Food Science, 73*(5), 67–80. https://doi.org/10.1111/j.1750-3841.2008.00766.x.

Krall, E., Hayes, C., & Garcia, R. (1998). How dentition status and masticatory function affect nutrient intake. *The Journal of the American Dental Association, 129*(9), 1261–1269. https://doi.org/10.14219/jada.archive.1998.0423.

Laguna, L., Hetherington, M. M., Chen, J., Artigas, G., & Sarkar, A. (2016). Measuring eating capability, liking and difficulty perception of older adults: A textural consideration. *Food Quality and Preference, 53*, 47–56. https://doi.org/10.1016/j.foodqual.2016.05.013.

Laguna, L., Sarkar, A., Artigas, G., & Chen, J. (2015). A quantitative assessment of the eating capability in the elderly individuals. *Physiology & Behavior, 147*, 274–281. https://doi.org/10.1016/j.physbeh.2015.04.052.

Le Bleis, F., Chaunier, L., Della Valle, G., Panouillé, M., & Réguerre, A. L. (2013). Physical assessment of bread destructuration during chewing. *Food Research International, 50*(1), 308–317. https://doi.org/10.1016/j.foodres.2012.10.042.

Le Bleis, F., Chaunier, L., Montigaud, P., & Della Valle, G. (2016). Destructuration mechanisms of bread enriched with fibers during mastication. *Food Research International, 80*, 1–11. https://doi.org/10.1016/j.foodres.2015.12.008.

Lillford, P. J. (2017). Food structure, its formation and breakdown. *Microscopy and Microanalysis, 23*(S1), 1122–1123. https://doi.org/10.1017/S1431927617006274.

Lonnie, M., Hooker, E., Brunstrom, J. M., Corfe, B. M., Green, M. A., Watson, A. W., et al. (2018). Protein for life: Review of optimal protein intake, sustainable dietary sources and the effect on appetite in ageing adults. *Nutrients, 10*(3), 1–18. https://doi.org/10.3390/nu10030360.

Lucas, P. W. (2004). *Dental functional morphology: How teeth work*. Cambridge University Press.

Lucas, P. W., & Luke, D. A. (1983a). Computer simulation of the breakdown of carrot particles during human mastication. *Archives of Oral Biology, 28*(9), 821–826. https://doi.org/10.1016/0003-9969(83)90038-9.

Lucas, P. W., & Luke, D. A. (1983b). Methods for analyzing the breakdown of food in human mastication. *Archives of Oral Biology*, *28*(9), 813–819. https://doi.org/10.1016/0003-9969(83)90037-7.

Lucas, P. W., Prinz, J. F., Agrawal, K. R., & Bruce, I. C. (2002). Food physics and oral physiology. *Food Quality and Preference*, *13*(4), 203–213. https://doi.org/10.1016/S0950-3293(00)00036-7.

Lutz, W., Sanderson, W., & Scherbov, S. (2008). The coming acceleration of global population ageing. *Nature*, *451*(7179), 716–719. https://doi.org/10.1038/nature06516.

Magne, H., Savary-Auzeloux, I., Rémond, D., & Dardevet, D. (2013). Nutritional strategies to counteract muscle atrophy caused by disuse and to improve recovery. *Nutrition Research Reviews*, *26*(02), 149–165. https://doi.org/10.1017/S0954422413000115.

Majzoobi, M., Ghiasi, F., Habibi, M., Hedayati, S., & Farahnaky, A. (2014). Influence of soy protein isolate on the quality of batter and sponge cake. *Journal of Food Processing and Preservation*, *38*(3), 1164–1170. https://doi.org/10.1111/jfpp.12076.

Malagelada, J. R., Go, V. L., & Summerskill, W. H. (1979). Different gastric, pancreatic, and biliary responses to solid-liquid or homogenized meals. Digestive Diseases and Sciences, 24(2), 101–110. Retrieved from http://www.ncbi.nlm.nih.gov/pubmed/371939.

Mathieu, V., Monnet, A., Jourdren, S., Panouillé, M., Chappard, C., & Souchon, I. (2016). Kinetics of bread crumb hydration as related to porous microstructure. *Food & Function*, 7(8), 3577–3589. https://doi.org/10.1039/C6FO00522E.

McCloskey, D. N. (2008). *The industrial revolution and liberty*. Retrieved January 18, 2019, from The Handbook of Libertarianism website. http://www.deirdremccloskey.com/articles/revolution.php#one.

Mese, H., & Matsuo, R. (2007). Salivary secretion, taste and hyposalivation. *Journal of Oral Rehabilitation*, *34*(10), 711–723. https://doi.org/10.1111/j.1365-2842.2007.01794.x.

Mestres, M., Moran, N., Jordan, A., & Buettner, A. (2005). aroma release and retronasal perception during and after consumption of flavored whey protein gels with different textures. 1. In vivo release analysis. *Journal of Agricultural and Food Chemistry*, *53*(2), 403–409. https://doi.org/10.1021/jf048596n.

Methven, L., Jiménez-Pranteda, M. L., & Lawlor, J. B. (2016). Sensory and consumer science methods used with older adults: A review of current methods and recommendations for the future. *Food Quality and Preference*, *48*, 333–344. https://doi.org/10.1016/J.FOODQUAL.2015.07.001.

Minekus, M., Alminger, M., Alvito, P., Ballance, S., Bohn, T., Bourlieu, C., et al. (2014). A standardised static in vitro digestion method suitable for food—An international consensus. *Food & Function*, *5*(6), 1113. https://doi.org/10.1039/c3fo60702j.

Mioche, L., Bourdiol, P., Monier, S., Martin, J.-F., & Cormier, D. (2004). Changes in jaw muscles activity with age: Effects on food bolus properties. *Physiology & Behavior*, *82*(4), 621–627. https://doi.org/10.1016/j.physbeh.2004.05.012.

Mohammed, I., Ahmed, A. R., & Senge, B. (2012). Dough rheology and bread quality of wheat–chickpea flour blends. *Industrial Crops and Products*, *36*(1), 196–202. https://doi.org/10.1016/j.indcrop.2011.09.006.

Monici, M. (2005). Cell and tissue autofluorescence research and diagnostic applications. *Biotechnology Annual Review*, *11*(SUPPL), 227–256. https://doi.org/10.1016/S1387-2656(05)11007-2.

Moore, D. R., Churchward-Venne, T. A., Witard, O., Breen, L., Burd, N. A., Tipton, K. D., et al. (2015). Protein ingestion to stimulate myofibrillar protein synthesis requires greater relative protein intakes in healthy older versus younger men. *The Journals of Gerontology Series A: Biological Sciences and Medical Sciences*, *70*(1), 57–62. https://doi.org/10.1093/gerona/glu103.

Mosca, A. C., van de Velde, F., Bult, J. H. F., van Boekel, M. A. J. S., & Stieger, M. (2015). Taste enhancement in food gels: Effect of fracture properties on oral breakdown, bolus

formation and sweetness intensity. *Food Hydrocolloids*, *43*, 794–802. https://doi.org/10.1016/j.foodhyd.2014.08.009.
Muñoz-González, C., Vandenberghe-Descamps, M., Feron, G., Canon, F., Labouré, H., & Sulmont-Rossé, C. (2018). association between salivary hypofunction and food consumption in the elderlies. A systematic literature review. *The Journal of Nutrition, Health & Aging*, *22*(3), 407–419. https://doi.org/10.1007/s12603-017-0960-x.
Nagler, R. M. (2004). salivary glands and the aging process: Mechanistic aspects, health-status and medicinal-efficacy monitoring. *Biogerontology*, *5*(4), 223–233. https://doi.org/10.1023/B:BGEN.0000038023.36727.50.
Nagler, R. M., & Hershkovich, O. (2005). Relationships between age, drugs, oral sensorial complaints and salivary profile. *Archives of Oral Biology*, *50*(1), 7–16. https://doi.org/10.1016/j.archoralbio.2004.07.012.
Neyraud, E., Palicki, O., Schwartz, C., Nicklaus, S., Feron, G., Chapman, K., et al. (2012). Variability of human saliva composition: Possible relationships with fat perception and liking. *Archives of Oral Biology*, *57*(5), 556–566. https://doi.org/10.1016/j.archoralbio.2011.09.016.
Nicholls, R. J., Appelqvist, I. A. M., Davies, A. P., Ingman, S. J., & Lillford, P. J. (1995). Glass transitions and the fracture behaviour of gluten and starches within the glassy state. *Journal of Cereal Science*, *21*(1), 25–36. https://doi.org/10.1016/S0733-5210(95)80005-0.
NIDCR. (2004). Tooth loss in seniors (age 65 and over). Retrieved April 27, 2017, from https://www.nidcr.nih.gov/DataStatistics/FindDataByTopic/ToothLoss/ToothLossSeniors65andOlder.htm#Table1.
Panouillé, M., Saint-Eve, A., Déléris, I., Le Bleis, F., & Souchon, I. (2014). Oral processing and bolus properties drive the dynamics of salty and texture perceptions of bread. *Food Research International*, *62*, 238–246. https://doi.org/10.1016/j.foodres.2014.02.031.
Panouillé, M., Saint-Eve, A., & Souchon, I. (2016). Instrumental methods for bolus characterization during oral processing to understand food perceptions. *Current Opinion in Food Science*, *9*, 42–49. https://doi.org/10.1016/j.cofs.2016.05.002.
Papas, A. S., Palmer, C. A., Rounds, M. C., Herman, J., McGandy, R. B., Hartz, S. C., et al. (1989). longitudinal relationships between nutrition and oral health. *Annals of the New York Academy of Sciences*, *561*(1), 124–142. https://doi.org/10.1111/j.1749-6632.1989.tb20976.x.
Pasini, G., Simonato, B., Giannattasio, M., Peruffo, A. D. B., & Curioni, A. (2001). Modifications of wheat flour proteins during in vitro digestion of bread dough, crumb, and crust: An electrophoretic and immunological study. *Journal of Agricultural and Food Chemistry*, *49*(5), 2254–2261. https://doi.org/10.1021/jf0014260.
Peleg, M. (1993). Tailoring texture for the elderly: Theoretical aspects and technological options. *Critical Reviews in Food Science and Nutrition*, *33*(1), 45–55. https://doi.org/10.1080/10408399309527611.
Peltzer, K., Hewlett, S., Yawson, A., Moynihan, P., Preet, R., Wu, F., et al. (2014). Prevalence of loss of all teeth (edentulism) and associated factors in older adults in China, Ghana, India, Mexico, Russia and South Africa. *International Journal of Environmental Research and Public Health*, *11*(11), 11308–11324. https://doi.org/10.3390/ijerph111111308.
Pentikäinen, S., Sozer, N., Närväinen, J., Ylätalo, S., Teppola, P., Jurvelin, J., et al. (2014). Effects of wheat and rye bread structure on mastication process and bolus properties. *Food Research International*, *66*, 356–364. https://doi.org/10.1016/j.foodres.2014.09.034.
Petitot, M., & Micard, V. (2010). Legume-fortified pasta. impact of drying and precooking treatments on pasta structure and inherent in vitro starch digestibility. *Food Biophysics*, *5*(4), 309–320. https://doi.org/10.1007/s11483-010-9180-1.
Peyron, M. A., Blanc, O., Lund, J. P., & Woda, A. (2004). Influence of age on adaptability of human mastication. *Journal of Neurophysiology*, *92*(2), 773–779. https://doi.org/10.1152/jn.01122.2003.

Peyron, M. A., Gierczynski, I., Hartmann, C., Loret, C., Dardevet, D., Martin, N., et al. (2011). Role of physical bolus properties as sensory inputs in the trigger of swallowing. *PLoS One*, *6*(6), e21167. https://doi.org/10.1371/journal.pone.0021167.

Peyron, M. A., Sayd, T., Sicard, J., Mirade, P.-S., Pinguet, J., Chambon, C., et al. (2021). Deciphering the protein digestion of meat products for the elderly by in vitro food oral processing and gastric dynamic digestion, peptidome analysis and modeling. *Food & Function*, *12*(16), 7283–7297. https://doi.org/10.1039/D1FO00969A.

Peyron, M. A., Veyrune, J. L., & Woda, A. (2013). Alimentation des séniors, comprendre le rôle de la mastication. *Innovations Agronomiques*, *33*, 93–103.

Peyron, M. A., Woda, A., Bourdiol, P., & Hennequin, M. (2017). Age-related changes in mastication. *Journal of Oral Rehabilitation*, *44*(4), 299–312. https://doi.org/10.1111/joor.12478.

Phillips, S. M. (2017). Current concepts and unresolved questions in dietary protein requirements and supplements in adults. *Frontiers in Nutrition*, *4*(May), 13. https://doi.org/10.3389/fnut.2017.00013.

Prinz, J. F., & Lucas, P. W. (1997). An optimization model for mastication and swallowing in mammals. *Proceedings of the Royal Society of London B: Biological Sciences*, *264*(1389), 1715–1721.

Rémond, D., Machebeuf, M., Yven, C., Buffière, C., Mioche, L., Mosoni, L., et al. (2007). Postprandial whole-body protein metabolism after a meat meal is influenced by chewing efficiency in elderly subjects. *American Journal of Clinical Nutrition*, *85*(5), 1286–1292. https://doi.org/10.1093/ajcn/85.5.1286.

Rémond, D., Shahar, D. R., Gille, D., Pinto, P., Kachal, J., Peyron, M.-A., et al. (2015). Understanding the gastrointestinal tract of the elderly to develop dietary solutions that prevent malnutrition. *Oncotarget*, *6*(16), 13858–13898. https://doi.org/10.18632/oncotarget.4030.

Repoux, M., Labouré, H., Courcoux, P., Andriot, I., Sémon, É., Yven, C., et al. (2012). Combined effect of cheese characteristics and food oral processing on in vivo aroma release. *Flavour and Fragrance Journal*, *27*(6), 414–423. https://doi.org/10.1002/ffj.3110.

Rodríguez-García, J., Puig, A., Salvador, A., & Hernando, I. (2012). Optimization of a sponge cake formulation with inulin as fat replacer: structure, physicochemical, and sensory properties. *Journal of Food Science*, 77(2). https://doi.org/10.1111/j.1750-3841.2011.02546.x.

Rose, M. R. (1991). *Evolutionary biology of aging*. Oxford University Press.

Roser, R. H., & Ortiz-Ospina, E. (2013). *World population growth—Our world in data*. Retrieved September 1, 2021, from https://ourworldindata.org/world-population-growth#licence.

Sahu, J. K., & Niranjan, K. (2009). Gas-liquid mixing. In P. J. Cullen (Ed.), *Food mixing: Principles and applications* (pp. 230–249). Wiley-Blackwell.

Salles, C., Chagnon, M.-C., Feron, G., Guichard, E., Laboure, H., Morzel, M., et al. (2011). In-mouth mechanisms leading to flavor release and perception. *Critical Reviews in Food Science and Nutrition*, *51*(1), 67–90. https://doi.org/10.1080/10408390903044693.

Saulnier, L., Micard, V., & Della Valle, G. (2014). Structure du pain et index glycémique. *Cahiers de Nutrition et de Dietetique*, *49*(2), 61–66. https://doi.org/10.1016/j.cnd.2014.01.004.

Schoeman, L., Williams, P., du Plessis, A., & Manley, M. (2016). X-ray micro-computed tomography (μCT) for non-destructive characterisation of food microstructure. *Trends in Food Science & Technology*, *47*, 10–24. https://doi.org/10.1016/J.TIFS.2015.10.016.

Schwartz, C., Vandenberghe-Descamps, M., Sulmont-Rossé, C., Tournier, C., & Feron, G. (2017). Behavioral and physiological determinants of food choice and consumption at sensitive periods of the life span, a focus on infants and elderly. *Innovative Food Science and Emerging Technologies*, *46*(June 2017), 91–106. https://doi.org/10.1016/j.ifset.2017.09.008.

Skamniotis, C. G. (2017). *Characterising and modelling fracture in functional pet foods*. London: Imperial College.

Skamniotis, C. G., Patel, Y., Charalambides, M. N., & Elliott, M. (2016). Fracture investigation in starch-based foods. *Interface Focus, 6*(3).

Slade, G. D., Akinkugbe, A. A., & Sanders, A. E. (2014). Projections of U.S. edentulism prevalence following 5 decades of decline. *Journal of Dental Research, 93*(10), 959–965. https://doi.org/10.1177/0022034514546165.

Szczesniak, A. S. (2002). Texture is a sensory property. *Food Quality and Preference, 13*(4), 215–225. https://doi.org/10.1016/S0950-3293(01)00039-8.

Tarrega, A., Yven, C., Sémon, E., & Salles, C. (2008). Aroma release and chewing activity during eating different model cheeses. *International Dairy Journal, 18*(8), 849–857. https://doi.org/10.1016/J.IDAIRYJ.2007.09.008.

Tournier, C., Grass, M., Septier, C., Bertrand, D., & Salles, C. (2014). The impact of mastication, salivation and food bolus formation on salt release during bread consumption. *Food & Function, 5*(11), 2969–2980. https://doi.org/10.1039/c4fo00446a.

Tournier, C., Grass, M., Zope, D., Salles, C., & Bertrand, D. (2012). Characterization of bread breakdown during mastication by image texture analysis. *Journal of Food Engineering, 113*(4), 615–622. https://doi.org/10.1016/j.jfoodeng.2012.07.015.

Turbin-Orger, A., Babin, P., Boller, E., Chaunier, L., Chiron, H., Della Valle, G., et al. (2015). Growth and setting of gas bubbles in a viscoelastic matrix imaged by X-ray microtomography: The evolution of cellular structures in fermenting wheat flour dough. *Soft Matter, 11*(17), 3373–3384. https://doi.org/10.1039/C5SM00100E.

Turbin-Orger, A., Boller, E., Chaunier, L., Chiron, H., Della Valle, G., & Réguerre, A. L. (2012). Kinetics of bubble growth in wheat flour dough during proofing studied by computed X-ray micro-tomography. *Journal of Cereal Science, 56*(3), 676–683. https://doi.org/10.1016/j.jcs.2012.08.008.

UN. (2017a). *World population ageing 2017—Highlights*. United Nations Publications.

UN. (2017b). *World population prospects: The 2017 revision*. No. ESA/P/WP. 241. Retrieved from https://esa.un.org/unpd/wpp/publications/files/key_findings_wpp_2015.pdf.

van der Bilt, A., Olthoff, L. W., Bosman, F., & Oosterhaven, S. P. (1993). The effect of missing postcanine teeth on chewing performance in man. *Archives of Oral Biology, 38*(5), 423–429. https://doi.org/10.1016/0003-9969(93)90214-7.

van der Bilt, A., Olthoff, L. W., van der Glas, H. W., van der Weelen, K., & Bosman, F. (1987). A mathematical description of the comminution of food in human mastication. *Archives of Oral Biology, 32*(8), 579–588.

van Eck, A., Hardeman, N., Karatza, N., Fogliano, V., Scholten, E., & Stieger, M. (2019). Oral processing behavior and dynamic sensory perception of composite foods: Toppings assist saliva in bolus formation. *Food Quality and Preference, 71*, 497–509. https://doi.org/10.1016/J.FOODQUAL.2018.05.009.

van Vliet, T. (2014). *Rheology and fracture mechanics of foods*. CRC Press.

Vandenberghe-Descamps, M., Labouré, H., Prot, A., Septier, C., Tournier, C., Feron, G., et al. (2016). Salivary flow decreases in healthy elderly people independently of dental status and drug intake. *Journal of Texture Studies, 47*(4), 353–360. https://doi.org/10.1111/jtxs.12191.

Vandenberghe-Descamps, M., Labouré, H., Septier, C., Feron, G., & Sulmont-Rossé, C. (2017). Oral comfort: A new concept to understand elderly people's expectations in terms of food sensory characteristics. *Food Quality and Preference, February*, 1–11. https://doi.org/10.1016/j.foodqual.2017.08.009.

Vandenberghe-Descamps, M., Sulmont-Rossé, C., Septier, C., Feron, G., & Labouré, H. (2017). Using food comfortability to compare food's sensory characteristics expectations of elderly people with or without oral health problems. *Journal of Texture Studies, 48*(4), 280–287. https://doi.org/10.1111/jtxs.12250.

Veyrune, J. L., & Mioche, L. (2000). Complete denture wearers: electromyography of mastication and texture perception whilst eating meat. *European Journal of Oral Sciences*, *108*(2), 83–92. https://doi.org/10.1034/j.1600-0722.2000.90780.x.
Vinyard, C. J., & Fiszman, S. (2016). Using electromyography as a research tool in food science. *Current Opinion in Food Science*, *9*, 50–55. https://doi.org/10.1016/j.cofs.2016.06.003.
Wang, X., & Chen, J. (2017). Food oral processing: Recent developments and challenges. *Current Opinion in Colloid & Interface Science*, *28*, 22–30. https://doi.org/10.1016/j.cocis.2017.01.001.
WHO, FAO, & UNU. (2007). Protein and aminoacid requirements in human nutrition: Report of a joint FAO/WHO/UNU expert consultation. In *WHO technical report series: Vol. 935*. World Health Organization.
Wilderjans, E., Luyts, A., Goesaert, H., Brijs, K., & Delcour, J. A. (2010). A model approach to starch and protein functionality in a pound cake system. *Food Chemistry*, *120*(1), 44–51. https://doi.org/10.1016/J.FOODCHEM.2009.09.067.
Wilson, C. E., & Brown, W. E. (1997). Influence of food matrix structure and oral breakdown during mastication on temporal perception of flavor. *Journal of Sensory Studies*, *12*(1), 69–86. https://doi.org/10.1111/j.1745-459X.1997.tb00054.x.
Witt, T., & Stokes, J. R. (2015). Physics of food structure breakdown and bolus formation during oral processing of hard and soft solids. *Current Opinion in Food Science*, *3*, 110–117. https://doi.org/10.1016/j.cofs.2015.06.011.
Wright, K. M., Sprunt, J., Smith, A. C., & Hills, B. P. (2003). Modelling flavour release from a chewed bolus in the mouth: Part II. The release kinetics. *International Journal of Food Science & Technology*, *38*(3), 351–360. https://doi.org/10.1046/j.1365-2621.2003.00681.x.
Young, A. K., Cheong, J. N., Hedderley, D. I., Morgenstern, M. P., & James, B. J. (2013). Understanding the link between bolus properties and perceived texture. *Journal of Texture Studies*, *44*(5), 376–386. https://doi.org/10.1111/jtxs.12025.
Zhu, J. F., Mizunuma, H., & Michiwaki, Y. (2014). Determination of characteristic shear rate of a liquid bolus through the pharynx during swallowing. *Journal of Texture Studies*, *45*(6), 430–439. https://doi.org/10.1111/jtxs.12094.

CHAPTER FIVE

Gluten-free bakery products: Ingredients and processes

Manuel Gómez*
Food Technology Area, College of Agricultural Engineering, University of Valladolid, Palencia, Spain
*Corresponding author: e-mail address: mgpallares@uva.es

Contents

Abstract

There is an increasing demand for gluten-free products around the world because certain groups of people, which have increased in the last decades, need to eliminate gluten from their diet. A growing number of people consider gluten-free products to be healthier. However, making gluten-free products such as bread is a technological challenge due to the important role of the gluten network in their development. However, other products, such as cakes and cookies usually made with wheat flour, can easily be made with gluten-free starches or flours since gluten does not play an essential role in their production. To replace wheat flour in these elaborations it is necessary to resort to gluten-free starches and/or flours and to gluten substitutes. Additionally, it can be convenient to incorporate other ingredients such as proteins, fibers, sugars or oils, as well as to modify their quantities in wheat flour formulations. Regarding gluten-free flours, it will also be necessary to know the parameters that influence their functionality in order to obtain regular products. These problems have originated a lower availability of gluten-free products which have a worse texture and are less tasty and more expensive than their homologues with gluten. These problems have been partially solved thanks to research on these types of products, their ingredients and their production methods. In recent years, studies about the nutritional improvement of these products have increased. This chapter delves into the main ingredients used in the production of gluten-free products, the processes for making gluten-free breads, cakes and cookies, and the nutritional quality of these products.

Advances in Food and Nutrition Research, Volume 99
ISSN 1043-4526
https://doi.org/10.1016/bs.afnr.2021.11.005

1. Introduction

The development of gluten-free products has been a challenge for researchers and companies due to the important role of gluten in the elaboration of some products. Research in this area began in the second half of the 20th century but has increased significantly in the first decades of the 21st century. These research works were initially aimed at finding the best gluten substitutes. Subsequently, emphasis was placed on the suitability of the different starch ingredients, on the effects of other ingredients, such as proteins or fibers, and on some aspects of processing. In recent years, more research has focused on the nutritional improvement of these products. However, there is still room for quality improvement in those gluten-free products available on the market, such as breads and pastries, which still have a lower organoleptic quality than their gluten-containing homologues and are commercialized at a much higher price. Thus, celiac patients manifest to be unsatisfied with the taste, texture, price, and availability of gluten-free products (Alencar, de Araújo, Faggian, de Araújo, & Capriles, 2021).

Studies on the development of gluten-free products and the improvement of their quality have been accompanied by a growth in the consumption of this type of products partly due to a wider availability and lower prices and, also, to certain beliefs on the part of consumers (Savarese, Wismer, & Graffigna, 2021). There are people who need to avoid gluten-containing products because these can have negative consequences for their health. Within this group, wheat allergens and celiac disease sufferers are particularly affected. There also seems to exist a non-celiac gluten intolerance and other gluten-related diseases, such as irritable bowel syndrome (IBS) and dermatitis herpetiformis (di Liberto et al., 2020). However, there is no certainty about how much gluten these individuals can tolerate; this seems to depend very much on each person and their pathology, so gluten consumption should be reduced as much as possible. Yet, there is also a growing segment of the population who considers that gluten is bad for their general health and that products without gluten are healthier. In fact, the percentage of the population consuming gluten-free products in countries such as the United States is much higher than the number of people with diagnosed allergies or intolerances. However, in general, gluten-free products are nutritionally worse than their gluten-containing homologues, as products without gluten tend to incorporate more simple sugars and fats (Miranda, Lasa, Bustamante, Churruca, & Simon, 2014).

In most Western countries, there is a "wheat culture" and thus, wheat flour is used to make a multitude of products, such as breads, cakes, cookies, pastries, sauces, breading, etc. It is important to distinguish that not in all of them gluten is essential. In fact, in many of these products the gluten network does not develop. This protein network, responsible for giving cohesion, extensibility and elasticity to doughs, only develops when wheat proteins are hydrated and subjected to mechanical work (Anjum et al., 2007). Mixing is minimized in some cookies to reduce the formation of the gluten network. Other products are made from whipped, very liquid batters, in which the lack of friction in mixing reduces the formation of the gluten network. Besides, some products are elaborated incorporating boiling water (scalded doughs) and in this operation, flour proteins are denatured and lose their classic functionality. In these elaborations, the substitution of wheat flour for another type of flour will be facilitated and, on many occasions, it will only be necessary to substitute this flour for an adequate combination of flours and/or starches. Nevertheless, in products where the gluten network develops, this substitution is more complex and, in addition to selecting an adequate combination of flours and/or starches, it will be necessary to resort to the so-called "gluten substitutes," which for the most part are included in the hydrocolloid group. Furthermore, it may be necessary to study other modifications both in formulation and in processing. Fig. 1 shows the ingredients typically used, in combination, to imitate the functionality of wheat flour in the production of gluten-free breads.

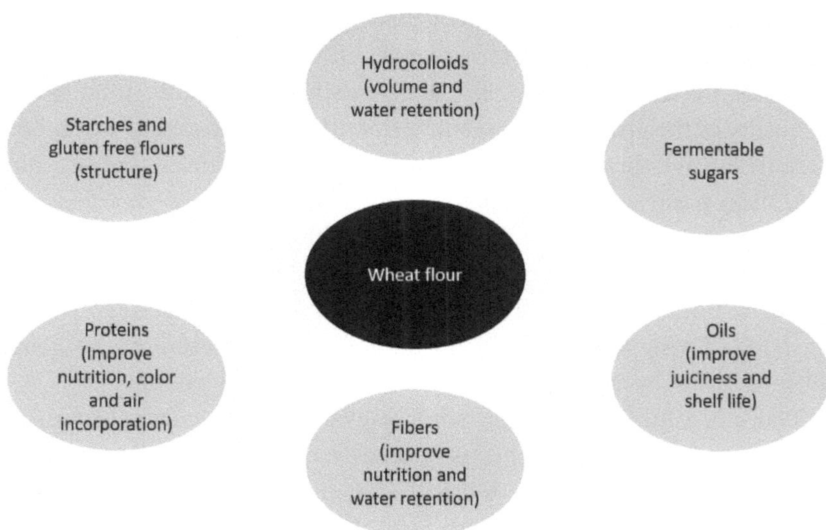

Fig. 1 Useful ingredients to replace wheat flourin gluten-free breadmaking.

2. Ingredients

2.1 Starchy products

Starch is the main component of wheat flour. Although more attention has generally been paid to its proteins, this component plays an essential role in the development of baked goods, such as breads, cakes or cookies. In the case of bread, starch is the main source of fermentable sugars. Thus, starch granules damaged in the milling process must be hydrolyzed by amylases, generating maltose and, to a lesser extent, glucose. This process begins during kneading and continues until baking. Starch also helps achieve a certain consistency of the dough and, during baking, it gelatinizes, increasing dough consistency and providing structure to the crumb. Starch gelatinization marks the end of dough expansion during baking. The processes of gelatinization and subsequent retrogradation will influence the texture of these products. In the case of the crust, the process is different since there is a rapid drying of the external part of the product. When the amount of water is reduced, the gelatinization phenomenon does not occur in the same way and the starch granules do not swell and break, which affects the texture of this part of the product in a different way (Martínez, Román, & Gómez, 2018; Vanin, Lucas, & Trystram, 2009).

Wheat flour starch is relatively uniform. Even when we can find various proportions of starch and various amounts of damaged starch among the different types of wheat flour, the characteristics of these starches, their amylose/amylopectin ratio, or their pasting behavior will be very similar. Nevertheless, when using alternative starch sources, the possibilities and variations are much greater, and these can have a definite influence on the quality of the products obtained. Furthermore, in the case of wheat flour, starch is always surrounded by proteins forming particles that are quite uniform in particle size. Wheat is a soft cereal and usually produces a fine flour after milling. In general, the grains used for gluten-free flours are harder than wheat and naturally produce coarser flours. Finer flours can be obtained from these flours or they can be subjected to wet milling processes to separate the starch. These variations will also influence the structure of the doughs and their ability to retain gas during fermentation and baking.

There are numerous starch sources available to produce gluten-free products, either native starches or gluten-free flours. The origin and the way these products are processed will influence their functionality, as well as the quality of the gluten-free products obtained. In addition, as described by Román, Gómez, and Martínez (2021), the particle size and water absorption capacity of these starches affected the rheology of the doughs and their internal structure. Consequently, there will be changes in their ability to expand during fermentation and baking and this will have an impact on the volume of the breads and, albeit indirectly, their texture. The proportion of amylose and amylopectin, and their fine structure will influence their pasting behavior and, therefore, their gelling and retrogradation processes, and thus crumb texture.

A study of world commercial gluten-free breads shows that the ingredients mostly used in their elaboration are maize starch and rice flour, followed by tapioca starch. In addition to these, other starches and flours can be used: potato or rice starch; flours from other gluten-free cereals, such as maize, sorghum, millet or teff- or from pseudocereals, such as quinoa, amaranth or buckwheat, or even flours from other sources, such as pulses, banana or chestnuts. Fig. 2 shows a summary of the starches and the gluten-free flours most commonly used in gluten-free breadmaking, and their main advantages.

2.1.1 Maize starch

Maize starch is obtained in starch companies where the maize grain is subjected to wet milling processes to separate the starch from the remaining

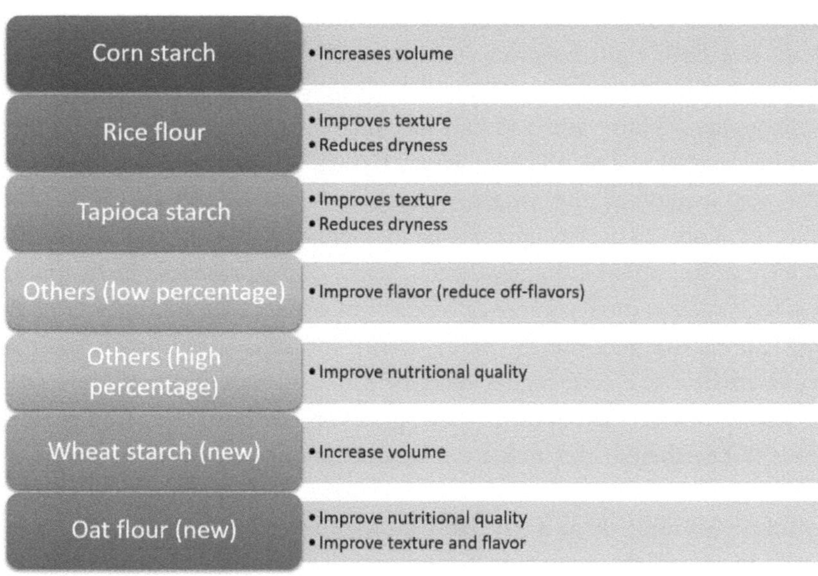

Fig. 2 Functionalities of different gluten-free flours and starches in gluten-free breadmaking.

components. Native maize starch is the most widely used ingredient in the production of gluten-free products. This is partly due to its great availability and low cost, the safety it has traditionally provided in terms of the absence of cross-contamination, and its regularity. However, the great advantage of maize starch lies in the possibility of obtaining large volume products, when compared to other alternative starch sources (Belorio & Gómez, 2020a; Martínez & Gómez, 2017), although the differences with other starch sources will depend on the gluten substitute used, differences are greater with hydroxypropyl methylcellulose (HPMC), and on the general formulation. These advantages notwithstanding, products made with maize starch have a drier texture that is not appreciated by consumers (Mancebo, Merino, Martínez, & Gómez, 2015), as well as a poor nutritional quality (less protein, fiber, minerals and vitamins) compared to other alternatives.

2.1.2 Rice flour

Rice flour is the second most widely used ingredient in the world to produce gluten-free products. Rice flour produces breads of smaller volume than maize starch, but reduces the dry texture of these breads and improves their acceptability (Mancebo, Merino, et al., 2015), so it is often mixed with starch. In general, rice flour is obtained from grains broken in rice factories in the process of separating husk and bran. These broken grains are by-products and, therefore, their cost is lower than whole grains. But, if there are not enough broken grains, rice flour must be produced from whole grains. Despite having the advantage of its bland taste and hypoallergenic properties, rice flour can lack homogeneity. Thus, aspects such as milling, particle size (de la Hera, Martínez, & Gómez, 2013; de la Hera, Rosell, & Gómez, 2014; Park et al., 2014) or variety at the origin (Cornejo & Rosell, 2015; Han, Cho, Kang, & Koh, 2012; Sivaramakrishnan, Senge, & Chattopadhyay, 2004), can have an important influence on the quality of the breads obtained, and also on aspects such as their volume or texture. In general, these studies agree that those aspects that increase the water absorption capacity of flours, such as a smaller particle size or a higher level of damaged starch, which depend on the variety of rice used and the milling system, reduced the volume of the breads. Nevertheless, this reduction can be minimized or eliminated if the water content of the formulation is increased. As observed by Mancebo, Martínez, Merino, de la Hera, and Gómez (2017) that gluten-free bread doughs improved their expansion capacity by lowering their consistency, until they reached a limit point where if the rheology is too fluid, a sufficiently firm structure is not possible and the doughs fall during fermentation or baking.

Therefore, when the ingredients have a higher water absorption capacity, which results in more consistent doughs, the water content in the formulation should be increased. Aspects such as the amylose/amylopectin content of these flours (Aoki, Kataoka, & Nishiba, 2020) or others related to the fine structure, such as the length of the amylose molecules (Román, Reguilón, Gómez, & Martínez, 2020), which influence the pasting properties, can modify the texture of the breads and their tendency to staling.

2.1.3 Root starch

The third most used ingredient to produce gluten-free products is root starch, such as tapioca or potato starch (Román, Belorio, & Gómez, 2019). These starches alone generate breads with lower volume than those obtained with maize starch. The smaller volume has been attributed to the larger size of the root starch granules, which prevents the formation of a more closed starch-hydrocolloid matrix (Martínez & Gómez, 2017). In fact, although in the case of wheat starch, the smaller granules allow breads to have more closed alveoli and a lower hardness, a combination of smaller and larger granules allows breads to have a higher specific volume (Román, de la Cal, Gómez, & Martínez, 2018). The major advantage of root starches, especially tapioca, is the improvement in texture compared to breads made with maize starch, which is why many research studies recommend mixing both types of starches (McCarthy, Gallagher, Gormley, Schober, & Arendt, 2005; Sanchez, Osella, & de la Torre, 2002; Sciarini, Ribotta, León, & Pérez, 2012). This improvement has been attributed to their lower retrogradation capacity. In fact, waxy starches or flours, with lower retrogradation capacity, can help improve the texture of breads by reducing their hardness, when used in low doses (10–15%) (Witczak, Korus, Ziobro, & Juszczak, 2019). However, when high levels are used, they significantly reduce the volume of breads and increase their hardness (Román, Reguilón, Gómez, & Martínez, 2020).

2.1.4 Wheat starch

Regarding starches, although it is possible to use any gluten-free starch, one of the starches with the greatest potential is wheat starch. In the process of obtaining wheat starch, all the protein is eliminated and, consequently, the components that can cause problems to people with celiac disease. However, the companies that commercialize this product as gluten-free, need to test it and confirm the absence of gluten. Wheat starch makes it possible to obtain breads with a slightly higher specific volume than maize (Martínez & Gómez, 2017), but the texture of these breads presents similar problems to those made with maize starch. Texture can be improved with the addition of rice flour in the formula (Mancebo, Merino, et al., 2015) or other ingredients that reduce dryness and lack of cohesiveness. In addition to these factors, the final taste of the breads, which will be slightly modified by using wheat starch, and their acceptability should also be analyzed. However, in most countries the use of wheat starch is uncommon because, in the past, it was not allowed for the production of gluten-free products and there is still some uncertainty on the part of consumers.

2.1.5 Modified starch

It is also possible to use physically or chemically modified starches. Within this group, there is a wide variety, depending on the origin of the starch and the type of modification. Despite the wide possibilities, research on the incorporation of these starches is scarce. The limited research agrees that the effects of using modified starches will depend on the type of modification they have been subjected to (Witczak, Juszczak, Ziobro, & Korus, 2012; Ziobro, Korus, Witczak, & Juszczak, 2012). However, some cross-linking starches can improve the volume and texture of breads, reducing their hardness and increasing their cohesiveness (Román, Reguilón, Gómez, & Martínez, 2020; Ziobro et al., 2012). These changes are the result of modifications in the pasting properties of starches which -after gelatinization and retrogradation- contribute to crumb texture (Román et al., 2021). In the case of using these starches, it may also be necessary to reconsider the hydration of the doughs as the changes may also influence dough rheology in cold. In addition, it should be noted that these starches must be declared as additives in most countries, so some manufacturers may prefer to use other ingredients.

2.1.6 Oat flour

One of the most promising flours for the production of gluten-free products is oat flour. Oats have a very interesting nutritional composition, as they contain a large amount of beta-glucans, which have been linked to important nutritional benefits, such as the prevention, treatment, and control of diabetes and cardiovascular diseases (Daou & Zhang, 2012; Lazaridou & Biliaderis, 2007). Unlike other cereals, oats have a significant amount of lipids in their endosperm, so the grain must be thermally treated to inactivate enzymes that can accelerate rancidity before milling. The taste of products made with oat flour is pleasant, but its use in the production of gluten-free breads needs to be further studied. Oats have traditionally been left aside in the production of gluten-free items because they were believed not to be tolerated by celiacs. It is now known that this intolerance was caused by contamination problems with other grains and that certified gluten-free oats can be obtained if extreme care is taken to avoid cross-contamination (Hüttner & Arendt, 2010; Poley, 2017). However, in many countries there is still a reluctance on the part of people affected with celiac disease to consume products made with oats. Changes in regulations accepting oat flour for the production of gluten-free products are still recent. Over time and with the increase in the supply of these products, a change in this mentality is foreseeable. In fact, in countries with a longer tradition of oat consumption, such as Finland, the use of oat flour in gluten-free products is more common than in other countries.

2.1.7 Other flours

Research on the use of other gluten-free flours to produce bread and other gluten-free products is very numerous. There are even literature reviews on these topics: the use of flours from pulses, such as chickpeas, lentils, beans, or lupine (Foschia, Horstmann, Arendt, & Zannini, 2017; Melini, Melini, Luziatelli, & Ruzzi, 2017), pseudocereals, such as quinoa, amaranth, or buckwheat (Alvarez-Jubete, Arendt, & Gallagher, 2010; Giménez-Bastida, Piskuła, & Zieliński, 2015), minor cereals such as millet (Taylor, Schober, & Bean, 2006) or teff (Zhu, 2018) and chestnuts (Zhu, 2017). Other flours have also been proposed, such as banana flour (Román, Gómez, Hamaker, & Martínez, 2019; Sarawong, Gutiérrez, Berghofer, & Schoenlechner, 2014). The fundamental reason for incorporating these flours is usually the potential nutritional improvement that can be achieved. Many of these grains not only have higher protein, fiber, mineral and vitamin contents than starches or rice flour, but also a higher content of antioxidant substances (Torres, Arufe, Chenlo, & Moreira, 2017). This improvement will depend on the type of grain source and on whether flours are whole-grain or not. However, commercial products do not usually contain these flours or incorporate them in very small proportions, typically less than 10%. Therefore, these additions do not substantially modify the nutritional characteristics of the processed products, but they can have a better "marketing" or image. On the contrary, the main limitation for the use of these flours is the flavor they provide to processed products, which is not very well accepted by most consumers. Research should be oriented to reduce or mask this unpleasant flavor without jeopardizing nutritional quality, which is quite complex. As indicated in the reviews discussed, thermal treatments, germination or fermentation have been put forth to improve the flavor of products made with these flours. Although they still do not reach the acceptability that products made with other flours or starches have, in small percentages, the flavor of these products can be improved (Pico, Antolín, Román, Bernal, & Gómez, 2019). Maize (Brites, Trigo, Santos, Collar, & Rosell, 2010; de la Hera, Talegón, Caballero, & Gómez, 2013) or sorghum (Schober, Messerschmidt, Bean, Park, & Arendt, 2005) flours can also be used, depending on their availability. In countries with a long tradition using maize or sorghum flours in baking or other preparations, consumers are more used to their taste. But, as with other gluten-free grains, the variety of the grain, the milling process, or the treatments that the flours receive will have an impact on the quality of the breads and, therefore, they should be studied. Other starch-rich flours, such as chestnut and banana flours, can provide interesting flavors and aromas when incorporated in small proportions, but in larger quantities, they give very strong flavors and tend to deteriorate the volume or texture of the breads.

2.2 Gluten substitutes

The gluten network, formed when wheat proteins are hydrated and mechanical work is applied, is essential for the development of some products such as bread. This network provides cohesion, elasticity and extensibility to doughs. It is also responsible for the dough ability to retain the gases produced during fermentation and the early stages of baking (Delcour et al., 2012). Therefore, when developing gluten-free products, we must find an ingredient that provides these characteristics. In fact, gluten substitution has been one of the most studied topics in the development of gluten-free breads, and there are specific reviews on the subject (Anton & Artfield, 2008; Mir, Shah, Naik, & Zargar, 2016; Zoghi, Mirmahdi, & Mohammadi, 2020). It should be noted that even when no ingredient has been found with the characteristics of the gluten network, which gives the dough elasticity, cohesiveness and extensibility, there are certain products that allow improving the retention of gases during fermentation and baking.

The production of gluten-free bread follows a similar process to the production of cakes and other products based on mixed doughs. Gluten-free bread doughs are less viscous and less elastic (less consistent), and their lower consistency makes a higher specific volume possible. They can even be mixed with the wire whip accessory, instead of the dough hook or flat beater mixer, depending on the formulation. In fact, the first gluten substitutes put forward were emulsifiers, which in principle could help the incorporation of air into the batter and could increase the size of these bubbles with the gases produced during fermentation. Today, the most widely used gluten substitutes are from the hydrocolloid family. In general, the hydrocolloids improve the dough structure, the tenderness of the breads and their shelf life, due to their hydrophilic nature, but the effects of the different hydrocolloids on the characteristics of the breads can be very different. Román, Belorio, and Gómez (2019) in their research on commercial gluten-free products, identified that hydroxypropyl methylcellulose (HPMC) is the most commonly used gluten substitute, followed by xanthan gum, psyllium and guar gum. The advantages and disadvantages of these gluten replacers are shown in Fig. 3. Xanthan gum and HPMC are also widely used in scientific studies (Masure, Fierens, & Delcour, 2016; Mir et al., 2016). However, in contrast to most research works, it is very common to combine different gluten substitutes in commercial products. This indicates that there is no perfect substitute and that each may provide some beneficial effects.

Unfortunately, studies comparing the functionality of different hydrocolloids are scarce and the conclusions they obtain are often not replicated (Akin & Miller, 2017; Lazaridou, Duta, Papageorgiou, Belc, & Biliaderis, 2007; Liu et al., 2018; Sabanis & Tzia, 2011). Such discrepancies in studies on gluten-free products are very common, as the results depend very much on the base formula used, the dosage of the ingredients and the hydration of the doughs. As mentioned previously, dough rheology is essential in product development and some ingredients with high water retention capacity,

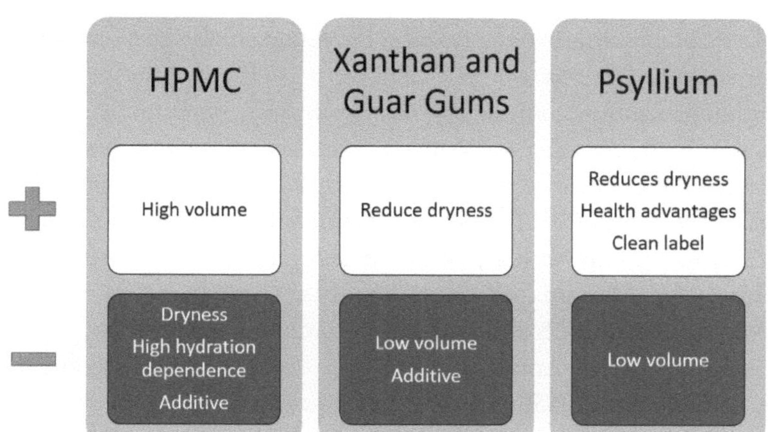

Fig. 3 Advantages and disadvantages of main gluten replacers.

which increases dough consistency, can generate breads with lower volumes and higher hardness; this could be reversed by increasing dough hydration. However, studies generally agree that HPMC is the gluten substitute that produces breads with a higher specific volume. This effect has been attributed to its ability to form a thermo-reversible gel which strengthens when heated and reverts back to weak entanglement after cooling (Grover, 1986). Thus, this temperature dependence stabilizes the gelatinizing crumb structure during baking (Crockett, Ie, & Vodovotz, 2011b). Although these breads tend to have lower hardness values, possibly due to their higher specific volume, their crumb is less cohesive and more fragile than the crumb of breads with gluten.

The differences in the use of HPMC, xanthan gum or psyllium, and the influence of dough hydration can be appreciated very clearly in the study by Belorio and Gómez (2020a). They showed that the optimal hydration to achieve a higher specific volume was greater with psyllium or xanthan gum than with HPMC. In addition, the influence of hydration on specific volume was much more pronounced when HPMC was used, with very significant reductions in specific volume when the hydration was decreased. This study also shows that HPMC contributes to achieving breads with a much higher specific volume than xanthan gum or psyllium. These differences were between 50% and 100%, depending on whether maize starch or rice flour is used as a base. However, if hydration is corrected to produce breads with a similar specific volume, breads with HPMC had a much harder, less cohesive and less elastic texture than breads with xanthan or psyllium. These changes are related to the phenomena that HPMC experiences during baking, and they reduce the final quality of the breads. For this reason, it is very common in commercial elaborations to mix HPMC with some other hydrocolloid that helps improve the texture, providing cohesiveness, elasticity and tenderness to the breads.

Although xanthan gum and psyllium are the most commonly used gums, it is also possible to use other hydrocolloids with high water absorption capacity, such as guar gum or locust bean gum, or other cellulose derivatives (Román, Belorio, & Gómez, 2019). Therefore, some authors propose a mixture of HPMC and psyllium to improve the quality of gluten-free breads (Cappa, Lucisano, & Mariotti, 2013; Haque & Morris, 1994; Mancebo, San Miguel, Martínez, & Gómez, 2015); to do this it is necessary to correct the hydration by increasing it as the doses of psyllium are incremented. Hager and Arendt (2013) proposed mixtures of xanthan gum and HPMC. The use of beta-glucans in combination with HPMC has also been put forth

(Kittisuban, Ritthiruangdej, & Suphantharika, 2014; Pérez-Quirce, Lazaridou, Biliaderis, & Ronda, 2017; Ronda, Perez-Quirce, Lazaridou, & Biliaderis, 2015). This gum has interesting nutritional advantages and, just like xanthan gum or psyllium, it has a high water absorption capacity, so it will also be necessary to consider dough hydration. However, due to their higher cost or lower availability, beta-glucans are rarely used in commercial formulations.

In the case of using HPMC, it should be noted that the name HPMC includes a host of products, which vary according to their molecular weight and structure (percentage of hydroxypropyl and methyl groups). Although most commercial products and scientific studies use HPMC 4KM with a low percentage of methoxyl groups and high thickening power, some researchers have reported higher specific volume using other variants with a higher molecular weight (Morreale, Garzón, & Rosell, 2018) or with a higher percentage of hydroxyl groups (Crockett et al., 2011b; Pérez-Quirce, Collar, & Ronda, 2014). However, these differences are much less important than those observed when replacing HPMC with another hydrocolloid, and cannot be extrapolated to all types of bread, as they will depend on the formulation, the process and the correction of hydration required. In addition, they may cause other problems such as crumb holes.

Although hydrocolloids are used to improve the organoleptic quality of gluten-free breads, it should not be forgotten that they can also improve the nutritional quality of these breads. Hydrocolloids belong to the group of soluble fibers (Viebke, Al-Assaf, & Phillips, 2014) and, as discussed later, the celiac population, as well as the majority of the population, consumes less fiber than necessary. Most studies on the nutritional advantages of hydrocolloids have focused on aspects such as their potential to reduce cholesterol or the glycemic index of products, their prebiotic effect, increasing satiety, reducing constipation, and even preventing certain types of cancer, such as colon cancer (Li & Nie, 2016). In fact, some health claims related to food hydrocolloids have been approved by European Food Safety Authority (EFSA) and are included under article 13.1, especially those pertaining to the maintenance of normal blood cholesterol concentrations (β-glucans, guar, konjac gum, pectins, HPMC) and the reduction of post-prandial glycemic response (β-glucans, HPMC, pectins) (Viebke et al., 2014). In addition, health claims are allowed in several countries when certain amounts of psyllium are incorporated in food products (Belorio & Gómez, 2020c). These claims are related to psyllium's ability to reduce the risk of heart disease (USA) or to help reduce/lower cholesterol (Canada). In this regard, the use of psyllium, which has very similar characteristics to the more traditionally used xanthan gum (Belorio, Marcondes, & Gómez, 2020), has increased significantly in recent years. Since it is not considered an additive, psyllium is suitable for making clean label products, it improves the quality of gluten-free breads made with HPMC, as mentioned before, and it has interesting nutritional properties. The nutritional benefits of psyllium have been discussed in several reviews and include effects on constipation, diarrhea, irritable bowel syndrome (IBS), inflammatory bowel disease–ulcerative colitis, colon cancer, diabetes and hypercholesterolemia (Franco, Sanches-Silva, Ribeiro-Santos, & de Melo, 2020; Wärnberg, Marcos, Bueno, & Moreno, 2009; Zhang, Wen, Zhang, & Duan, 2019). These properties have resulted in an increased use of psyllium and its proportion in the elaboration of gluten-free breads in recent years.

2.3 Other ingredients

2.3.1 Proteins

Gluten-free products generally have a lower protein content than their gluten-containing analogues (Conte, Fadda, Piga, & Collar, 2016; Miranda et al., 2014). This is because the gluten-free flours and starches most widely used to replace wheat flour in gluten-free formulations have a lower protein content. Thus, the incorporation of protein in these products may be an attempt to reproduce the nutritional quality of gluten-containing products. If this is the case, since cereal proteins are not of high nutritional quality, almost any protein source can be valid. However, if the aim is to provide products with a high protein content and quality, the nutritional quality of the protein source must be considered. Nevertheless, it should be noted that gluten-free diets do not usually have a lower protein intake than traditional diets, as celiacs and people who follow a gluten-free diet usually ingest protein from other sources.

Perhaps, the functionality of the proteins is more important than the nutritional properties in the elaboration of gluten-free products. Even though none of the suitable proteins to produce gluten-free products can achieve the gluten functionality, they do have certain properties that may be of interest in the development of gluten-free products. Nevertheless, the functionality of the different proteins is very different and must be evaluated. Thus, when deciding whether to incorporate a protein source in the production of gluten-free products, aspects such as its nutritional quality, functionality, possible allergenicity, origin (which can be a problem for vegans) as well as its cost, must be taken into account. Some important characteristics of the proteins most frequently used in the production of gluten-free products are shown in Fig. 4.

For this reason, the incorporation of protein in gluten-free breads has been addressed in various studies for several years, both in starch-based breads (Horstmann, Foschia, & Arendt, 2017; Kittisuban et al., 2014; Krupa-Kozak, Baczek, & Rosell, 2013; Witczak, Juszczak, Ziobro, & Korus, 2017; Ziobro, Juszczak, Witczak, & Korus, 2016; Ziobro, Witczak, Juszczak, & Korus, 2013) and in rice flour-based breads (Marco & Rosell, 2008; Nozawa, Ito, & Arai, 2016; Phongthai, D'Amico, Schoenlechner, & Rawdkuen, 2016; Shin, Gang, & Song, 2010; Storck et al., 2013), or in mixtures of both (Aprodu, Badiu, & Banu, 2016; Crockett, Ie, & Vodovotz, 2011a; Gallagher, Gormley, & Arendt, 2003; Smerdel et al., 2012). Most of these studies have a similar finding, which is the improvement of bread color, as proteins participate in Maillard reactions generating darker colors, comparable to those of wheat breads. Thus, Pico, Reguilón, Bernal, and Gómez (2019) observed that any protein source can generate slightly darker breads, but the most effective ones in this regard were dairy proteins, as confirmed by other studies (Gallagher et al., 2003; Krupa-Kozak et al., 2013). The improvement in the color of breads achieved by the addition of protein is important, as color is a parameter that influences the consumers purchasing opinion (Castro, Oblitas, Chuquizuta, & Avila-George, 2017). Gluten-free breads without added protein often have a lighter color, and consumers prefer

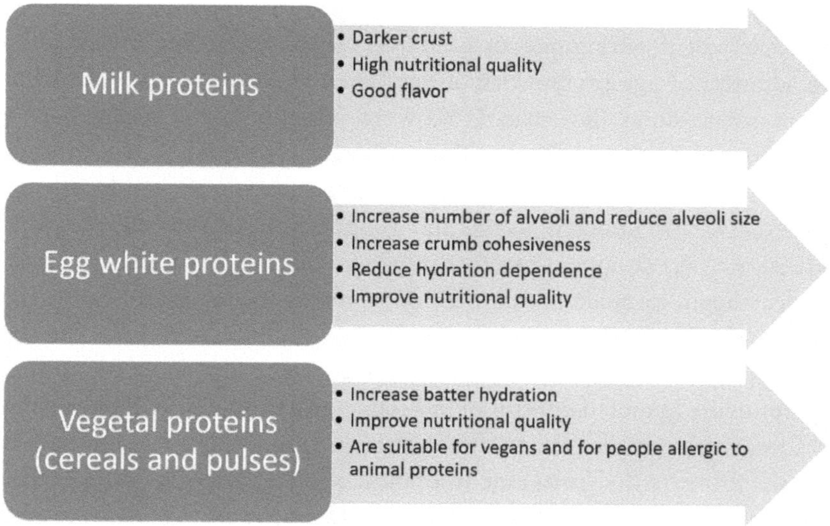

Fig. 4 Functionalities of different protein sources in gluten-free breadmaking.

somewhat darker colors (Carini et al., 2015). Pico, Antolín, et al. (2019) observed that the type of protein also influenced the aromas generated, through its participation in Maillard reactions. Among the proteins studied, rice and whey proteins were the ones that generated volatile substances (2-acetyl-1-pyrroline and pyrazines) related to favorable sensory characteristics. The results relating to the influence of proteins on other parameters such as dough rheology, volume or texture are contradictory. This is due to the distinct functionality of the different proteins, as well as to the percentages used, the different bread formulations, and the fact that in most cases the hydration is not modified when proteins are incorporated.

A study about the incorporation of high amounts of protein (30%) in gluten-free breads found that this incorporation significantly affected the optimal hydration of doughs made with maize starch and HPMC (Sahagún & Gómez, 2018a). Thus, although any bread enriched with protein did not reach the specific volume of the control, breads with vegetable proteins (rice and pea), which have a higher hydration capacity than animal proteins (whey and egg white), required a much higher amount of water in the formulation to reach the maximum specific volume. The hardness of these breads was also lower than that generated with animal proteins. In the case of whey protein, the effect was due to the lower specific volume of the breads. But, in the case of egg proteins, this was related to protein coagulation during baking, leading to an increase in crumb hardness. Nevertheless, this protein also significantly increased cohesiveness, which is one of the aspects that need to be addressed in this type of breads. The improvement in springiness or cohesiveness of gluten-free breads due to the addition of egg protein had already been observed in previous studies where lower doses (less than 10%) were incorporated (Phongthai et al., 2016; Storck et al., 2013; Ziobro, Witczak, et al., 2013) without significantly increasing bread hardness. In addition, egg proteins produced breads with more closed and uniform alveoli and a higher tolerance to changes in the hydration of the doughs. Therefore, when modifying the optimal hydration, the decrease in specific volume was not as pronounced as in the control or in the breads with vegetable protein. Finally, it should be noted that the optimally hydrated dough of the egg protein breads was much more liquid than the optimally hydrated control or vegetable protein doughs. These changes achieved with egg protein have been attributed to its high foaming capacity and, therefore, to the greater incorporation of air, in the form of smaller bubbles, during the mixing of the ingredients. However, changes in the production process of the egg proteins and in their final properties can slightly modify the effects of egg proteins on the quality of gluten-free breads (Masure, Wouters, Fierens, & Delcour, 2019) and, as a consequence, should therefore also be considered.

In addition to the usual vegetable proteins from cereals, pulses or potatoes, the use of other sources such as rapeseed (Korus, Chmielewska, Witczak, Ziobro, & Juszczak, 2021) or sunflower (Zorzi, Garske, Flôres, & Thys, 2020) proteins has also been proposed. In any case, factors such as the cost of these proteins and their availability on the market, as well as their functionality, must always be taken into consideration. Regarding animal proteins, besides those from dairy or egg proteins, in recent years, the use of cricket powder has been proposed as a protein ingredient to produce gluten-free breads (da Machado & Thys, 2019; Nissen, Samaei, Babini, & Gianotti, 2020). The properties of this ingredient are similar to those of vegetable proteins due to its high water absorption capacity. However, if the dough hydration is not modified, it generates breads with lower volume and greater hardness than the control, as well as darker crumbs. It also modifies the flavor profile significantly. In addition, the potential rejection by large numbers of consumers (Onwezen, Bouwman, Reinders, & Dagevos, 2021) currently makes the cricket powder an unsuitable option for companies producing this type of product.

It has been shown that the effect of protein can be positive in some cases and negative in others, depending on the percentage of protein incorporated. In some situations, these effects can be compensated for by combining different protein sources. Thus, to achieve a greater incorporation of proteins, for the elaboration of products with high protein levels, the most appropriate solution may be to combine different protein sources. In this way, for instance, the advantages of egg proteins (greater cohesiveness, finer and more uniform alveoli, and protein quality) can be used and the negative effect they have on hardness can be minimized with the addition of pulse proteins (Bravo-Núñez, Sahagún, & Gómez, 2019). Another interesting effect of the addition of high percentages of protein is the possibility of reducing starch digestibility (Sahagún, Benavent-Gil, Rosell, & Gómez, 2020), but this effect also depends on the hydration of the doughs and the specific volume of the breads and, therefore, needs to be studied in detail.

Regardless of the reason why proteins may be incorporated in the production of gluten-free breads, the reality is that the vast majority of commercial gluten-free breads contain some types of protein (Román, Belorio, & Gómez, 2019). Despite the positive results obtained with dairy proteins in improving bread color and aroma, they are the least used. This is due to the fact that many manufacturers offer gluten-free and lactose-free breads, or dairy products, on account of the increased lactose intolerance associated with celiac disease (Ojetti et al., 2005).

Numerous investigations have aimed to obtain a protein network similar to the gluten network from other protein sources. Some of them have attempted to achieve this through the action of transglutaminase (Marco & Rosell, 2008; Moore, Heinbockel, Dockery, Ulmer, & Arendt, 2006; Renzetti, dal Bello, & Arendt, 2008; Storck et al., 2013), an enzyme capable of binding lysine and glutamine and creating bonds between protein chains. Nevertheless, the characteristics of the gluten network are much more complex, and the incorporation of transglutaminase does not provide significant improvements in the production of gluten-free breads.

The use of mesostructured whey proteins has also been tested (van Riemsdijk, Pelgrom, van der Goot, Boom, & Hamer, 2011; van Riemsdijk, van der Goot, & Hamer, 2011; van Riemsdijk, Van der Goot, Hamer, & Boom, 2011). Although the authors claim that it is possible to obtain similar doughs to those obtained with wheat flour, as well as breads with similar volumes, the manufacturing process must be modified by eliminating the sheeting phase. Furthermore, these studies do not evaluate critical issues regarding the quality of gluten-free breads, such as their texture, staling and organoleptic quality (through sensorial tests). In fact today, ten years after these studies were carried out, no company is using this technique, so it can be assumed that the use of this protein presents significant problems that were not detected in the published articles.

In the last few years, the use of zein (maize protein) to produce gluten-free products has been further developed. Interesting results have been obtained, but as with mesostructured whey protein, they have not been transferred to the commercial production of gluten-free breads. The first study on the use of zein in doughs found that it was possible to obtain cohesive and extensible doughs (similar to gluten-containing doughs) from starch and maize protein (Schober, Bean, Boyle, & Park, 2008). However, to produce these doughs, it was necessary to keep them above the glass transition temperature of zein (40 °C). It is known that the functionality of this protein is more linked to hydrophobic interactions rather than disulfide bonds, as in gluten doughs, and that the use of α-zeins is preferred over other types of zeins (β or γ), which depends on the isolation procedures (Schober, Bean, Tilley, Smith, & Ioerger, 2011). It is also known that the mechanical work applied during kneading is key to the development of fibrillar structures, as in the case of gluten-containing doughs (Andersson, Öhgren, Johansson, Kniola, & Stading, 2011). Although there have been attempts to improve the results by reducing the glass transition temperature with citric acid (Berta, Gmoser, Krona, & Stading, 2015) or by applying a previous heat treatment on zein (Federici, Selling, Campanella, & Jones, 2021), they are not enough for industrial application. It also seems that processing can be improved with other diluted acids such as acetic or lactic acid, both present in sourdoughs (Sly, Taylor, & Taylor, 2014). However, commercial bread production with these proteins is not possible because of the problems arising from the need to keep the temperatures, and other aspects that also need to be evaluated. Thus, when the doughs are processed at the usual temperatures, zein fibers do not form, and gluten-free breads made with zein can be improved with the addition of HPMC, like other gluten-free breads made with other proteins (Berta, Koelewijn, Öhgren, & Stading, 2019). Another problem with these doughs is the negative effect of salt on dough rheology and bread volume (Smith, Bean, Selling, Sessa, & Aramouni, 2017). These investigations have resulted in reviews on the subject where further information can be found (Erickson, Campanella, & Hamaker, 2012; Taylor, Taylor, Campanella, & Hamaker, 2016). Studies about doughs with a similar functionality using sorghum proteins exhibited comparable results (Oom, Pettersson, Taylor, & Stading, 2008; Schober et al., 2011).

2.3.2 Fibers

The benefits of fiber consumption are widely known and include better control of blood glucose and cholesterol levels, protection against cardiovascular diseases, regulation of intestinal function, promotion of gut health and protection against colon cancer (Ktenioudaki & Gallagher, 2012). Partially because of these benefits, it has been confirmed that higher fiber consumption would be one of the dietary measures that could most reduce mortality in individuals, along with a reduced salt intake and higher consumption of whole-grain products (Afshin et al., 2019; Reynolds et al., 2019). Although the non-celiac population should also consume more fiber and whole-grain products, people following a strict gluten-free diet have a lower fiber intake than the rest of the population, which is below the levels recommended by various international organizations (Hager, Axel, & Arendt, 2011; Martin, Geisel, & Maresch, 2013; Shepherd & Gibson, 2013). Unlike what might be thought, the fiber content of gluten-free breads, even though quite diverse, is not usually lower than that of wheat breads made with white flour, and it ranges from 1.3% to 7.2%, although values between 3% and 6% are more frequent (Matos & Rosell, 2011; Mazzeo, Cauzzi, Brighenti, & Pellegrini, 2015). This is explained because, despite the fiber content of gluten-free starches and flours being usually lower than wheat flours, the hydrocolloids used as gluten substitutes are also fibers. One of the most recommended sources of fiber are whole-grain ingredients, but the supply of gluten-free whole-grain products is very scarce, due to the problems we have discussed. Therefore, it may be appropriate to increase the fiber content of gluten-free breads by incorporating foreign fibers, or ingredients with a high fiber content, for nutritional reasons. In fact, almost half of commercial breads include these ingredients in their compositions (Román, Belorio, & Gómez, 2019). In some labels, only the words "vegetable fiber" are indicated, but in most of them, the type of fiber (inulin, apple fiber, beet fiber, corn fiber, or bamboo, as well as rice or soy bran) is specified. The inclusion of fibers and their effect on the quality of gluten-free products has also been addressed in different reviews (Arslan, Rakha, Xiaobo, & Mahmood, 2019; Tsatsaragkou, Protonotariou, & Mandala, 2016).

Among the products that can increase fiber content, whole-grain gluten-free flours may be highlighted. These works are discussed in the section about flours and starches. Another option is the incorporation of cereal bran or pseudocereals, such as rice (Phimolsiripol, Mukprasirt, & Schoenlechner, 2012) or quinoa (Föste et al., 2014). It is also possible to resort to the use of fiber-rich by-products from the food industry, such as orange pomace (O'Shea, Doran, Auty, Arendt, & Gallagher, 2013; O'Shea, Rößle, Arendt, & Gallagher, 2015), apple pomace (Rocha-Parra, Ribotta, & Ferrero, 2015) or tiger-nut by-product (Aguilar, Albanell, Miñarro, Guamis, & Capellas, 2015). These products have the advantage that they can provide other interesting nutrients in addition to fiber, and their incorporation gives added value to some by-products. However, they have significant disadvantages, as they provide strange colors and flavors, in some cases not much appreciated by consumers, so they must be added in small percentages and a large increase in fiber content is not achieved (Gómez & Martinez, 2018). The scarce market availability of these products with homogeneous properties over time should also be considered. For this reason, the incorporation of these sources of fiber in commercial products is slim, and the use of purified fibers is more common. These by-products have different hydration properties, and therefore modify the doughs in different ways depending on their origin and the treatments they have received. It will therefore be essential to modify the hydration of the doughs to optimize the specific volume of the breads obtained (Föste, Verheyen, Jekle, & Becker, 2020). But this problem is common in the case of incorporating purified fibers.

There are numerous studies on the inclusion of fibers in gluten-free breads, although most of them only evaluate one type of fiber. Thus, there are research works about the incorporation of inulin and other fructooligosaccharides (Capriles & Arêas, 2013; Korus, Grzelak, Achremowicz, & Sabat, 2006; Ziobro, Korus, Juszczak, & Witczak, 2013), resistant starch (Korus, Witczak, Ziobro, & Juszczak, 2009), citrus fiber (Korus, Juszczak, Witczak, & Ziobro, 2020), or cereal fibers (Sabanis, Lebesi, & Tzia, 2009a, 2009b). As in the case of proteins, the functionality of all these fibers is different and must be considered when choosing them. Martínez, Díaz, and Gómez (2014) compared various types of fibers and observed very different results in terms of their influence on doughs and breads. Therefore, soluble fibers with low thickening power, such as nutriose or polydextrose, reduced the G* value of the doughs, increasing the specific volume of the breads and reducing their hardness. These fibers also produced breads with a thinner and more uniform alveoli and a darker color. The effect on the alveoli was attributed to soluble fibers combining with the hydrocolloid in the aqueous phase, favoring the creation of a film that coated starch granules and flour particles, giving more stability to the structure. The effect on color has also been proven with the addition of inulin (Capriles & Arêas, 2013; Korus et al., 2006) and can be attributed to a partial hydrolysis of these fibers, composed of glucose or fructose, in fermentation, which generates the mono- and disaccharides that can participate in Maillard and sugar caramelization reactions. This hydrolysis can be minimized by choosing a yeast with low invertase activity that increases the amount of inulin in the final bread (Morreale, Benavent-Gil, & Rosell, 2019).

The effect of fibers on volume may be related to lower G* values, since more fluid doughs generate breads with higher volume up to a certain limit, as already mentioned. Consequently, when fibers are incorporated, their water absorption capacity and the way they modify the rheology of the doughs should be considered, while their hydration should be adapted to achieve the best results. For this reason, fructooligosaccharides with lower molecular weight increase the consistency of doughs, less than those with higher molecular weight, and this has an influence on bread volume and texture (Ziobro, Korus, et al., 2013). On the contrary, resistant starch

(Korus et al., 2009) or citrus fibers (Korus et al., 2020) increase dough consistency by reducing bread volume if dough hydration is not modified. A similar process occurs with celluloses of different origins, shapes and sizes, but it must be considered that the shape and particle size also influence the rheology of the doughs, their internal structure and their final volume (Martínez et al., 2014; Ren, Linter, & Foster, 2020; Sabanis et al., 2009b). Thus, fine insoluble fibers produced breads with higher specific volumes and lower hardness than controls; however, coarser insoluble fibers gave rise to breads with lower specific volumes and greater hardness than controls. In general, the addition of fiber reduces the hardness of breads if the specific volume is maintained or if the specific volume does not change much (Korus et al., 2020; Sciarini et al., 2017). This result is partly related to the reduction of starch in the formula; after gelatinization during baking and subsequent retrogradation, starch is mostly responsible for bread texture. It is also necessary to consider the various interactions between starch and fibers, depending on their typology. Kiumarsi et al. (2019) observed that breads with inulin had lower firmness values than those containing resistant starch, which they attributed to this type of interactions. The higher water holding capacity of fibers may also contribute to reducing both water migrations during storage and staling phenomena, as citrus fibers do (Korus et al., 2020), but this will depend on the characteristics of each fiber.

In addition to the variety in their functionality, it is necessary to note the effect of fibers on health, which also differs depending on their solubility, thickening power and other characteristics of each fiber (Padayachee, Day, Howell, & Gidley, 2017). Thus, for example, the efficacy of psyllium in preventing constipation is much higher than other fibers (Gélinas, 2013). At the same time, the prebiotic effects of inulin have not been demonstrated for other fibers (Shah, Li, Al Sabbah, Xu, & Mráz, 2020). Furthermore, inulin can improve calcium bioavailability in calcium-enriched breads (Krupa-Kozak, Świątecka, Bączek, & Brzóska, 2016). Therefore, the incorporation of fibers is usually beneficial in the elaboration of gluten-free breads, but the reason for their addition must be considered (health properties or improvement of the organoleptic characteristics of the breads). Depending on this reason, fiber type and quantity should be properly selected and some processing parameters such as hydration should be changed. At the commercial level, it should also be considered that some celluloses obtained from wheat and oats, despite being gluten-free, often generate uncertainty among people who follow a gluten-free diet.

3. Processes

3.1 Bread

In the elaboration of gluten-free breads, it is necessary to consider some aspects that differentiate them from breads made with wheat flour (Table 1). However, research has been much more focused on the evaluation of formulations than on the analysis of the processes.

First of all, for the development of gluten-free products, it is necessary to mix the ingredients, but not to give mechanical work to develop the gluten network as when using wheat flour. This phase can therefore be much shorter than the kneading in gluten-containing doughs. However, there is also an aeration of the dough and an adaptation of the yeast in this phase. If it takes a little longer, the yeast can adapt better and the subsequent fermentation occurs faster (Gómez, Talegón, & de la Hera, 2013).

Doughs for making gluten-free products are usually much less elastic and stickier than those made with gluten, so some of the machinery adapted for gluten products cannot be used. In many cases, gluten-free doughs are deposited by extrusion inside a mold. In other cases, they are formed by rolling and cutting, as in breads with high hydration such as ciabatta.

The fermentation in these breads is also usually shorter than in gluten breads. In fact, most research articles use fermentation times of 30–40 min, as opposed to the typical 90–120 min for wheat breads. The reason for this

Table 1 Main changes in gluten-free breadmaking versus wheat flour breadmaking.

Breadmaking	Problem	Solution/change
Mixing	No need to develop the gluten network	Less time Other equipment
Fermentation	Less stability	Less time
Fermentation	Low level of damaged starch or amylases (sugar)	Add fermentable sugars
Baking	More water Lighter crust	More time Increase temperature
Final Bread	Dryness Staling	Increase oil content
Final Bread	No wheat bread flavor/off-flavor	Add flavorings

is the low stability of this type of dough, as well as the dangers of overfermentation, which can result in a drop in volume. Likewise, a longer fermentation time can both generate larger alveoli, which is not always positive, and modify the mouthfeel (Puerta et al., 2021), but how it influences acceptability has not been studied. Longer fermentation can bring beneficial effects on the shelf life of breads (Gómez, Oliete, Pando, Ronda, & Caballero, 2008). In the case of wheat breads, it is possible to provide doughs with more stability during fermentation with the use of additives such as emulsifiers (Gómez et al., 2004). However, in the case of gluten-free breads, the risk of volume reduction and possible loss of quality with longer fermentation is usually more important.

An important aspect of fermentation is the need for fermentable sugars in the medium. Although some gluten-containing products add sugars to accelerate or enhance fermentation, in many other cases, these sugars come from the hydrolysis of damaged starch by amylases. On the contrary, in many gluten-free elaborations this hydrolysis does not occur, or it occurs to a lesser extent, due to the low level of damaged starch or the small amount of amylases. This lack of fermentable sugars is especially important in starch-based formulations. Starches are obtained by wet milling, while most flours are extracted through dry milling, a process that results in a higher damaged starch (Heo, Lee, Shim, Yoo, & Lee, 2013). In addition, hydrolytic enzymes are eliminated in this process. Sugars can also help improve the aroma of gluten-free breads through Maillard reactions, depending on the type of sugar and provided that there are adequate amino acids for these reactions (Pacyński, Zawirska-Wojtasiak, & Mildner-Szkudlarz, 2015). For this reason, the incorporation of added sugars in gluten-free breads is much more common than in gluten-containing breads (Román, Belorio, & Gómez, 2019). In general, sucrose is used, but other sources of added sugars are also suitable, such as honey, molasses, caramel, or syrups of different origins, which are used in commercial preparations (Román, Belorio, & Gómez, 2019).

Finally, baking must be modified. This seems obvious because of the higher moisture content of gluten-free doughs, which must be eliminated to a large degree, and the lighter colors they may have depending on the ingredients. This last deficiency may be due not only to the minor presence of proteins in some formulas, but also to the lower content of hydrolytic enzymes, which are present in wheat flour. Another possibility is the fact that these enzymes have had a shorter time to act, due to the reduced fermentation time. Nevertheless, baking has received little attention among research groups and it is hardly covered in published scientific articles. Baking temperature and time have been found to influence bread color (Salah, Olkhovatov, & Aïder, 2019) and thus, it may be of interest to increase either baking temperature or time to improve dark colors, or water loss. However, losses of volatiles are more pronounced in gluten-free breads than in gluten-containing breads (Pico, Khomenko, Capozzi, Navarini, & Biasioli, 2020), which is a problem in breads that tend to have a softer aroma. It would be necessary to study possible temperature gradients or how other baking parameters influence the quality of gluten-free breads.

The slight aroma of gluten-free breads may account for the incorporation of flavorings in almost one-third of commercial gluten-free breads (Román, Belorio, & Gómez, 2019). These flavorings may be aimed at obtaining a product with an aroma similar to that of wheat breads, as gluten-free breads have different aromas, both in the crumb (Pico, Bernal, & Gómez, 2017; Pico, Hansen, & Petersen, 2017) and the crust (Pico, Antolín, Román, Gómez, & Bernal, 2018). But the function of aromas can also be to mask some foreign aromas by replacing them with more pleasant ones.

Another ingredient that is common in the preparation of gluten-free breads is oil (or fats). Thus, according to the study by Román, Belorio, and Gómez (2019), almost all commercial gluten-free breads contained this type of ingredient. It is known that oil can help stabilize bubbles, improve the specific volume of breads (Houben, Höchstötter, & Becker, 2012), and reduce starch retrogradation and crumb hardening phenomena (Smith & Johansson, 2004). Moreover, as oils help reduce dough consistency (Moreira, Chenlo, & Torres, 2012), they also help generate breads with a higher specific volume (Mancebo et al., 2017). But perhaps the main reason for including oils in gluten-free bread formulas is to provide a moist texture to breads that otherwise tend to be very dry.

A final aspect of gluten-free bread processing is the use of sourdoughs. Sourdoughs have unquestionable benefits for the elaboration of wheat breads, including both improved bread aroma and shelf life, as well as better nutritional characteristics (Arora et al., 2021; Fernández-Peláez, Paesani, & Gómez, 2020). In fact, the use of sourdoughs in commercial gluten-free breads has grown in recent years and about one-fifth incorporate them (Román, Belorio, & Gómez, 2019). However, it is not possible to know the elaboration method of these sourdoughs or fermented flours. The incorporation of sourdoughs has also been addressed in scientific articles, in much larger proportion than any other aspect. In fact, there are already several literature reviews on the subject (Moroni, dal Bello, & Arendt, 2009; Nionelli & Rizzello, 2016; Zannini, Pontonio, Waters, & Arendt, 2012). It is difficult to draw general conclusions since results depend on the way the sourdoughs are obtained and on the formula used to make the gluten-free bread. In addition, important aspects such as the aroma of the breads or their acceptability are not usually included in the research. What does seem to be demonstrated is the ability of sourdoughs to reduce the glycemic index of these breads (Novotni et al., 2012), which is usually higher than that of wheat breads. This effect may be related to lactic (Östman, Nilsson, Liljeberg Elmståhl, Molin, & Björck, 2002) and acetic (Liljeberg & Björck, 1998) acids, generated in these fermentations (Gobbetti, de Angelis, Corsetti, & di Cagno, 2005), which have the capacity to reduce the glycemic index.

In addition to sourdough incorporation, changes in the formulation, the inclusion of certain fibers, or the use of certain gluten-free flours, can help reduce the glycemic index of gluten-free breads (Capriles & Arêas, 2016). Even the use of flours with a larger particle size can help in this reduction (de la Hera et al., 2014) due to the difficulty for starch hydration in coarser particles or in particles resulting from the milling of harder grains, or their harder parts (Román, Gómez, Li, Hamaker, & Martínez, 2017).

3.2 Cakes

There is a multiplicity of formulas for making cakes and the functionality or importance of each ingredient can vary between them. According to Pyler and Gorton (2009), cakes can be divided into two groups: emulsified shortening cakes, or batter cakes, and egg foam cakes. The first includes cakes that present a significant amount of fat or oil in their formulation and, the second, those that have either a very low proportion or no fat at all. Pound cakes or layer cakes are found among the first ones, in which the fatty substances cover the air bubbles present in the batter, giving them stability. These cakes have a less aerated structure (fewer and larger crumb pores) and they usually need the help of an improver to increase their volume. They also are often smoother than the rest, which is due to the effect of fats. Foam cakes incorporate a large amount of air bubbles in the batter, due to the foaming action of the egg white. These bubbles are stabilized by the egg white and they grow during baking by the simple expansion of the gases with heating. The cakes obtained are drier, but with a finer pore structure and a higher specific volume. For further knowledge of these types of cake and the functionality of their ingredients, the reviews by Godefroidt, Ooms, Pareyt, Brijs, and Delcour (2019) and Wilderjans, Luyts, Brijs, and Delcour (2013) can be consulted. In general, in cake elaboration, the incorporation of air during mixing must be considered, and not just its quantity. It is important that the air bubbles be as small as possible, since this generates stability, and minimizes the phenomena of coalescence, mobility of the bubbles, and the subsequent loss of volume. The viscosity of the batter is also important, because if the phase surrounding the bubbles is too liquid, bubble mobility increases and stability is lost; but if it is too thick, it can interfere by limiting the expansion of the bubbles during baking.

In cakes, the development of the gluten network is very limited given the low resistance that the mixer/whisk accessories found in the dough or batter. Therefore, little mechanical work is applied and the gluten network does not develop, or it develops minimally. The presence of some ingredients such as fats and sugar do not contribute to the development of the gluten network either (Cauvain & Young, 2006; Donelson & Wilson, 1960). In fact, as it will be demonstrated, it is possible to obtain gluten-free cakes with very similar specific volumes to cakes made with wheat flour. In the case of some more viscous batter, which offers more resistance to the mixing accessories, a greater difference can be noted between gluten and gluten-free cakes.

As it has been indicated, gluten does not play an essential role in this type of elaborations and products with acceptable volume can be obtained by simply substituting wheat flour for another gluten-free flour or starch. In this sense, flours most widely used, both at the industrial level and the research level, are rice flours. But, although in many articles the characterization of these flours is quite scarce, aspects such as particle size or damaged starch can greatly influence the quality of the final product (Kim & Shin, 2014). Particularly, in the case of rice flours, those with a smaller particle size are preferable, this characteristic being more important and relevant in sponge cakes (de la Hera, Martínez, Oliete, & Gómez, 2013). It is observed that the larger the particle size, the larger the size of the bubbles present in the batter, and the lower the bubble stability that reduces the final volume of the cakes and increases their hardness. Changes in the percentage of damaged starch can influence the viscosity of the batter, as well as starch and protein contents (Wilderjans, Luyts, Goesaert, Brijs, & Delcour, 2010). The effects of these factors in wheat cakes is already known (Gaines, 1985; Yamamoto, Worthington, Hou, & Ng, 1996). The influence of particle size is noticed even in mixtures of gluten-free flours, such as soybean flour with maize starch, where the gluten-free flour accounts for less than 50% (Dhen et al., 2015).

In fact, in commercial cakes it is common to combine these flours with some type of starch, mostly maize or potato, which can help minimize this effect (Belorio & Gómez, 2020b). However, it has been proven that the total substitution of wheat flour for starch can produce a collapse in the cake structure during the cooling process, generating a cake with smaller volume and coarser crumbs (Wilderjans, Pareyt, Goesaert, Brijs, & Delcour, 2008), a problem that can be solved with the addition of proteins (Ronda, Oliete, Gómez, Caballero, & Pando, 2011). However, the lack of knowledge and regularity of rice and other gluten-free grain flours may lead some manufacturers to prefer to use more regular starch blends to minimize the influence of flours. It should be noted that most gluten-free grains are harder than wheat, so in order to produce fine flours they need more energy; besides, they generate a greater amount of damaged starch, which would increase the water absorption capacity of the flours. Another possible alternative is to separate the finer flours that are generated in the milling process, without forcing it. But in that case, the flour producer should also find a market for the coarser flours.

To improve the nutritional value of the cakes, the use of whole wheat flours has been investigated. Sometimes these flours must undergo heat treatment to minimize the rancidity phenomenon. In general, the use of these parts of the grain usually reduces the volume and acceptability of the cakes, due, at least in part, to their larger particle size. Yet, the type of heat treatment received can reduce these problems, as Paesani, Bravo-Núñez, and Gómez (2021) found with maize flours. In general, a smoother treatment is preferred in order not to modify too much batter rheology, even when, sometimes, it may be interesting to modify it, so the results will depend on the formula and the flour used. To improve the nutritional quality of these products, the incorporation of fibers has also been used (Gularte, de la Hera, Gómez, & Rosell, 2012; Tsatsaragkou, Papantoniou, & Mandala, 2015), but the effects are very different depending on the type of fiber, its solubility and its thickening character. Other authors have proposed the use of pulse flours (Gularte, Gómez, & Rosell, 2012), other grains (Marston, Khouryieh, & Aramouni, 2016; Rothschild et al., 2015), or the incorporation of agro-industrial by-products (Drabińska, Ciska, Szmatowicz, & Krupa-Kozak, 2018; Kırbaş, Kumcuoglu, & Tavman, 2019; Majzoobi, Poor, Jamalian, & Farahnaky, 2016). It is difficult to draw general conclusions about the use of these ingredients as they will depend on the formula used and the characteristics of the flours or products incorporated. However, in general, these flours reduce cake volume and modify their organoleptic characteristics, especially their flavor. The way aspects such as flour composition, milling methods or subsequent treatments affect the quality of the products obtained, should always be known. In this way, it would be possible to obtain flours with homogeneous quality. However, it should be kept in mind that cakes are not consumed for their nutritional benefits and that their organoleptic quality should be a priority while developing them, since it usually decreases with this type of nutritional improvements.

As mentioned previously, gluten-free cakes can develop a similar volume to wheat cakes, but in many cases a low cohesiveness (crumbling in the mouth) is noted (de la Hera, Martínez, Oliete, & Gómez, 2013; Dhen et al., 2015). This can be a problem as it makes them brittle during handling, or creates a swallowing sensation typical of gluten-free bakery products (Gallagher et al., 2003). To overcome this problem, the egg white content can be increased, as already discussed in the case of gluten-free breads. In fact, most industrial gluten-free muffins incorporate a greater amount of egg in their formulations than muffins made with wheat flour. Thus, the liquid batter turns into a solid foam, because of the starch gelatinization and the egg protein coagulation that take place during the last phases of the baking process, and influences the final texture (Wilderjans et al., 2013). Different studies have shown that an increase in the egg protein content can help generate less dense batters (with greater incorporation of air in the form of small particles) and cakes with greater specific volume and more cohesive and elastic texture (Matos, Sanz, & Rosell, 2014; Ronda, Gómez, Caballero, Oliete, & Blanco, 2009; Sahagún, Bravo-Núñez, Báscones, & Gómez, 2018). However, if the amount of egg protein is too high, the final texture of the cakes can be excessively hard (Sahagún et al., 2018). The incorporation of whey protein can also help produce cakes with greater volume and better texture (Ammar et al., 2021; Sahagún et al., 2018). In cakes with high protein incorporations, it may be interesting to mix proteins from various sources, mostly whey protein since others, such as egg or vegetable proteins, have negative effects when incorporated in large proportions (Bravo-Núñez, Sahagún, Bravo-Núñez, & Gómez, 2020).

The presence of some additives, such as hydrocolloids and emulsifiers, that can help stabilize the batter or increase the shelf life of the cakes is more common in commercial gluten-free muffins than in those made with gluten (Belorio & Gómez, 2020b). The effectiveness of hydrocolloids, and especially of xanthan gum, in regulating the viscosity of batters and in increasing the specific volume and shelf life of cakes due to their higher water absorption capacity, has already been proven both in wheat cakes (Gómez, Ronda, Caballero, Blanco, & Rosell, 2007) and in gluten-free cakes (Andrade et al., 2018; Preichardt, Vendruscolo, Gularte, & Moreira, 2011; Ronda et al., 2009; Sumnu, Koksel, Sahin, Basman, & Meda, 2010; Turabi, Sumnu, & Sahin, 2010). However, the amount of hydrocolloids required is much lower than in the production of gluten-free breads. Emulsifiers, on the other hand, can stabilize the air bubbles incorporated in the batter, generating cakes with greater volume and a finer structure and reducing the staling phenomenon. These effects have been proven both in wheat cakes (Jyotsna, Prabhasankar, Indrani, & Venkateswara-Rao, 2004; Sakiyan, Sumnu, Sahin, & Bayram, 2004) and in cakes made with rice flour (Mohamed & Hamid, 1998; Ronda et al., 2009). In general, it is possible to obtain quality cakes without the presence of additives, but to solve some problems that arise depending on the formulation or the type of gluten-free flour/starch used, it may be convenient to use them.

3.3 Cookies

The term cookie, depending on the language used, includes a number of products made with a mixture of flour (usually wheat), sugar and fat as the main ingredients, and with a low moisture content (usually less than 5%). Within this group, we can find sweet and salty cookies, elaborated with different procedures and various formulations. It is therefore very difficult to provide general recommendations. But it should be noted that in most cookies the gluten network is not formed and, even when made with wheat flour, an effort to avoid the formation of this network is required. This is due to the high fat and sugar contents, and the low flour content of some of these cookies. But it is also due to the less mechanical work required during the process, since doughs are very liquid or mixing times are reduced. In all these cases it will be quite simple to produce gluten-free products and it may be enough to replace wheat flour with other suitable gluten-free flour, or with their mixture. In these cases, most of the recommendations regarding ingredients and processes will be valid to produce gluten-free cookies. The reviews by Pareyt and Delcour (2008) can be very useful in this context. In other elaborations, such as sweet cookies (Maria type) or crackers with a higher flour content in the formulation, the gluten network is developed and thus the aspects related to gluten quality are important when selecting the most suitable wheat flours (Kweon, Slade, Levine, & Gannon, 2014). In these cases, it may be necessary to modify the formulation, incorporate other ingredients or additives, or change production methods. We should not forget that these products are not excessively healthy, due to their high fat and sugar contents, so their organoleptic quality and consumer acceptability should be a priority in their development.

The majority of gluten-free cookie studies are based on the development of sugar snap cookies, or similar formulations. In these investigations, the cookies are produced by sheeting and cutting the pieces. These pieces tend to expand during baking, reducing their hardness. In this type of elaborations, the water absorption capacity of the flours is usually important, since the higher the water absorption capacity, the higher the G' and G'' values of the doughs, and the lower the expansion in the oven, all of which increase the hardness of the final product. Thus, Pareyt and Delcour (2008) stated that even in wheat flours, aspects such as protein and damaged starch content, which affect the water absorption capacity, influenced the characteristics of the cookies. Kweon et al. (2014) proposed that the SRC (solvent retention capacity) test is suitable for the quality control of flours. To obtain this type of gluten-free cookies, flours of different origins and particle sizes or with different pretreatments can be used (Mancebo, Picón, & Gómez, 2015). But as we mentioned before, flours with a higher water absorption capacity, such as the finest ones, or those where part of the starch is gelatinized, generate harder cookies with a smaller diameter. Adedara and Taylor (2021) also stated that gluten-free cookies can be obtained (with sorghum flour) very similar to those made with wheat flour and note that the starch does not gelatinize in the cookie making process, due to the low moisture content. However, incorporating pregelatinized starch in the formula reduces the breaking strength of the sorghum cookies. An opposite effect was observed when the water or sugar contents in the doughs were increased.

As mentioned previously, in contrast to the finer wheat flours, gluten-free flours have very different particle sizes, and it is possible to obtain different particle sizes by more aggressive milling or by sieving. Belorio, Sahagún, and Gómez (2019)'s study with maize flours, showed that, although a larger particle size was preferred to get a greater cookie expansion and reduce its hardness, a percentage of fine flours was also necessary to help provide cohesiveness to the dough. Without this percentage, the doughs were excessively brittle and difficult to handle.

As in other gluten-free products, it is also possible to use combinations with starches or proteins. In general, starches have a lower water absorption capacity than flours, and flours have a lower water absorption capacity than insoluble proteins, so the use of starch can increase dough expansion, while proteins can reduce it (Mancebo, Rodríguez, & Gómez, 2016). However, animal proteins, such as dairy or egg white proteins, are soluble proteins and if they replace part of the flour, dough consistency was reduced and expansion increased. Besides having a direct effect on the nutritional composition of the cookies, the addition of proteins makes darker cookies possible, due to their participation in the Maillard reactions, and can also modify the texture. Thus, for example, the inclusion of egg white in high percentages can increase cookie hardness due to the coagulation processes of this protein when heated (Sahagún & Gómez, 2018b). But the incorporation of egg can also contribute to giving cohesion to excessively brittle cookies. Something similar happens with the incorporation of fibers. Thus, while soluble fibers, such as inulin, reduce the consistency of doughs when they replace flour, insoluble fibers tend to increase the consistency, to a greater extent as their water absorption capacity increased (Mancebo, Rodríguez, Martínez, & Gómez, 2018). In these cases, aspects such as the particle size of the fibers can influence the final result.

Studies about the use of different types of gluten-free flours in the production of cookies are very numerous, as well as those about the incorporation of industry by-products rich in fiber and other nutrients (Gómez & Martinez, 2018; Xu, Zhang, Wang, & Li, 2020). In this sense, only between 2019 and May 2021, several research works have been published on the inclusion of sorghum flours (Adedara & Taylor, 2021), pulses (Cappa, Kelly, & Ng, 2020; di Cairano et al., 2020; Torra et al., 2021), pseudocereals (di Cairano et al., 2020, 2021; Sabbione, Suárez, Añón, & Scilingo, 2019), chestnut (Torra et al., 2021), alfalfa (Rocchetti et al., 2019), Agaricus bisporus (Sulieman et al., 2019), fruit by-products (Gagneten et al., 2021), and banana starch (Román, Sahagún, Gómez, & Martínez, 2019), among others. As in other cases, efforts are made to improve the nutritional quality of these products, and the investigations usually focus on the composition of the final product or the reduction of starch digestibility, although sometimes they include analyses of acceptability and cookie shape. Some studies also address the presence of antioxidant components or a possible prebiotic effect. In general, if we substitute nutritionally deficient flours or starches by flours or products with a better nutritional content, the final cookies will have improved nutritional quality, since nutrients such as fiber, protein or minerals are not usually modified in the elaboration process. However, in most of these studies, aspects of the milling processes or the treatment of the flours are not usually considered. Paesani, Bravo-Núñez, and Gómez (2020) observed that the way in which wholemeal flours, in this case maize, are stabilized to reduce the risk of rancidity, hasd an important influence on cookie quality, as their water absorption capacity was modified. Cappa et al. (2020) verified that certain characteristics of bean flours, such as particle size or water retention capacity, had an important influence on the characteristics of the cookies. As mentioned before, it is very difficult to provide general recommendations, but to produce cookies with regular characteristics, it is necessary to know in each case which flour factors influence these characteristics. If the intention is to improve the nutritional quality of the cookies, it is better to start with the healthiest possible formulas and to bear in mind that consumers are not usually very tolerant to a reduction of the organoleptic quality in this type of products.

4. Nutritional quality of gluten-free products

The elimination of gluten from the diet is a necessary practice for gluten allergic and gluten intolerant people (celiacs), but it has also been recommended for other people suffering from gluten-related disorders (GRD), such as irritable bowel syndrome (IBS), Dermatitis Herpetiformis, non-celiac gluten sensitivity, and even diabetes. An analysis of the advantages of this diet for these groups can be found in the review by Di Liberto et al. (2020). However, in some countries, the consumption of gluten-free products has experienced a very significant increase, based on the belief that gluten is bad for people's health (Savarese et al., 2021). Some influencers have spread these statements, without any scientific evidence, and they have propagated on social media. The reasons why consumers choose to eat a gluten-free diet are being studied, so that this knowledge may be later used to design certain policies or public actions aiming to avoid adverse effects (Norwood, 2021; Xhakollari, Canavari, & Osman, 2021).

It is complicated to evaluate the advantages of a gluten-free diet since, as in the case of a gluten diet, they can be more or less healthy. But the elimination of gluten out of a health concern, may bring about more profound dietary changes and these may come along with an increased consumption of fruits and vegetables, less salt, sugars and saturated fats, and better lifestyle habits. Thus, Christoph, Larson, Hootman, Miller, and Neumark-Sztainer (2018) observed that people who, without suffering from GRD, did not consume gluten because they believed it was less healthy, engaged in healthier behaviors and had a better dietary intake, but were also more likely to engage in unhealthy weight control behaviors. It is also possible that a placebo effect may manifest when the gluten intake is stopped, or a nocebo effect may manifest when gluten products are ingested and the person thinks they may be harmful for her (Enck & Klosterhalfen, 2021). Similarly, the problem for some people may not be gluten but other components, such as FODMAPs or amylase trypsin inhibitors (ATIs), present in wheat (Koumbi, Giouleme, & Vassilopoulou, 2020; Mumolo et al., 2020). However, a gluten-free diet may have adverse effects on non-celiac disease patients (Niland & Cash, 2018).

The nutritional quality of gluten-free products is often inferior to that of their gluten-containing counterparts. Some of the solutions proposed to improve this quality, as well as their drawbacks they present, are shown in Table 2. If we compare the nutritional quality of gluten-free products with their gluten-containing equivalents, all studies agree that gluten-free

Table 2 Some solutions to improve the nutritional quality of gluten-free products.

Nutritional aspects to improve	Solution	Problems
Low protein (products)	Add proteins	Depend on quantity and kind of protein
	Use protein rich flours (e.g., pulses)	Off-flavors; lower volume
Low fiber (diet)	Add fiber	Depend on quantity and kind of fiber
	Use fiber rich flours (e.g., whole grain flours)	Off-flavors; lower volume
Low minerals and vitamins (diet)	Add mineral and vitamins	
	Use mineral and vitamin rich flours (e.g., whole grain flours)	Off-flavors; lower volume
High fat (products)	Reduce fats	Increases dryness and reduce shelf life
	Improve the nutritional quality of fats (less saturated fats)	
	Replace fats/oil by hydrocolloids	Reduce shelf life and lower acceptability
High sugars (products)	Reduce sugars	Need fermentable sugars
High GI	Use sourdoughs or acids	

breads have a lower protein content than their gluten-containing homologues (Kulai & Rashid, 2014; Matos & Rosell, 2011; Mazzeo et al., 2015;Miranda et al., 2014; Román, Belorio, & Gómez, 2019). This is because starch is widely used in most formulas and the protein percentage in most commonly used gluten-free flours, such as rice flour, is lower than in wheat flour. These studies also tend to agree on the fact that gluten-free breads have a higher percentage of fat, typically monounsaturated and polyunsaturated, than wheat breads (Kulai & Rashid, 2014; Matos & Rosell, 2011; Mazzeo et al., 2015; Miranda et al., 2014; Román, Belorio, & Gómez, 2019). This difference is due to the use of oil and, -to a lesser degree, fat to provide tenderness to gluten-free breads. Although gluten-free starches and flours had a lower fiber content than wheat flour, the fiber content of gluten-free breads compared to wheat breads was either not significantly different (Kulai & Rashid, 2014; Missbach et al., 2015) or was even slightly higher (Román, Belorio, & Gómez, 2019). This is because gluten-free breads contain hydrocolloids as substitutes for gluten and, nutritionally, they belong to the fiber group.

Salt content is not very different either in gluten-free and gluten-containing breads: while in gluten-free breads some studies find lower contents (Román, Belorio, & Gómez, 2019) but with a great variability, others observe a higher contents (Miranda et al., 2014; Tres et al., 2020). It therefore appears that these results depend on the countries analyzed and that there are gluten-free breads on the market with both lower and higher sodium content than wheat breads. Additionally, no major differences in the content of simple sugars were observed, the percentages in gluten-free breads being between 4.5 and 5% (Miranda et al., 2014; Román, Belorio, & Gómez, 2019). In any case, it should be noted that gluten-free breads present a very wide range of compositions, since the variety of formulations is much broader than in wheat breads (Román, Belorio, & Gómez, 2019). In addition, Tres et al. (2020) observed that the deviations between declared and analyzed composition (in fat and salt contents) in gluten-free breads was much larger than in gluten breads, and most studies comparing the composition of gluten-free and gluten-containing products are based on the declared composition.

There are numerous studies that analyze the composition of a wide range of gluten and gluten-free products (Cornicelli, Saba, Machello, Silano, & Neuhold, 2018; do Nascimento, Fiates, dos Anjos, & Teixeira, 2014; Fry, Madden, & Fallaize, 2018; Mazzeo et al., 2015). However, within each category, a vast variety of products is included and, consequently, it is not possible to reach conclusions about the differences in the composition of a single product, but of a large group; so that the market supply, which is much larger in products with gluten, also influences. In general, there has been an increase in the supply of gluten-free products over the years, and they also have a higher cost. However, the nutritional differences between gluten-free and gluten-containing products other than bread may agree with those mentioned above. Thus, in a study about muffins, gluten-free products only showed significant differences (lower content) in total carbohydrates and sugars compared to those made with wheat flour (Belorio & Gómez, 2020b). Hence, the lower protein content of gluten-free flours was compensated for by the larger proportion of egg in the formula, and in these muffins, it is not necessary to increase the oil content as in gluten-free breads. This study agrees on the greater variability of the composition of gluten-free products, which indicates different ways companies or countries address the problems presented by gluten-free products.

Studies that have compared gluten and gluten-free diets state that, despite the lower protein content of gluten-free starches and flours, the gluten-free diet is not less rich in protein, indicating changes in the consumption habits (products consumed) of people following a gluten-free diet. Nevertheless, it is observed that the gluten-free diet is deficient in fiber micronutrients, such as Vit D, Vit B12 and folate, and in some minerals, such as iron, zinc, magnesium and calcium; while it is rich in products with a high glycemic index (Di Nardo et al., 2019; Vici, Belli, Biondi, & Polzonetti, 2016). In recent years, the changes that the gluten-free diet have produced on the microbiome are also being investigated (Caio et al., 2020; Reddel, Putignani, & del Chierico, 2019). Ways to improve the nutritional quality of gluten-free products have already been discussed in this chapter and there are some reviews on the subject (Capriles & Arêas, 2014; Matos & Rosell, 2015; Melini et al., 2017). In general, there is a consensus that the inclusion of whole-grain flours, or pseudocereals and pulses could improve the nutritional quality of gluten-free products. Much research has been published on this subject, but this practice has not been very successful in the development of commercial products, partly because of the higher cost of these raw materials, and partly because of their possible lack of regularity. However, the main issue is related to the effects of these flours on the organoleptic quality of the products and their acceptability. It should be noted that one of the complaints of people following a gluten-free diet, in addition to the cost and lack of variety of products on offer, is the low organoleptic quality of the products, especially when compared to products with gluten. For this reason, companies focus on obtaining products with the highest possible organoleptic quality.

Another alternative to improve the nutritional quality of gluten-free products would be the fortification or the supplementation of gluten-free flours by adding certain minerals and vitamins. These additions have been detected in gluten-free breads, depending on the country of origin and -in some cases- on the legislation on the subject (Román, Belorio, & Gómez, 2019).

5. Conclusions

Gluten does not play an essential role in all baked products. In these cases, as in cakes and many types of cookies, making gluten-free products is relatively simple. The most important factor is a proper choice of gluten-free flour that should replace wheat flour. However, in products such as bread, gluten plays an essential role and it is not possible to completely replace it and obtain products similar to those obtained with wheat flour. In these cases, to obtain products that are as similar as possible to those made with wheat, we must resort to complex mixtures of ingredients where each one contributes its functionality. The products on the market have improved a lot, but there is still room for improvement, both in their organoleptic quality and in their nutritional quality.

References

Adedara, O. A., & Taylor, J. R. N. (2021). Roles of protein, starch and sugar in the texture of sorghum biscuits. *LWT - Food Science and Technology, 136*, 110323.

Afshin, A., Sur, P. J., Fay, K. A., Cornaby, L., Ferrara, G., Salama, J. S., et al. (2019). Health effects of dietary risks in 195 countries, 1990–2017: A systematic analysis for the Global Burden of Disease Study 2017. *The Lancet, 393*, 1958–1972.

Aguilar, N., Albanell, E., Miñarro, B., Guamis, B., & Capellas, M. (2015). Effect of tiger nut-derived products in gluten-free batter and bread. *Food Science and Technology International, 21*, 323–331.

Akin, P. A., & Miller, R. A. (2017). Starch-hydrocolloid interaction in chemically leavened gluten-free sorghum bread. *Cereal Chemistry, 94*, 897–902.

Alencar, N. M. M., de Araújo, V. A., Faggian, L., de Araújo, M. B. S., & Capriles, V. D. (2021). What about gluten-free products? An insight on celiac consumers' opinions and expectations. *Journal of Sensory Studies, 36*, e12664.

Alvarez-Jubete, L., Arendt, E. K., & Gallagher, E. (2010). Nutritive value of pseudocereals and their increasing use as functional gluten-free ingredients. *Trends in Food Science and Technology, 21*, 106–113.

Ammar, I., Gharsallah, H., Brahim, A. B., Attia, H., Ayadi, M. A., Hadrich, B., et al. (2021). Optimization of gluten-free sponge cake fortified with whey protein concentrate using mixture design methodology. *Food Chemistry, 343*, 128457.

Andersson, H., Öhgren, C., Johansson, D., Kniola, M., & Stading, M. (2011). Extensional flow, viscoelasticity and baking performance of gluten-free zein-starch doughs supplemented with hydrocolloids. *Food Hydrocolloids, 25*, 1587–1595.

Andrade, F. J. E. T., Albuquerque, P. B. S., Moraes, G. M. D., Farias, M. D. P., Teixeira-Sá, D. M. A., Vicente, A. A., et al. (2018). Influence of hydrocolloids (galactomannan and xanthan gum) on the physicochemical and sensory characteristics of gluten-free cakes based on fava beans (Phaseolus lunatus). *Food & Function, 9*, 6369–6379.

Anjum, F. M., Khan, M. R., Din, A., Saeed, M., Pasha, I., & Arshad, M. U. (2007). Wheat gluten: High molecular weight glutenin subunits - structure, genetics, and relation to dough elasticity. *Journal of Food Science, 72*, 56–63.

Anton, A. A., & Artfield, S. D. (2008). Hydrocolloids in gluten-free breads: A review. *International Journal of Food Sciences and Nutrition, 59*, 11–23.

Aoki, N., Kataoka, T., & Nishiba, Y. (2020). Crucial role of amylose in the rising of gluten- and additive-free rice bread. *Journal of Cereal Science, 92*, 102905.

Aprodu, I., Badiu, E. A., & Banu, I. (2016). Influence of protein and water addition on gluten-free dough properties and bread quality. *International Journal of Food Engineering, 12*, 355–363.

Arora, K., Ameur, H., Polo, A., Di Cagno, R., Rizzello, C. G., & Gobbetti, M. (2021). Thirty years of knowledge on sourdough fermentation: A systematic review. *Trends in Food Science and Technology, 108*, 71–83.

Arslan, M., Rakha, A., Xiaobo, Z., & Mahmood, M. A. (2019). Complimenting gluten free bakery products with dietary fiber: Opportunities and constraints. *Trends in Food Science and Technology, 83*, 194–202.

Belorio, M., & Gómez, M. (2020a). Effect of hydration on gluten-free breads made with hydroxypropyl methylcellulose in comparison with psyllium and xanthan gum. *Food, 9*, 1548.

Belorio, M., & Gómez, M. (2020b). Gluten-free muffins versus gluten containing muffins: Ingredients and nutritional differences. *Trends in Food Science and Technology, 102*, 249–253.

Belorio, M., & Gómez, M. (2020c). Psyllium: A useful functional ingredient in food systems. *Critical Reviews in Food Science and Nutrition*. https://doi.org/10.1080/10408398.2020.1822276.

Belorio, M., Marcondes, G., & Gómez, M. (2020). Influence of psyllium versus xanthan gum in starch properties. *Food Hydrocolloids, 105*, 105843.

Belorio, M., Sahagún, M., & Gómez, M. (2019). Influence of flour particle size distribution on the quality of maize gluten-free cookies. *Food, 8*, 83.

Berta, M., Gmoser, R., Krona, A., & Stading, M. (2015). Effect of viscoelasticity on foam development in zein-starch dough. *LWT - Food Science and Technology, 63*, 1229–1235.

Berta, M., Koelewijn, I., Öhgren, C., & Stading, M. (2019). Effect of zein protein and hydroxypropyl methylcellulose on the texture of model gluten-free bread. *Journal of Texture Studies, 50*, 341–349.

Bravo-Núñez, Á., Sahagún, M., Bravo-Núñez, A., & Gómez, M. (2020). Optimisation of protein-enriched gluten-free layer cakes using a mixture design. *International Journal of Food Science and Technology, 55*, 2171–2178.

Bravo-Núñez, Á., Sahagún, M., & Gómez, M. (2019). Assessing the importance of protein interactions and hydration level on protein-enriched gluten-free breads: A novel approach. *Food and Bioprocess Technology, 12*, 820–828.

Brites, C., Trigo, M. J., Santos, C., Collar, C., & Rosell, C. M. (2010). Maize-based gluten-free bread: Influence of processing parameters on sensory and instrumental quality. *Food and Bioprocess Technology, 3*, 707–715.

Caio, G., Lungaro, L., Segata, N., Guarino, M., Zoli, G., Volta, U., et al. (2020). Effect of gluten-free diet on gut microbiota composition in patients with celiac disease and non-celiac gluten/wheat sensivity. *Nutrients, 12*, 1832.

Cappa, C., Kelly, J. D., & Ng, P. K. W. (2020). Baking performance of 25 edible dry bean powders: Correlation between cookie quality and rapid test indices. *Food Chemistry, 302*, 125338.

Cappa, C., Lucisano, M., & Mariotti, M. (2013). Influence of Psyllium, sugar beet fibre and water on gluten-free dough properties and bread quality. *Carbohydrate Polymers, 98*, 1657–1666.
Capriles, V. D., & Arêas, J. A. G. (2013). Effects of prebiotic inulin-type fructans on structure, quality, sensory acceptance and glycemic response of gluten-free breads. *Food & Function, 4*, 104–110.
Capriles, V. D., & Arêas, J. A. G. (2014). Novel approaches in gluten-free breadmaking: Interface between food science, nutrition, and health. *Comprehensive Reviews in Food Science and Food Safety, 13*, 871–890.
Capriles, V. D., & Arêas, J. A. G. (2016). Approaches to reduce the glycemic response of gluten-free products: In vivo and in vitro studies. *Food & Function, 7*, 1266–1272.
Carini, E., Scazzina, F., Curti, E., Fattori, F., Mazzeo, T., & Vittadini, E. (2015). Physicochemical, sensory properties and starch in vitro digestion of gluten-free breads. *International Journal of Food Sciences and Nutrition, 66*, 867–872.
Castro, W., Oblitas, J., Chuquizuta, T., & Avila-George, H. (2017). Application of image analysis to optimization of the bread-making process based on the acceptability of the crust color. *Journal of Cereal Science, 74*, 194–199.
Cauvain, S. P., & Young, L. S. (2006). *Baked products: Science, technology and practice*. Oxford: Blackwell Publishing.
Christoph, M. J., Larson, N., Hootman, K. C., Miller, J. M., & Neumark-Sztainer, D. (2018). Who values gluten-free? Dietary intake, behaviors, and sociodemographic characteristics of young adults who value gluten-free food. *Journal of the Academy of Nutrition and Dietetics, 118*, 1389–1398.
Conte, P., Fadda, C., Piga, A., & Collar, C. (2016). Techno-functional and nutritional performance of commercial breads available in Europe. *Food Science and Technology International, 22*, 621–633.
Cornejo, F., & Rosell, C. M. (2015). Physicochemical properties of long rice grain varieties in relation to gluten free bread quality. *LWT - Food Science and Technology, 62*, 1203–1210.
Cornicelli, M., Saba, M., Machello, N., Silano, M., & Neuhold, S. (2018). Nutritional composition of gluten-free food versus regular food sold in the Italian market. *Digestive and Liver Disease, 50*, 1305–1308.
Crockett, R., Ie, P., & Vodovotz, Y. (2011a). Effects of soy protein isolate and egg white solids on the physicochemical properties of gluten-free bread. *Food Chemistry, 129*, 84–91.
Crockett, R., Ie, P., & Vodovotz, Y. (2011b). How do xanthan and hydroxypropyl methylcellulose individually affect the physicochemical properties in a model gluten-free dough? *Journal of Food Science, 76*, E274–E282.
da Machado, C. R., & Thys, R. C. S. (2019). Cricket powder (Gryllus assimilis) as a new alternative protein source for gluten-free breads. *Innovative Food Science and Emerging Technologies, 56*, 102180.
Daou, C., & Zhang, H. (2012). Oat beta-glucan: Its role in health promotion and prevention of diseases. *Comprehensive Reviews in Food Science and Food Safety, 11*, 355–365.
de la Hera, E., Martínez, M. M., & Gómez, M. (2013). Influence of flour particle size on quality of gluten-free rice bread. *LWT - Food Science and Technology, 54*, 199–206.
de la Hera, E., Martínez, M. M., Oliete, B., & Gómez, M. (2013). Influence of flour particle size on quality of gluten-free rice cakes. *Food and Bioprocess Technology, 6*, 2280–2288.
de la Hera, E., Rosell, C. M., & Gómez, M. (2014). Effect of water content and flour particle size on gluten-free bread quality and digestibility. *Food Chemistry, 151*, 526–531.
de la Hera, E., Talegón, M., Caballero, P., & Gómez, M. (2013). Influence of maize flour particle size on gluten-free breadmaking. *Journal of the Science of Food and Agriculture, 93*, 924–932.

Delcour, J. A., Joye, I. J., Pareyt, B., Wilderjans, E., Brijs, K., & Lagrain, B. (2012). Wheat gluten functionality as a quality determinant in cereal-based food products. *Annual Review of Food Science and Technology*, *3*, 469–492.

Dhen, N., Román, L., Ben Rejeb, I., Martínez, M. M., Garogouri, M., & Gómez, M. (2015). Particle size distribution of soy flour affecting the quality of enriched gluten-free cakes. *LWT - Food Science and Technology*, *66*, 179–185.

di Cairano, M., Condelli, N., Caruso, M. C., Cela, N., Tolve, R., & Galgano, F. (2021). Use of underexploited flours for the reduction of glycaemic index of gluten-free biscuits: Physicochemical and sensory characterization. *Food and bioprocess technology*, *14*, 1490–1502.

di Cairano, M., Condelli, N., Caruso, M. C., Marti, A., Cela, N., & Galgano, F. (2020). Functional properties and predicted glycemic index of gluten free cereal, pseudocereal and legume flours. *LWT - Food Science and Technology*, *133*, 109860.

di Liberto, D., Carlisi, D., D'anneo, A., Emanuele, S., Giuliano, M., De Blasio, A., et al. (2020). Gluten free diet for the management of non celiac diseases: The two sides of the coin. *Healthcare*, *8*, 400.

Di Nardo, G., Villa, M. P., Conti, L., Ranucci, G., Pacchiarotti, C., Principessa, L., et al. (2019). Nutritional deficiencies in children with celiac disease resulting from a gluten-free diet: A systematic review. *Nutrients*, *11*, 1588.

do Nascimento, A. B., Fiates, G. M. R., dos Anjos, A., & Teixeira, E. (2014). Availability, cost and nutritional composition of gluten-free products. *British Food Journal*, *116*, 1842–1852.

Donelson, D. H., & Wilson, J. T. (1960). Effect of the relative quality of flour fractions on cake quality. *Cereal Chemistry*, *37*, 241–262.

Drabińska, N., Ciska, E., Szmatowicz, B., & Krupa-Kozak, U. (2018). Broccoli by-products improve the nutraceutical potential of gluten-free mini sponge cakes. *Food Chemistry*, *267*, 170–177.

Enck, P., & Klosterhalfen, S. (2021). The placebo and nocebo responses in clinical trials in inflammatory bowel diseases. *Frontiers in Pharmacology*, *12*, 641436.

Erickson, D. P., Campanella, O. H., & Hamaker, B. R. (2012). Functionalizing maize zein in viscoelastic dough systems through fibrous, beta-sheet-rich protein networks: An alternative, physicochemical approach to breadmaking. *Trends in Food Science & Technology*, *24*, 74–81.

Federici, E., Selling, G. W., Campanella, O. H., & Jones, O. G. (2021). Thermal treatment of dry zein to improve rheological properties in gluten-free dough. *Food Hydrocolloids*, *115*, 106629.

Fernández-Peláez, J., Paesani, C., & Gómez, M. (2020). Sourdough technology as a tool for the development of healthier grain-based products: An update. *Agronomy*, *10*, 1962.

Foschia, M., Horstmann, S. W., Arendt, E. K., & Zannini, E. (2017). Legumes as functional ingredients in gluten-free bakery and pasta products. *Annual Review of Food Science and Technology*, *8*, 75–96.

Föste, M., Nordlohne, S. D., Elgeti, D., Linden, M. H., Heinz, V., Jekle, M., et al. (2014). Impact of quinoa bran on gluten-free dough and bread characteristics. *European Food Research and Technology*, *239*, 767–775.

Föste, M., Verheyen, C., Jekle, M., & Becker, T. (2020). Fibres of milling and fruit processing by-products in gluten-free bread making: A review of hydration properties, dough formation and quality-improving strategies. *Food Chemistry*, *306*, 125451.

Franco, E. A. N., Sanches-Silva, A., Ribeiro-Santos, R., & de Melo, N. R. (2020). Psyllium (Plantago ovata Forsk): From evidence of health bene fi ts to its food application. *Trends in Food Science & Technology*, *96*, 166–175.

Fry, L., Madden, A. M., & Fallaize, R. (2018). An investigation into the nutritional composition and cost of gluten free versus non-gluten free food productos in the UK. *Journal of Human Nutrition and Dietetics*, *31*, 108–120.

Gagneten, M., Archaina, D. A., Salas, M. P., Leiva, G. E., Salvatori, D. M., & Schebor, C. (2021). Gluten-free cookies added with fibre and bioactive compounds from blackcurrant residue. *International Journal of Food Science and Technology, 56*, 1734–1740.

Gaines, C. S. (1985). Associations among soft wheat flour particle size, protein content, chlorine response, kernel hardness, milling quality, white layer cake volume, and sugar-snap cookie spread. *Cereal Chemistry, 62*, 290–292.

Gallagher, E., Gormley, T. R., & Arendt, E. K. (2003). Crust and crumb characteristics of gluten free breads. *Journal of Food Engineering, 56*, 153–161.

Gélinas, P. (2013). Preventing constipation: A review of the laxative potential of food ingredients. *International Journal of Food Science and Technology, 48*, 445–467.

Giménez-Bastida, J. A., Piskuła, M., & Zieliński, H. (2015). Recent advances in development of gluten-free buckwheat products. *Trends in Food Science and Technology, 44*, 58–65.

Gobbetti, M., de Angelis, M., Corsetti, A., & di Cagno, R. (2005). Biochemistry and physiology of sourdough lactic acid bacteria. *Trends in Food Science and Technology, 16*, 57–69.

Godefroidt, T., Ooms, N., Pareyt, B., Brijs, K., & Delcour, J. A. (2019). Ingredient functionality during foam-type cake making: A review. *Comprehensive Reviews in Food Science and Food Safety, 18*, 1550–1562.

Gómez, M., del Real, S., Rosell, C. M., Ronda, F., Blanco, C. A., & Caballero, P. A. (2004). Functionality of different emulsifiers on the performance of breadmaking and wheat bread quality. *European Food Research and Technology, 219*, 145–150.

Gómez, M., & Martinez, M. M. (2018). Fruit and vegetable by-products as novel ingredients to improve the nutritional quality of baked goods. *Critical Reviews in Food Science and Nutrition, 58*, 2119–2135.

Gómez, M., Oliete, B., Pando, V., Ronda, F., & Caballero, P. A. (2008). Effect of fermentation conditions on bread staling kinetics. *European Food Research and Technology, 226*, 1379–1387.

Gómez, M., Ronda, F., Caballero, P. A., Blanco, C. A., & Rosell, C. M. (2007). Functionality of different hydrocolloids on the quality and shelf-life of yellow layer cakes. *Food Hydrocolloids, 21*, 167–173.

Gómez, M., Talegón, M., & de la Hera, E. (2013). Influence of mixing on quality of gluten-free bread. *Journal of Food Quality, 36*, 139–145.

Grover, J. A. (1986). Methylcellulose (MC) and hydroxypropylmethylcellulose (HPMC). In M. Glicksmen (Ed.), *Food Hydrocolloids* (pp. 121–154). Boca Raton: CRC Press.

Gularte, M. A., de la Hera, E., Gómez, M., & Rosell, C. M. (2012). Effect of different fibers on the enrichment of gluten-free layer cake. *LWT - Food Science and Technology, 48*, 209–2014.

Gularte, M. A., Gómez, M., & Rosell, C. M. (2012). Impact of legume flours on quality and in vitro digestibility of starch and protein from gluten-free cakes. *Food and Bioprocess Technology, 5*, 3142–3150.

Hager, A., & Arendt, E. K. (2013). Influence of hydroxypropylmethylcellulose (HPMC), xanthan gum and their combination on loaf specific volume, crumb hardness and crumb grain characteristics of gluten-free breads based on rice, maize, teff and buckwheat. *Food Hydrocolloids, 32*, 195–203.

Hager, A. S., Axel, C., & Arendt, E. K. (2011). Status of carbohydrates and dietary fiber in gluten-free diets. *Cereal Foods World, 56*, 109–114.

Han, H. M., Cho, J. H., Kang, H. W., & Koh, B. K. (2012). Rice varieties in relation to rice bread quality. *Journal of the Science of Food and Agriculture, 92*, 1462–1467.

Haque, A., & Morris, E. R. (1994). Combined use of ispaghula and HPMC to replace or augment gluten in breadmaking. *Food Research International, 27*, 379–393.

Heo, S., Lee, S. M., Shim, J. H., Yoo, S. H., & Lee, S. (2013). Effect of dry- and wet-milled rice flours on the quality attributes of gluten-free dough and noodles. *Journal of Food Engineering, 116*, 213–217.

Horstmann, S. W., Foschia, M., & Arendt, E. K. (2017). Correlation analysis of protein quality characteristics with gluten-free bread properties. *Food & Function, 8*, 2465–2474.
Houben, A., Höchstötter, A., & Becker, T. (2012). Possibilities to increase the quality in gluten-free bread production: An overview. *European Food Research and Technology, 235*, 195–208.
Hüttner, E. K., & Arendt, E. K. (2010). Recent advances in gluten-free baking and the current status of oats. *Trends in Food Science and Technology, 21*, 303–312.
Jyotsna, R., Prabhasankar, P., Indrani, D., & Venkateswara-Rao, G. (2004). Improvement of rheological and baking properties of cake batters with emulsifier gels. *Journal of Food Science, 69*, 16–19.
Kim, J.-M., & Shin, M. (2014). Effects of particle size distributions of rice flour on the quality of gluten-free rice cupcakes. *LWT - Food Science and Technology, 59*, 526–532.
Kırbaş, Z., Kumcuoglu, S., & Tavman, S. (2019). Effects of apple, orange and carrot pomace powders on gluten-free batter rheology and cake properties. *Journal of Food Science and Technology, 56*, 914–926.
Kittisuban, P., Ritthiruangdej, P., & Suphantharika, M. (2014). Optimization of hydroxypropylmethylcellulose, yeast β-glucan, and whey protein levels based on physical properties of gluten-free rice bread using response surface methodology. *LWT - Food Science and Technology, 57*, 738–748.
Kiumarsi, M., Shahbazi, M., Yeganehzad, S., Majchrzak, D., Lieleg, O., & Winkeljann, B. (2019). Relation between structural, mechanical and sensory properties of gluten-free bread as affected by modified dietary fibers. *Food Chemistry, 277*, 664–673.
Korus, J., Chmielewska, A., Witczak, M., Ziobro, R., & Juszczak, L. (2021). Rapeseed protein as a novel ingredient of gluten-free bread. *European Food Research and Technology, 247*, 2015–2025.
Korus, J., Grzelak, K., Achremowicz, K., & Sabat, R. (2006). Influence of prebiotic additions on the quality of gluten-free bread and on the content of inulin and fructooligosaccharides. *Food Science and Technology International, 12*, 489–495.
Korus, J., Juszczak, L., Witczak, M., & Ziobro, R. (2020). Effect of citrus fiber on the rheological properties of dough and quality of the gluten-free bread. *Applied Sciences, 10*, 6633.
Korus, J., Witczak, M., Ziobro, R., & Juszczak, L. (2009). The impact of resistant starch on characteristics of gluten-free dough and bread. *Food Hydrocolloids, 23*, 988–995.
Koumbi, L., Giouleme, O., & Vassilopoulou, E. (2020). Non-celiac gluten sensitivity and irritable bowel disease: Looking for the culprits. *Current Developments in Nutrition, 4*, nzaa176.
Krupa-Kozak, U., Baczek, N., & Rosell, C. M. (2013). Application of dairy proteins as technological and nutritional improvers of calcium-supplemented gluten-free bread. *Nutrients, 5*, 4503–4520.
Krupa-Kozak, U., Świątecka, D., Bączek, N., & Brzóska, M. M. (2016). Inulin and fructooligosaccharide affect in vitro calcium uptake and absorption from calcium-enriched gluten-free bread. *Food & Function*, 7, 1950–1958.
Ktenioudaki, A., & Gallagher, E. (2012). Recent advances in the development of high-fibre baked products. *Trends in Food Science and Technology, 28*, 4–14.
Kulai, T., & Rashid, M. (2014). Assessment of nutritional adequacy of packaged gluten-free food products. *Canadian Journal of Dietetic Practise and Research, 75*, 186–190.
Kweon, M., Slade, L., Levine, H., & Gannon, D. (2014). Cookie- versus cracker-baking-what's the difference? Flour functionality requirements explored by SRC and alveography. *Critical Reviews in Food Science and Nutrition, 54*, 115–138.
Lazaridou, A., & Biliaderis, C. G. (2007). Molecular aspects of cereal β-glucan functionality: Physical properties, technological applications and physiological effects. *Journal of Cereal Science, 46*, 101–118.

Lazaridou, A., Duta, D., Papageorgiou, M., Belc, N., & Biliaderis, C. G. (2007). Effects of hydrocolloids on dough rheology and bread quality parameters in gluten-free formulations. *Journal of Food Engineering*, *79*, 1033–1047.
Li, J. M., & Nie, S. P. (2016). The functional and nutritional aspects of hydrocolloids in foods. *Food Hydrocolloids*, *53*, 46–61.
Liljeberg, H., & Björck, I. (1998). Delayed gastric emptying rate may explain improved glycaemia in healthy subjects to a starchy meal with added vinegar. *European Journal of Clinical Nutrition*, *52*, 368–371.
Liu, X., Mu, T., Sun, H., Zhang, M., Chen, J., & Fauconnier, M. L. (2018). Influence of different hydrocolloids on dough thermo-mechanical properties and in vitro starch digestibility of gluten-free steamed bread based on potato flour. *Food Chemistry*, *239*, 1064–1074.
Majzoobi, M., Poor, Z. V., Jamalian, J., & Farahnaky, A. (2016). Improvement of the quality of gluten-free sponge cake using different levels and particle sizes of carrot pomace powder. *International Journal of Food Science and Technology*, *51*, 1369–1377.
Mancebo, C. M., Martínez, M. M., Merino, C., de la Hera, E., & Gómez, M. (2017). Effect of oil and shortening in rice bread quality: Relationship between dough rheology and quality characteristics. *Journal of Texture Studies*, *48*, 597–606.
Mancebo, C. M., Merino, C., Martínez, M. M., & Gómez, M. (2015). Mixture design of rice flour, maize starch and wheat starch for optimization of gluten free bread quality. *Journal of Food Science and Technology*, *52*, 6323–6333.
Mancebo, C. M., Picón, J., & Gómez, M. (2015). Effect of flour properties on the quality characteristics of gluten free sugar-snap cookies. *LWT - Food Science and Technology*, *64*, 264–269.
Mancebo, C. M., Rodríguez, P., & Gómez, M. (2016). Assessing rice flour-starch-protein mixtures to produce gluten free sugar-snap cookies. *LWT - Food Science and Technology*, *67*, 127–132.
Mancebo, C. M., Rodríguez, P., Martínez, M. M., & Gómez, M. (2018). Effect of the addition of soluble (nutriose, inulin and polydextrose) and insoluble (bamboo, potato and pea) fibres on the quality of sugar-snap cookies. *International Journal of Food Science and Technology*, *53*, 129–136.
Mancebo, C. M., San Miguel, M.Á., Martínez, M. M., & Gómez, M. (2015). Optimisation of rheological properties of gluten-free doughs with HPMC, psyllium and different levels of water. *Journal of Cereal Science*, *61*, 8–15.
Marco, C., & Rosell, C. M. (2008). Breadmaking performance of protein enriched, gluten-free breads. *European Food Research and Technology*, *227*, 1205–1213.
Marston, K., Khouryieh, H., & Aramouni, F. (2016). Effect of heat treatment of sorghum flour on the functional properties of gluten-free bread and cake. *LWT - Food Science and Technology*, *65*, 637–644.
Martin, J., Geisel, T., & Maresch, C. (2013). Inadequate nutrient intake in patients with celiac disease: Results from a German dietary survey. *Digestion*, *87*, 240–246.
Martínez, M. M., Díaz, Á., & Gómez, M. (2014). Effect of different microstructural features of soluble and insoluble fibres on gluten-free dough rheology and bread-making. *Journal of Food Engineering*, *142*, 49–56.
Martínez, M. M., & Gómez, M. (2017). Rheological and microstructural evolution of the most common gluten-free flours and starches during bread fermentation and baking. *Journal of Food Engineering*, *197*, 78–86.
Martínez, M. M., Román, L., & Gómez, M. (2018). Implications of hydration depletion in the in vitro starch digestibility of white bread crumb and crust. *Food Chemistry*, *239*, 295–303.
Masure, H. G., Fierens, E., & Delcour, J. A. (2016). Current and forward looking experimental approaches in gluten-free bread making research. *Journal of Cereal Science*, *67*, 92–111.

Masure, H. G., Wouters, A. G. B., Fierens, E., & Delcour, J. A. (2019). Impact of egg white and soy proteins on structure formation and crumb firming in gluten-free breads. *Food Hydrocolloids*, *95*, 406–417.

Matos, M. E., & Rosell, C. M. (2011). Chemical composition and starch digestibility of different gluten-free breads. *Plant Foods for Human Nutrition*, *66*, 224–230.

Matos, M. E., & Rosell, C. M. (2015). Understanding gluten-free dough for reaching breads with physical quality and nutritional balance. *Journal of the Science of Food and Agriculture*, *95*, 653–661.

Matos, M. E., Sanz, T., & Rosell, C. M. (2014). Establishing the function of proteins on the rheological and quality properties of rice-based gluten free muffins. *Food Hydrocolloids*, *35*, 150–158.

Mazzeo, T., Cauzzi, S., Brighenti, F., & Pellegrini, N. (2015). The development of a composition database of gluten-free products. *Public Health Nutrition*, *18*, 1353–1357.

McCarthy, D. F., Gallagher, E., Gormley, T. R., Schober, T. J., & Arendt, E. K. (2005). Application of response surface methodology in the development of gluten-free bread. *Cereal Chemistry*, *82*, 609–615.

Melini, F., Melini, V., Luziatelli, F., & Ruzzi, M. (2017). Current and forward-looking approaches to technological and nutritional improvements of gluten-free bread with legume flours: A critical review. *Comprehensive Reviews in Food Science and Food Safety*, *16*, 1101–1122.

Mir, S. A., Shah, M. A., Naik, H. R., & Zargar, I. A. (2016). Influence of hydrocolloids on dough handling and technological properties of gluten-free breads. *Trends in Food Science and Technology*, *51*, 49–57.

Miranda, J., Lasa, A., Bustamante, M. A., Churruca, I., & Simon, E. (2014). Erratum to: Nutritional differences between a gluten-free diet and a diet containing equivalent products with gluten. *Plant Foods for Human Nutrition*, *69*, 182–187.

Missbach, B., Schwingshackl, L., Billmann, A., Mystek, A., Hickelsberger, M., Bauer, G., et al. (2015). Gluten-free food database: The nutritional quality and cost of packaged gluten-free foods. *PeerJ*, *3*.

Mohamed, S., & Hamid, N. A. (1998). Effects of ingredients on the characteristics of rice cakes. *Journal of the Science of Food and Agriculture*, *76*, 464–468.

Moore, M. M., Heinbockel, M., Dockery, P., Ulmer, H. M., & Arendt, E. K. (2006). Network formation in gluten-free bread with application of transglutaminase. *Cereal Chemistry*, *83*, 28–36.

Moreira, R., Chenlo, F., & Torres, M. D. (2012). Effect of shortenings on the rheology of gluten-free doughs: Study of chestnut flour with chia flour, olive and sunflower oils. *Journal of Texture Studies*, *43*, 375–383.

Moroni, A. V., dal Bello, F., & Arendt, E. K. (2009). Sourdough in gluten-free bread-making: An ancient technology to solve a novel issue? *Food Microbiology*, *26*, 676–684.

Morreale, F., Benavent-Gil, Y., & Rosell, C. M. (2019). Inulin enrichment of gluten free breads: Interaction between inulin and yeast. *Food Chemistry*, *278*, 545–551.

Morreale, F., Garzón, R., & Rosell, C. M. (2018). Understanding the role of hydrocolloids viscosity and hydration in developing gluten-free bread. A study with hydroxypropylmethylcellulose. *Food Hydrocolloids*, 77, 629–635.

Mumolo, M. G., Rettura, F., Melissari, S., Costa, F., Ricchiuti, A., Ceccarelli, L., et al. (2020). Is gluten the only culprit for non-celiac. *Nutrients*, *12*, 3785.

Niland, B., & Cash, B. D. (2018). Health benefits and adverse effects of a gluten-free diet in non-celiac disease patients. *Gastroenterology and Hepatology*, *14*, 82–91.

Nionelli, L., & Rizzello, C. G. (2016). Sourdough-based biotechnologies for the production of gluten-free foods. *Food*, *5*, 65.

Nissen, L., Samaei, S. P., Babini, E., & Gianotti, A. (2020). Gluten free sourdough bread enriched with cricket flour for protein fortification: Antioxidant improvement and volatilome characterization. *Food Chemistry*, *333*, 127410.

Norwood, F. B. (2021). Perceived impact of information signals on opinions about gluten-free diets. *PLoS One, 16,* e0248570. https://doi.org/10.1371/journal.pone.0248570.

Novotni, D., Čukelj, N., Smerdel, B., Bituh, M., Dujmić, F., & Ćurić, D. (2012). Glycemic index and firming kinetics of partially baked frozen gluten-free bread with sourdough. *Journal of Cereal Science, 55,* 120–125.

Nozawa, M., Ito, S., & Arai, E. (2016). Effect of ovalbumin on the quality of gluten-free rice flour bread made with soymilk. *LWT - Food Science and Technology, 66,* 598–605.

O'Shea, N., Doran, L., Auty, M., Arendt, E. K., & Gallagher, E. (2013). The rheology, microstructure and sensory characteristics of a gluten-free bread formulation enhanced with orange pomace. *Food & Function, 4,* 1856–1863.

O'Shea, N., Rößle, C., Arendt, E. K., & Gallagher, E. (2015). Modelling the effects of orange pomace using response surface design for gluten-free bread baking. *Food Chemistry, 166,* 223–230.

Ojetti, V., Nucera, G., Migneco, A., Gabrielli, M., Lauritano, C., Danese, S., et al. (2005). High prevalence of celiac disease in patients with lactose intolerance. *Digestion, 71,* 106–110.

Onwezen, M. C., Bouwman, E. P., Reinders, M. J., & Dagevos, H. (2021). A systematic review on consumer acceptance of alternative proteins: Pulses, algae, insects, plant-based meat alternatives, and cultured meat. *Appetite, 159,* 105058.

Oom, A., Pettersson, A., Taylor, J. R. N., & Stading, M. (2008). Rheological properties of kafirin and zein prolamins. *Journal of Cereal Science, 47,* 109–116.

Östman, E. M., Nilsson, M., Liljeberg Elmståhl, H. G. M., Molin, G., & Björck, I. M. E. (2002). On the effect of lactic acid on blood glucose and insulin responses to cereal products: Mechanistic studies in healthy subjects and in vitro. *Journal of Cereal Science, 36,* 339–346.

Pacyński, M., Zawirska-Wojtasiak, R., & Mildner-Szkudlarz, S. (2015). Improving the aroma of gluten-free bread. *LWT - Food Science and Technology, 63,* 706–713.

Padayachee, A., Day, L., Howell, K., & Gidley, M. J. (2017). Complexity and health functionality of plant cell wall fibers from fruits and vegetables. *Critical Reviews in Food Science and Nutrition, 57,* 59–81.

Paesani, C., Bravo-Núñez, Á., & Gómez, M. (2020). Effect of extrusion of whole-grain maize flour on the characteristics of gluten-free cookies. *LWT - Food Science and Technology, 132,* 109931.

Paesani, C., Bravo-Núñez, Á., & Gómez, M. (2021). Effect of stabilized wholegrain maize flours on the quality characteristics of gluten-free layer cakes. *LWT - Food Science and Technology, 135,* 109959.

Pareyt, B., & Delcour, J. A. (2008). The role of wheat flour constituents, sugar, and fat in low moisture cereal based products: A review on sugar-snap cookies. *Critical Reviews in Food Science and Nutrition, 48,* 824–839.

Park, J. H., Kim, D. C., Lee, S. E., Kim, O. W., Kim, H., Lim, S. T., et al. (2014). Effects of rice flour size fractions on gluten free rice bread. *Food Science and Biotechnology, 23,* 1875–1883.

Pérez-Quirce, S., Collar, C., & Ronda, F. (2014). Significance of healthy viscous dietary fibres on the performance of gluten-free rice-based formulated breads. *International Journal of Food Science and Technology, 49,* 1375–1382.

Pérez-Quirce, S., Lazaridou, A., Biliaderis, C. G., & Ronda, F. (2017). Effect of β-glucan molecular weight on rice flour dough rheology, quality parameters of breads and in vitro starch digestibility. *LWT - Food Science and Technology, 82,* 446–453.

Phimolsiripol, Y., Mukprasirt, A., & Schoenlechner, R. (2012). Quality improvement of rice-based gluten-free bread using different dietary fibre fractions of rice bran. *Journal of Cereal Science, 56,* 389–395.

Phongthai, S., D'Amico, S., Schoenlechner, R., & Rawdkuen, S. (2016). Comparative study of rice bran protein concentrate and egg albumin on gluten-free bread properties. *Journal of Cereal Science, 72*, 38–45.

Pico, J., Antolín, B., Román, L., Bernal, J., & Gómez, M. (2019). Selection of the most suitable mixture of flours and starches for the improvement of gluten-free breads through their volatile profiles. *European Food Research and Technology, 245*, 1755–1766.

Pico, J., Antolín, B., Román, L., Gómez, M., & Bernal, J. (2018). Analysis of volatile compounds in gluten-free bread crusts with an optimised and validated SPME-GC/QTOF methodology. *Food Research International, 106*, 686–695.

Pico, J., Bernal, J. L., & Gómez, M. (2017). Influence of different flours and starches on gluten-free bread aroma. *Journal of Food Science and Technology, 54*, 1433–1441.

Pico, J., Hansen, Å. S., & Petersen, M. A. (2017). Comparison of the volatile profiles of the crumb of gluten-free breads by DHE-GC/MS. *Journal of Cereal Science, 76*, 280–288.

Pico, J., Khomenko, I., Capozzi, V., Navarini, L., & Biasioli, F. (2020). Real-time monitoring of volatile compounds losses in the oven during baking and toasting of gluten-free bread doughs: A PTR-MS evidence. *Food, 9*, 1498.

Pico, J., Reguilón, M. P., Bernal, J., & Gómez, M. (2019). Effect of rice, pea, egg white and whey proteins on crust quality of rice flour-corn starch based gluten-free breads. *Journal of Cereal Science, 86*, 92–101.

Poley, J. R. (2017). The gluten-free diet: Can oats and wheat starch be part of it? *Journal of the American College of Nutrition, 36*, 1–8.

Preichardt, L. D., Vendruscolo, C. T., Gularte, M. A., & Moreira, A. D. S. (2011). The role of xanthan gum in the quality of gluten free cakes: Improved bakery products for coeliac patients. *International Journal of Food Science and Technology, 46*, 2591–2597.

Puerta, P., Garzón, R., Rosell, C. M., Fiszman, S., Laguna, L., & Tárrega, A. (2021). Modifying gluten-free bread's structure using different baking conditions: Impact on oral processing and texture perception. *LWT - Food Science and Technology, 140*, 110718.

Pyler, E. J., & Gorton, L. A. (2009). *Baking science and technology. Volume II: Formulation & production*. Kansas City: Sosland Publishing Company.

Reddel, S., Putignani, L., & del Chierico, F. (2019). The impact of low-FODMAPs, gluten-free, and ketogenic diets on gut microbiota modulation in pathological conditions. *Nutrients, 11*, 373.

Ren, Y., Linter, B. R., & Foster, T. J. (2020). Starch replacement in gluten free bread by cellulose and fibrillated cellulose. *Food Hydrocolloids, 107*, 105957.

Renzetti, S., dal Bello, F., & Arendt, E. K. (2008). Microstructure, fundamental rheology and baking characteristics of batters and breads from different gluten-free flours treated with a microbial transglutaminase. *Journal of Cereal Science, 48*, 33–45.

Reynolds, A., Mann, J., Cummings, J., Winter, N., Mete, E., & Te Morenga, L. (2019). Carbohydrate quality and human health: a series of systematic reviews and meta-analyses. *The Lancet, 393*, 434–445.

Rocchetti, G., Senizza, A., Gallo, A., Lucini, L., Giuberti, G., & Patrone, V. (2019). In vitro large intestine fermentation of gluten-free rice cookies containing alfalfa seed (Medicago sativa L.) flour: A combined metagenomic/metabolomic approach. *Food Research International, 120*, 312–321.

Rocha-Parra, A. F., Ribotta, P. D., & Ferrero, C. (2015). Apple pomace in gluten-free formulations: Effect on rheology and product quality. *International Journal of Food Science and Technology, 50*, 682–690.

Román, L., Belorio, M., & Gómez, M. (2019). Gluten-free breads: The gap between research and commercial reality. *Comprehensive Reviews in Food Science and Food Safety, 18*, 690–702.

Román, L., de la Cal, E., Gómez, M., & Martínez, M. M. (2018). Specific ratio of A-to B-type wheat starch granules improves the quality of gluten-free breads: Optimizing dough viscosity and pickering stabilization. *Food Hydrocolloids*, *82*, 510–518.

Román, L., Gómez, M., Hamaker, B. R., & Martínez, M. M. (2019). Banana starch and molecular shear fragmentation dramatically increase structurally driven slowly digestible starch in fully gelatinized bread crumb. *Food Chemistry*, *274*, 664–671.

Román, L., Gómez, M., Li, C., Hamaker, B. R., & Martínez, M. M. (2017). Biophysical features of cereal endosperm that decrease starch digestibility. *Carbohydrate Polymers*, *165*, 180–188.

Román, L., Gómez, M., & Martínez, M. M. (2021). Mesoscale structuring of gluten-free bread with starch. *Current Opinion in Food Science*, *38*, 189–195.

Román, L., Reguilón, M. P., Gómez, M., & Martínez, M. M. (2020). Intermediate length amylose increases the crumb hardness of rice flour gluten-free breads. *Food Hydrocolloids*, *100*, 105451.

Román, L., Sahagún, M., Gómez, M., & Martínez, M. M. (2019). Nutritional and physical characterization of sugar-snap cookies: Effect of banana starch in native and molten states. *Food & Function*, *10*, 616–624.

Ronda, F., Gómez, M., Caballero, P. A., Oliete, B., & Blanco, C. A. (2009). Improvement of quality of gluten-free layer cakes. *Food Science and Technology International*, *15*, 193–202.

Ronda, F., Oliete, B., Gómez, M., Caballero, P. A., & Pando, V. (2011). Rheological study of layer cake batters made with soybean protein isolate and different starch sources. *Journal of Food Engineering*, *102*, 272–277.

Ronda, F., Perez-Quirce, S., Lazaridou, A., & Biliaderis, C. G. (2015). Effect of barley and oat β-glucan concentrates on gluten-free rice-based doughs and bread characteristics. *Food Hydrocolloids*, *48*, 197–207.

Rothschild, J., Rosentrater, K. A., Onwulata, C., Singh, M., Menutti, L., Jambazian, P., et al. (2015). Influence of quinoa roasting on sensory and physicochemical properties of allergen-free, gluten-free cakes. *International Journal of Food Science and Technology*, *50*, 1873–1881.

Sabanis, D., Lebesi, D., & Tzia, C. (2009a). Development of fibre-enriched gluten-free bread: A response surface methodology study. *International Journal of Food Sciences and Nutrition*, *60*, 174–190.

Sabanis, D., Lebesi, D., & Tzia, C. (2009b). Effect of dietary fibre enrichment on selected properties of gluten-free bread. *LWT - Food Science and Technology*, *42*, 1380–1389.

Sabanis, D., & Tzia, C. (2011). Effect of hydrocolloids on selected properties of gluten-free dough and bread. *Food Science and Technology International*, *17*, 279–291.

Sabbione, A. C., Suárez, S., Añón, M. C., & Scilingo, A. (2019). Amaranth functional cookies exert potential antithrombotic and antihypertensive activities. *International Journal of Food Science and Technology*, *54*, 1506–1513.

Sahagún, M., Benavent-Gil, Y., Rosell, C. M., & Gómez, M. (2020). Modulation of in vitro digestibility and physical characteristics of protein enriched gluten free breads by defining hydration. *LWT - Food Science and Technology*, *117*, 108642.

Sahagún, M., Bravo-Núñez, Á., Báscones, G., & Gómez, M. (2018). Influence of protein source on the characteristics of gluten-free layer cakes. *LWT - Food Science and Technology*, *94*, 50–56.

Sahagún, M., & Gómez, M. (2018a). Assessing influence of protein source on characteristics of gluten-free breads optimising their hydration level. *Food and Bioprocess Technology*, *11*, 1686–1694.

Sahagún, M., & Gómez, M. (2018b). Influence of protein source on characteristics and quality of gluten-free cookies. *Journal of Food Science and Technology*, *55*, 4131–4138.

Sakiyan, O., Sumnu, G., Sahin, S., & Bayram, G. (2004). Influence of fat content and emulsifier type on the rheological properties of cake batter. *European Food Research and Technology, 219*, 635–638.

Salah, K., Olkhovatov, E. A., & Aïder, M. (2019). Effect of canola proteins on rice flour bread and mathematical modelling of the baking process. *Journal of Food Science and Technology, 56*, 3744–3753.

Sanchez, H. D., Osella, C. A., & de la Torre, M. A. (2002). Optimization of gluten-free bread prepared from cornstarch, rice flour, and cassava starch. *Journal of Food Science, 67*, 416–419.

Sarawong, C., Gutiérrez, Z. R., Berghofer, E., & Schoenlechner, R. (2014). Effect of green plantain flour addition to gluten-free bread on functional bread properties and resistant starch content. *International Journal of Food Science and Technology, 49*, 1825–1833.

Savarese, M., Wismer, W., & Graffigna, G. (2021). Conceptualizing "free-from" food consumption determinants: A systematic integrative literature review focused on gluten and lactose. *Food Quality and Preference, 90*, 104170.

Schober, T. J., Bean, S. R., Boyle, D. L., & Park, S. H. (2008). Improved viscoelastic zein-starch doughs for leavened gluten-free breads: Their rheology and microstructure. *Journal of Cereal Science, 48*, 755–767.

Schober, T. J., Bean, S. R., Tilley, M., Smith, B. M., & Ioerger, B. P. (2011). Impact of different isolation procedures on the functionality of zein and kafirin. *Journal of Cereal Science, 54*, 241–249.

Schober, T. J., Messerschmidt, M., Bean, S. R., Park, S. H., & Arendt, E. K. (2005). Gluten-free bread from sorghum: Quality differences among hybrids. *Cereal Chemistry, 82*, 394–404.

Sciarini, L. S., Bustos, M. C., Vignola, M. B., Paesani, C., Salinas, C. N., & Pérez, G. T. (2017). A study on fibre addition to gluten free bread: Its effects on bread quality and in vitro digestibility. *Journal of Food Science and Technology, 54*, 244–252.

Sciarini, L. S., Ribotta, P. D., León, A. E., & Pérez, G. T. (2012). Incorporation of several additives into gluten free breads: Effect on dough properties and bread quality. *Journal of Food Engineering, 111*, 590–597.

Shah, B. R., Li, B., Al Sabbah, H., Xu, W., & Mráz, J. (2020). Effects of prebiotic dietary fibers and probiotics on human health: With special focus on recent advancement in their encapsulated formulations. *Trends in Food Science and Technology, 102*, 178–192.

Shepherd, S. J., & Gibson, P. R. (2013). Nutritional inadequacies of the gluten-free diet in both recently-diagnosed and long-term patients with coeliac disease. *Journal of Human Nutrition and Dietetics, 26*, 349–358.

Shin, M., Gang, D. O., & Song, J. Y. (2010). Effects of protein and transglutaminase on the preparation of gluten-free rice bread. *Food Science and Biotechnology, 19*, 951–956.

Sivaramakrishnan, H. P., Senge, B., & Chattopadhyay, P. K. (2004). Rheological properties of rice dough for making rice bread. *Journal of Food Engineering, 62*, 37–45.

Sly, A. C., Taylor, J., & Taylor, J. R. N. (2014). Improvement of zein dough characteristics using dilute organic acids. *Journal of Cereal Science, 60*, 157–163.

Smerdel, B., Pollak, L., Novotni, D., Čukelj, N., Benković, M., Lušić, D., et al. (2012). Improvement of gluten-free bread quality using transglutaminase, various extruded flours and protein isolates. *Journal of Food and Nutrition Research, 51*, 242–253.

Smith, B. M., Bean, S. R., Selling, G., Sessa, D., & Aramouni, F. M. (2017). Effect of salt and ethanol addition on zein–starch dough and bread quality. *Journal of Food Science, 82*, 613–621.

Smith, P. R., & Johansson, J. (2004). Influences of the proportion of solid fat in a shortening on loaf volume and staling of bread. *Journal of Food Processing and Preservation, 28*, 359–367.

Storck, C. R., da Rosa Zavareze, E., Gularte, M. A., Elias, M. C., Rosell, C. M., & Guerra Dias, A. R. (2013). Protein enrichment and its effects on gluten-free bread characteristics. *LWT - Food Science and Technology, 53*, 346–354.

Sulieman, A. A., Zhu, K. X., Peng, W., Hassan, H. A., Obadi, M., Siddeeg, A., et al. (2019). Rheological and quality characteristics of composite gluten-free dough and biscuits supplemented with fermented and unfermented Agaricus bisporus polysaccharide flour. *Food Chemistry, 271*, 193–203.

Sumnu, G., Koksel, F., Sahin, S., Basman, A., & Meda, V. (2010). The effects of xanthan and guar gums on staling of gluten-free rice cakes baked in different ovens. *International Journal of Food Science and Technology, 45*, 87–93.

Taylor, J. R. N., Schober, T. J., & Bean, S. R. (2006). Novel food and non-food uses for sorghum and millets. *Journal of Cereal Science, 44*, 252–271.

Taylor, J. R. N., Taylor, J., Campanella, O. H., & Hamaker, B. R. (2016). Functionality of the storage proteins in gluten-free cereals and pseudocereals in dough systems. *Journal of Cereal Science, 67*, 22–34.

Torra, M., Belorio, M., Ayuso, M., Carocho, M., Ferreira, I. C. F. R., Barros, L., et al. (2021). Chickpea and chestnut flours as non-gluten alternatives in cookies. *Food, 10*, 911.

Torres, M. D., Arufe, S., Chenlo, F., & Moreira, R. (2017). Coeliacs cannot live by gluten-free bread alone—Every once in awhile they need antioxidants. *International Journal of Food Science and Technology, 52*, 81–90. https://doi.org/10.1111/ijfs.13287.

Tres, A., Tarnovska, N., Varona, E., Quintanilla-Casas, B., Vichi, S., Gibert, A., et al. (2020). Determination and comparison of the lipid profile and sodium content of gluten-free and gluten-containing breads from the spanish market. *Plant Foods for Human Nutrition, 75*, 344–354.

Tsatsaragkou, K., Papantoniou, M., & Mandala, I. (2015). Rheological, physical, and sensory attributes of gluten-free rice cakes containing resistant starch. *Journal of Food Science, 80*, E341–E348.

Tsatsaragkou, K., Protonotariou, S., & Mandala, I. (2016). Structural role of fibre addition to increase knowledge of non-gluten bread. *Journal of Cereal Science, 67*, 58–67.

Turabi, E., Sumnu, G., & Sahin, S. (2010). Quantitative analysis of macro and micro-structure of gluten-free rice cakes containing different types of gums baked in different ovens. *Food Hydrocolloids, 24*, 755–762.

van Riemsdijk, L. E., Pelgrom, P. J. M., van der Goot, A. J., Boom, R. M., & Hamer, R. J. (2011). A novel method to prepare gluten-free dough using a meso-structured whey protein particle system. *Journal of Cereal Science, 53*, 133–138.

van Riemsdijk, L. E., van der Goot, A. J., & Hamer, R. J. (2011). The use of whey protein particles in gluten-free bread production, the effect of particle stability. *Food Hydrocolloids, 25*, 1744–1750.

van Riemsdijk, L. E., Van der Goot, A. J., Hamer, R. J., & Boom, R. M. (2011). Preparation of gluten-free bread using a meso-structured whey protein particle system. *Journal of Cereal Science, 53*, 355–361.

Vanin, F. M., Lucas, T., & Trystram, G. (2009). Crust formation and its role during bread baking. *Trends in Food Science and Technology, 20*, 333–343.

Vici, G., Belli, L., Biondi, M., & Polzonetti, V. (2016). Gluten free diet and nutrient deficiencies: A review. *Clinical Nutrition, 35*, 1236–1241.

Viebke, C., Al-Assaf, S., & Phillips, G. O. (2014). Food hydrocolloids and health claims. *Bioactive Carbohydrates and Dietary Fibre, 4*, 101–114.

Wärnberg, J., Marcos, A., Bueno, G., & Moreno, L. A. (2009). Functional benefits of psyllium fiber supplementation. *Current Topics in Nutraceutical Research, 7*, 55–63.

Wilderjans, E., Luyts, A., Brijs, K., & Delcour, J. A. (2013). Ingredient functionality in batter type cake making. *Trends in Food Science and Technology, 30*, 6–15.

Wilderjans, E., Luyts, A., Goesaert, H., Brijs, K., & Delcour, J. A. (2010). A model approach to starch and protein functionality in a pound cake system. *Food Chemistry, 120*, 44–51.
Wilderjans, E., Pareyt, B., Goesaert, H., Brijs, K., & Delcour, J. A. (2008). The role of gluten in a pound cake system: A model approach based on gluten-starch blends. *Food Chemistry, 110*, 909–915.
Witczak, M., Juszczak, L., Ziobro, R., & Korus, J. (2012). Influence of modified starches on properties of gluten-free dough and bread. Part I: Rheological and thermal properties of gluten-free dough. *Food Hydrocolloids, 28*, 353–360.
Witczak, T., Juszczak, L., Ziobro, R., & Korus, J. (2017). Rheology of gluten-free dough and physical characteristics of bread with potato protein. *Journal of Food Process Engineering, 40*, e12491.
Witczak, M., Korus, J., Ziobro, R., & Juszczak, L. (2019). Waxy starch as dough component and anti-staling agent in gluten-free bread. *LWT - Food Science and Technology, 99*, 476–482.
Xhakollari, V., Canavari, M., & Osman, M. (2021). Why people follow a gluten-free diet? An application of health behaviour models. *Appetite, 161*, 105136.
Xu, J., Zhang, Y., Wang, W., & Li, Y. (2020). Advanced properties of gluten-free cookies, cakes, and crackers: A review. *Trends in Food Science and Technology, 103*, 200–213.
Yamamoto, H., Worthington, S. T., Hou, G., & Ng, P. K. W. (1996). Rheological properties and baking qualities of selected soft wheats grown in the United States. *Cereal Chemistry, 73*, 215–221.
Zannini, E., Pontonio, E., Waters, D. M., & Arendt, E. K. (2012). Applications of microbial fermentations for production of gluten-free products and perspectives. *Applied Microbiology and Biotechnology, 93*, 473–485.
Zhang, J., Wen, C., Zhang, H., & Duan, Y. (2019). Review of isolation, structural properties, chain conformation, and bioactivities of psyllium polysaccharides. *International Journal of Biological Macromolecules, 139*, 409–420.
Zhu, F. (2017). Properties and food uses of chestnut flour and starch. *Food and Bioprocess Technology, 10*, 1173–1191.
Zhu, F. (2018). Chemical composition and food uses of teff (Eragrostis tef). *Food Chemistry, 239*, 402–415.
Ziobro, R., Juszczak, L., Witczak, M., & Korus, J. (2016). Non-gluten proteins as structure forming agents in gluten free bread. *Journal of Food Science and Technology, 53*, 571–580.
Ziobro, R., Korus, J., Juszczak, L., & Witczak, T. (2013). Influence of inulin on physical characteristics and staling rate of gluten-free bread. *Journal of Food Engineering, 116*, 21–27.
Ziobro, R., Korus, J., Witczak, M., & Juszczak, L. (2012). Influence of modified starches on properties of gluten-free dough and bread. Part II: Quality and staling of gluten-free bread. *Food Hydrocolloids, 29*, 68–74.
Ziobro, R., Witczak, T., Juszczak, L., & Korus, J. (2013). Supplementation of gluten-free bread with non-gluten proteins. Effect on dough rheological properties and bread characteristic. *Food Hydrocolloids, 32*, 213–220.
Zoghi, A., Mirmahdi, R. S., & Mohammadi, M. (2020). The role of hydrocolloids in the development of gluten-free cereal-based products for coeliac patients: A review. *International Journal of Food Science and Technology*.
Zorzi, C. Z., Garske, R. P., Flôres, S. H., & Thys, R. C. S. (2020). Sunflower protein concentrate: A possible and beneficial ingredient for gluten-free bread. *Innovative Food Science and Emerging Technologies, 66*, 102539.

CHAPTER SIX

Enhancing health benefits of bakery products using phytochemicals

Jing Gao, Audrey Hui Si Koh, and Weibiao Zhou*
Department of Food Science and Technology, National University of Singapore, Singapore, Singapore
*Corresponding author: e-mail address: weibiao@nus.edu.sg

Contents

Abstract

There has been a growing interest in functional bakery products with enhanced health benefits, especially the prevention of some chronic diseases such as type 2 diabetes, cardiovascular diseases and neurodegenerative disorders. Fortification of wheat flour with phytochemicals, plant components with various bio-activities, is one of the promising approaches to improving public health with the ubiquitous consumption of baked

Advances in Food and Nutrition Research, Volume 99
ISSN 1043-4526
https://doi.org/10.1016/bs.afnr.2021.12.002

goods. This chapter reviews the current knowledge of several representative phytochemicals, mainly plant polyphenols, including catechins, anthocyanins, fucoidan and quercetin extracted from various plant resources, and their application in bakery products, regarding their stability, impact on product quality and potential health benefits.

Abbreviations

EC (-)-epicatechin
ECG (-)-epicatechin gallate
EGC (-)-epigallocatechin
EGCG (-)-epigallocatechin gallate
GI glycemic index
GPE grape pomace extract
GPT grape pomace powder
GTE green tea extract
GTP green tea powder
iAUC incremental area under the curve
TAC total anthocyanin content
TPC total phenolic content

1. Introduction

Functional food is defined as "a food or food ingredient that provides a health benefit beyond satisfying traditional nutritional requirement" (Sanders, 1998). There is an increasing demand of functional food products due to their healthier appeal to the consumers. Phytochemicals of plants are well known to have nutritional functions as well as medicinal benefits that help reduce the risk of numerous diseases. For the past decades, development of new bakery products enriched with health-promoting bioactive compounds, such as phytochemicals, offers new visions and opportunities for alternative strategies in food production, supply and consumption. However, the stability of phytochemicals as well as their interactions with the food matrix should be taken carefully into consideration for commercial application as functional ingredients. It is also important to show how functional ingredients and foods are actually digested and metabolized in order to substantiate their health benefits. There is a need for a multidimensional and multidisciplinary approach when it comes to connecting foods and their functionality. This chapter discusses the development of functional bakery products using four types of phytochemicals, namely green tea catechins, anthocyanins, fucoidan and quercetin, from the angles of their application, stability, as well as their impact on product quality and potential health benefits.

2. Catechins from green tea

For years, green tea (*Camellia sinensis*) has been widely known and consumed for its palatability and medicinal values. The health benefits of green tea, such as natural antioxidant, anti-inflammatory, antiarteriosclerotic and anticarcinogenic effects, are attributed to its high amount of polyphenolic compounds, among which catechins are the major ones (~75%) (Braicu, Ladomery, Chedea, Irimie, & Berindan-Neagoe, 2013; Khan & Mukhtar, 2013). Green tea extract (GTE) contains mainly four types of catechins, namely (-)-epigallocatechin (EGC), (-)-epigallocatechin gallate (EGCG), (-)-epicatechin (EC), and (-)-epicatechin gallate (ECG) (Wang & Zhou, 2004). As shown in Fig. 1, the presence of hydroxyl groups in catechin molecules gives them strong antioxidant activities. Among all the catechins, (-)-EGCG is the most abundant (7–74 mg/g) and with the most biological activities (Friedman, Levin, Choi, Kozukue, & Kozukue, 2006), and therefore, has been used as the quality indicator of green tea products.

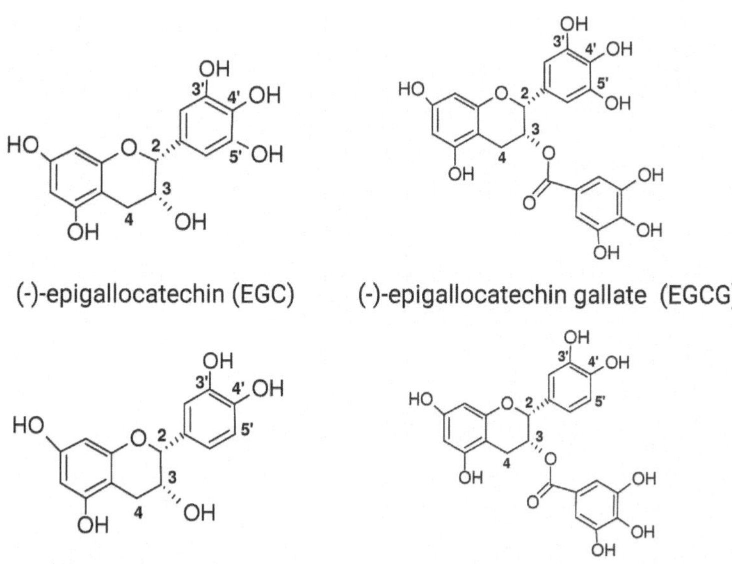

Fig. 1 Chemical structure of four major types of catechins in green tea extract.

2.1 Green tea catechins in bakery products

Green tea catechins are considered as Generally Recognized as Safe (GRAS) by the U.S. Food and Drug Administration (FDA) (Perumalla & Hettiarachchy, 2011). The addition of green tea catechins in bakery products combines the flavor of bakery goods and the functionality of tea. Both green tea powder (GTP) and green tea extract (GTE) have been used in developing tea-enhanced bakery products. GTP is prepared from dry green tea leaves, containing 8%–30% catechins and carbohydrates (e.g., cellulosic fiber) (Yilmaz, 2006). GTE is prepared by spray drying the strong green tea infusion, which contains 40%–50% solids, and the final GTE powder contains 29%–80% catechins (Wang, Provan, & Helliwell, 2000). The contents of (−)-EGCG and (−)-ECG are the indicator of the quality of GTE (Sharma & Zhou, 2011). GTE has a characteristic taste that combines bitterness, astringency, umaminess, sweetness and slight sourness (Namal Senanayake, 2013).

As shown in Fig. 2, is the fortification of GTP or GTE in bakery products leads to distinct differences in the color of final product. GTP imparts a greenish hue while GTE imparts a reddish hue to the final products due to the color of the powder itself. Table 1 lists the details of applications of tea catechins in white bread (Goh et al., 2015; Wang & Zhou, 2004), whole-wheat flour pan bread (Ning et al., 2017), rye bread (Bajerska et al., 2010), biscuit (Ahmad et al., 2015; Mildner-Szkudlarz et al., 2009; Phongnarisorn, Orfila, Holmes, & Marshall, 2018; Sharma & Zhou, 2011) and cake (Lu et al., 2010; Mau et al., 2015).

Lu et al. (2010) substituted 10%, 20% and 30% of wheat flour with GTP in sponge cake and found no difference in the hedonic sensory score up to 20% substitution. On the other hand, GTE powder containing 60% catechins could only be incorporated at 0.15%–0.5% of flour basis in baked bread and 0.15%–0.3% in biscuit, beyond which it caused deteriorations in product quality (Sharma & Zhou, 2011; Wang et al., 2007). The application of GTE was in a much lower dosage as compared to the GTP, which might be due to the higher content of catechins in GTE.

2.2 Stability of green tea catechins during baking

Tea catechins are thermo-sensitive and highly unstable under alkaline conditions (Massounga Bora, Ma, Li, & Liu, 2018). When added into bread dough, GTE was stable and 94% of (-)-EGCG and (-)-ECG were detected after a 9-week storage at −20 °C (Wang & Zhou, 2004). Moreover, there was no significant difference between unfrozen and frozen dough in terms of catechin retention up to 0.15% fortification on flour basis.

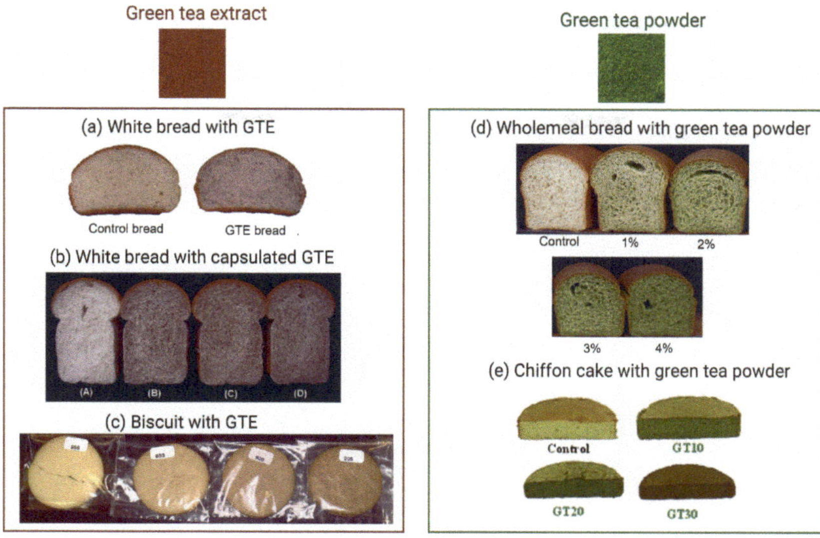

Fig. 2 Functional bakery products developed using green tea powder (GTP) and green tea extract (GTE): (a) white bread prepared using GTE powder at 0 and 0.15 g/100 g of flour (Wang, Zhou, & Isabelle, 2007); (b) white bread prepared (A) without any GTE powder, (B) bread with unencapsulated GTE, (C) bread with freeze dried encapsulated GTE, and (D) bread with spray dried encapsulated GTE (Pasrija, Ezhilarasi, Indrani, & Anandharamakrishnan, 2015); (c) biscuit dough and biscuit prepared 0, 0.2, 0.4, 0.6 g GTE/100 g flour (from left to right) (Amber, 2013; Sharma, 2013); (d) whole-wheat flour pan bread prepared using green tea powder at 1, 2, 3, and 4 g/100 g of flour (Ning, Hou, Sun, Wan, & Dubat, 2017); and (e) chiffon cake developed using green tea powder at the substitution level of 0 (control), 10 (GT10), 20 (GT20) and 30% (GT30) of flour (Mau et al., 2015).

GTE fortified bread (50–150 mg GTE/100 g flour) showed a 84% retention of catechin after baking at 215 °C for 11 min (Wang & Zhou, 2004). In another study of bread fortified with 450–2000 mg GTE/100 g flour, the retention level was 75%–89% after baking at 200 °C for 10 min (Goh et al., 2015). For biscuit, GTE was added at levels of 150, 200, and 300 mg per 100 g of flour. The retention of (–)-EGCG and (–)-ECG was 21% and 30%, respectively, after baking at 160 °C for 10 min (Sharma & Zhou, 2011). The overall stability order of tea catechins in bread was (–)-EC ≈ (–)-CG > (–)-ECG > (–)-GCG > (–)-EGCG > (–)-EGC (Wang & Zhou, 2004). Similarly, the order of stability in biscuit was (–)-CG > (–)-GCG > (–)-ECG > (–)-EGCG (Sharma & Zhou, 2011). In both bread and biscuit matrices, (–)-EGCG and (–)-EGC were more susceptible to degradation than (–)-ECG and (–)-EC.

Table 1 Fortification of green tea catechins in bakery products.

Bakery products	Fortification material	Fortification levels	Impact on quality/ sensory	Recommended addition level	References
White bread	GTE (60% total tea catechins)	0.15 and 0.5 g/100 g flour	• Brightness and sweetness of bread decreased • Hardness, stickiness and astringency increased	• <5 g/kg for astringency and sweetness • <1.5 g/kg for brightness, hardness and stickiness	Wang et al. (2007)
Wholemeal bread	Green tea powder (GTP)	1, 2, 3, and 4 g/100 g flour	• Bread volume, L, a, b values decreased • Pore size, hardness, chewiness increased	1/100 g flour	Ning et al. (2017)
Rye bread	Aqueous GTE (total catechin $\geq$15%, EGCG $\geq$9%)	0.5, 0.8, and 1.1 g/100 g flour	Sensory score for aroma, flavor and overall acceptability decreased significantly at 1.1% addition level	0.8%	Bajerska, Mildner-Szkudlarz, Jeszka, and Szwengiel (2010)

Biscuit	GTE (total catechin ≥50%, EGCG ≥ 25%)	0.2, 0.4, 0.6, and 0.8 g/100 g flour	• Hardness, stickiness and density increased • Fracturability and thickness decreased	N.A.	Sharma (2013)
	GTE (136 mg gallic acid equivalent/g dry leaves)	0.02%, 0.12% and 1% (lipid weigh basis)	• Increased lipid stability • Higher acceptance scores as compared to samples with BHA	1% for lipid stability	Mildner-Szkudlarz, Zawirska-Wojtasiak, Obuchowski, and Gośliński (2009)
	Grounded green tea leaves	1, 2, 4 g/100 g flour	• Reduced retrogradation, L and b values of color, and structure integrity • Reduced acceptability of color, aroma and taste	Less than 1% addition	Ahmad et al. (2015)
Sponge cake	Grounded green tea leaves	Replace 10%, 20% and 30% wheat flour with tea powder	• Increased cake hardness, chewiness, and crumb *a* value • Decreased volume, and crust L, a, b values	Up to 20% substitution	Lu, Lee, Mau, and Lin (2010)

The stability of catechins demonstrated a dosage-dependent behavior that a higher concentration of GTE led to a higher retention level after baking (Sharma & Zhou, 2011; Wang & Zhou, 2004). The loss of tea catechins during baking is mainly attributed to thermal degradation, oxidation and epimerization. The epimerization depends strongly on pH and temperature: the higher the pH and temperature, the less stable the tea catechins. The high temperature in bread crumb during baking (i.e., core temperature at 80–100 °C for 8–9 min when baked at 215 °C for 11 min) offered sufficient energy for the epimerization of tea catechins (Wang & Zhou, 2004). A similar epimerization reaction was found in biscuit during the biscuit-making process where the amount of (−)-CG in the biscuit was more than initially added in making the dough (Sharma & Zhou, 2011).

The low-acidic environment of bread dough (pH of 5–6) was favorable for the epimerization reaction. On the other hand, biscuit contains alkaline ingredients, such as sodium bicarbonate, ammonium bicarbonate and baking powder. The loss of (–)-EGCG was partly attributed to degradation in biscuit due to high pH, and not epimerization (Sharma & Zhou, 2011). The degradation of catechins due to the high pH in biscuit dough makes the fortification of catechins even more challenging. Moreover, biscuit is higher in oxidation potential as compared to the bread. It is well known that yeast immediately consumes oxygen during the proofing process, hence minimizing the opportunity of oxidation of catechins in bread. On the other hand, there is an ample amount of active oxygen for oxidation in biscuit due to the absence of yeast. Microencapsulation of green tea catechins could be a solution to increase their stability in biscuit and other similar bakery products.

A mathematical model combining both thermal degradation and epimerization was established to predict the stability of (−)-EGCE under various breadmaking conditions (Wang, Zhang, Zhang, & Li, 2008; Wang, Zhou, & Jiang, 2008). Epimerization and degradation of tea catechins in bread followed pseudo first-order reactions and the rate constants of reaction kinetics followed the Arrhenius equation. In the bread system, activation energy (*Ea*) for epimerization of (-)-EGCG to (-)-GCG and for (-)-GCG to (-)-EGCG was similar to those in aqueous solution, which were 105.1 and 84.3 kJ/mol respectively.

2.3 Impact of green tea catechins on product quality

2.3.1 Impact on product color

The fortification of either GTP or GTE causes a color change in the bakery products, which can be measured spectrophotometrically using Hunter L^*

a* b* values. The brightness of bread decreased in a dose-dependent manner with the addition of 0.05%–0.15% GTE, indicating a more intense color of the bread with a higher level of GTE (Wang et al., 2007). Similarly, the brightness of pan bread decreased from 80 to 45 with an addition of 4/100 g GTP (Ning et al., 2017). It is interesting to note that the brightness value of GTE bread was comparable to commercial wholemeal bread and fiber-enriched bread. Hence, consumer acceptance may not be an issue as regards the color of GTE fortified bread at the level up to 0.5%.

There was a significant increase in the red tone when the amount of GTE was increased from 0.15% to 0.50% of flour weight (Wang et al., 2007) (Fig. 2a). In comparison, Ning et al. (2017) observed a reduction of *a** value from 5.5 to almost 0 with an addition of 2 g GTP/100 g flour, which indicates that bread crumb became greenish in color (Fig. 2d). A similar result was observed in the study of chiffon cake fortified with GTP (Mau et al., 2015) (Fig. 2c). This greenish color was attributed to the 1 g chlorophyll found in every 100 g of green tea powder. Chlorophyll broke down during heating to magnesium chlorophyll which imparts a brown color. For cookies, there was no significant change in *L** a* b* values when GTP was added at the level of 1 to 2% of flour weight (Ahmad et al., 2015). Hence, color change in the final products depends on raw material (GTE or GTP), its addition level, and the food matrix applied.

Microencapsulates incorporated bread exhibited slightly improved color analysis results such as a higher *L** value and a lower *a** value than the GTE incorporated bread (Pasrija et al., 2015) (Fig. 2b). The encapsulation of GTE reduced the direct exposure of bread ingredients to GTE. Nevertheless, the color of bread containing microencapsulated GTE was still distinctly different from the control bread based on visual inspection.

2.3.2 Impact on product specific volume and texture

2.3.2.1 Bread

In white bread, fortification of GTE had a negative effect on bread volume from 0.05% dosage level onwards (Wang, Zhou, Yu, & Chow, 2006). Specially, the specific volume of bread was reduced by 16.6% when GTE was added at 0.15% of flour weight (Wang et al., 2006). Moreover, baked bread fortified with 0.1% GTE had a 27.7% increasement in firmness as compared to the control (Wang et al., 2006). A similar negative impact on the volume and texture of whole-wheat flour pan bread was also reported (Ning et al., 2017), where 2 g GTP/100 g flour addition led to a significant volume reduction. An addition of GTP at the level of 3 and 4 g/100 g flour caused an increase of 25% and 35%, respectively, in the hardness of whole-wheat bread (Ning et al., 2017). Microencapsulation of green tea catechins (1.36–2.31 mg/g powder) could minimize the impact of green tea polyphenol on the quality of bread, leading to no significant change in bread volume or crumb firmness when 1.5–2.5 g of microencapsulated powders were added per loaf (100 g flour) (Pasrija et al., 2015).

A well-developed gluten network is responsible for gas retention during proofing and baking. The disulfide bonds of the gluten network are formed through a SH-SS interchange reaction during dough mixing (Belton, 1999). Tea catechins interfere with the gluten network formation during dough mixing. As phenolic compounds, tea catechins can negatively affect the SH-SS interchange reaction by scavenging the free radical (GS*) initiated by the scission of disulfide bonds, which leads to excessive SH groups and inadequate disulfide bonds (Wang & Zhou, 2004).

Moreover, catechins can interact with other ingredients and cause the alternation of bread structure and quality. For example, tea polyphenols interacted with starch and proteins via hydrophobic interactions and hydrogen bonding to change the bread structure (Sivam, Sun-Waterhouse, Perera, & Waterhouse, 2013). Also, the increased cellulose content in dough, due to the incorporation of GTP, diluted the gluten and altered dough water absorption, resulting in reduced water availability to gluten and poor viscoelastic properties of dough (Ning et al., 2017). Nevertheless, the study of Ananingsih, Gao, and Zhou (2013) showed that the gassing power of yeast was not affected by GTE based on Rheofermentometer analysis. Hence, it is reasonable to suggest that volume reduction was more related to adverse changes in dough structure than gas production.

Overall, fortification of catechins showed adverse effects on the bread quality, producing a smaller bread with a firmer crumb texture (Ning et al., 2017; Wang et al., 2006). For GTE, a fortification level at less than 0.1/100 g flour would not cause significant changes in the volume and texture of bread (Wang et al., 2006). For GTP, bread quality was not significantly affected at the dosage of 1/100 g flour (Ning et al., 2017). Microencapsulation is a promising way to incorporate green tea catechins at a higher dosage without adversely affecting bread quality.

2.3.2.2 Cake

The batter's apparent viscosity (measured at 30 rpm in a rotational viscometer) of sponge cake and chiffon cake significantly increased with the addition of GTP, i.e., from 5308 to 22,206 cp for sponge cake and from 12,390 to 14,739 cp for chiffon cake both at 30% substitution level, respectively (Lu et al., 2010; Mau et al., 2015). The volume of the sponge cake decreased from 1872 to 1158 mL (a 38% reduction) (Lu et al., 2010) and that of the chiffon cake decreased from 2166 to 1592 mL (a 26% reduction) at 30% substitution (Mau et al., 2015). With the substitution of flour with GTP at the level of 30%, there was a 88% increase in hardness and a 21% reduction in cohesiveness for the sponge cake (Lu et al., 2010); and a 84% increase in the hardness and a 14% reduction in the cohesiveness of chiffon cake (Mau et al., 2015). Similar to bread, the texture of cake is closely related to its volume. The adverse effect of GTP might be explained by the dilution effect of cellulose on weakening the gluten matrix that is responsible for gas retention in bakery products.

2.3.2.3 Cookie/biscuit

Hardness and fracturability are the important quality characteristics of biscuits and cookies, which influence the texture perception during mastication. The hardness of cookie is closely linked to the interaction of protein and starch by hydrogen bonding (Hoseney & Rogers, 1994). There was a 38% reduction in the force to break/snap the cookies as the level of GTP fortification increased from 0% to 4% (Ahmad et al., 2015). Phongnarisorn et al. (2018) also reported an increase in biscuit hardness with the addition of matcha green tea powder (6/100 g flour). Moreover, the addition of GTE in biscuit (2/100 g flour) caused a significant reduction in the fracturability of the biscuit based on the three-point-bending test (Sharma, 2013). The fracturability of a biscuit is mainly determined by the recrystallization of sugar from the continuous phase after baking and cooling. Recrystallisation of sugar is affected by many factors, one being the level of impurities in the continuous phase. It could be postulated that catechins in GTE, being water-soluble, could act as an "impurity" in the continuous phase, inhibiting proper crystallization of sugar and thus lowering the fracturability of GTE fortified biscuits.

2.4 Impact on product sensory properties

2.4.1 Bread

The marketing success of a functional bakery product greatly depends on consumers' acceptance of its taste, aroma and texture. Both instrumental analysis and sensory evaluations with trained and untrained panels were conducted to quantify the changes in porosity, brightness, hardness, stickiness, astringency and sweetness of baked bread fortified with 0.15% and 0.50%

GTE (Wang et al., 2007). The reduction in bread brightness due to GTE fortification was perceived by both trained and untrained panels.

The untrained panel did not detect any changes in the bread texture, while the trained panel considered GTE fortified bread as harder and stickier in texture. Overall, threshold levels for the effect of GTE fortification on the bread sensory properties to be significant were identified by the trained panel to be 0.50% for astringency and sweetness, and 0.15% for brightness, hardness, and stickiness. Below the threshold values, the sensory quality of the bread fortified with GTE was not significantly compromised. This offers bread manufacturers a good guideline for producing GTE fortified white bread.

For rye bread, aroma, flavor and overall acceptability were evaluated on a 10-point hedonic scale (Bajerska et al., 2010). The addition of GTE up to 1.1% (flour base) did not affect the scores of color and texture attributes. The dark color of rye bread masked the color brought by green tea catechins. On the other hand, scores for the flavor and overall acceptability decreased significantly when the bread was prepared with 1.1% GTE. The recommended usage of GTE in rye bread was 0.8% without adversely affecting the sensory scores in aroma, color, texture, flavor and overall acceptability. The higher dosage applicable in rye bread as compared to white bread might be due to the masking effect of flavor bouquets produced during the sourdough fermentation process (Bajerska et al., 2010).

2.4.2 Cake

For sponge cake, 86 untrained consumers were recruited to rate their degree of liking on a 7-point hedonic scale (Lu et al., 2010). Substitution of wheat flour with GTP at 10% and 20% levels did not affect the scores of color, sweetness, flavor, texture and overall liking of the sponge cake. However, sponge cake formulated by replacing cake flour with 30% GTP had the lowest sensory score for color, sweetness, flavor and texture due to an increase in bitterness. Both tea catechins and caffeine of GTP contribute to the bitter taste. The authors suggested a maximum of 20% substitution of GTP in order to make the sponge cake with a satisfactory sensory profile. In a similar study, untrained consumers ($n = 54$) were used to study the overall preference of chiffon cake on a 7-point hedonic scale. No difference was found in hedonic sensory results for the control, 10% and 20% GTP substituted cakes, whereas 30% substituted cakes were rated lower in all sensory results (Mau et al., 2015). Hence, 20% substitution seems to be the highest level recommended for GTP fortification in cakes.

2.4.3 Cookie/biscuit

The addition of 4% (flour weight) GTP in cookie significantly increased the scores of color, aroma and taste as compared to the control in sensory analysis (Ahmad et al., 2015). The preferred flavor profile was likely due to the presence of green tea volatiles and polyphenols. On the other hand, the scores of appearance and texture reduced significantly with the addition of GTP even at the lowest level (1% flour weight). As the overall liking was the lowest for cookies with 4% GTP, this dosage was not recommended in formulating GTP enriched cookies.

Gramza-Michałowska et al. (2016) incorporated 5.5% grounded green tea leaves into cookie. The scores of taste, crispness and overall acceptance were lower for the fortified cookie as compared to the control. The green/-gray color and grassy aftertaste were the major concerns of the untrained consumer panel ($n=30$). Furthermore, Gómez-Mascaraque et al. (2017) studied the impact of adding free or microencapsulated GTE on biscuit using 214 untrained consumers. There was no difference in the overall liking, texture, or flavor liking of the control, free GTE and microencapsulated GTE biscuits. As there was only one dosage of GTE incorporation, no threshold could be established from such study. Nevertheless, it was suggested that GTE could be added without affecting the organoleptic qualities of biscuit.

In another study, the optimal combination of Matcha green tea powder (MGTP, 10% total catechin) and sugar content in biscuit was explored using the response surface method (Phongnarisorn et al., 2018). It was found that low MGTP and low sugar content received the highest acceptability. In particular, sweetness acceptability was significantly affected by the level of MGTP with greater acceptability at higher sugar level (30/100g flour) across all the MGTP levels (2, 4 and 6g/100g flour). The optimal combination was calculated to be 1.84g MGTP/100g flour with 27.73g sugar/100g flour.

In summary, color, bitterness/sweetness and texture are the most important determinants of the acceptability of green tea catechin fortified bakery products. The harder texture and the bitterness and astringency brought by green tea polyphenols such as catechins represent a major challenge for their incorporation into bakery products. Therefore, determining the threshold levels for fortification, below which adverse effects on organoleptic properties become undetectable, holds the key to commercial success of the fortified bakery products.

2.5 Health benefits of fortifying green tea catechins in bakery products

2.5.1 Antioxidant activity

Catechins are known of their antioxidant activity. The presence of O-dihydroxy and O-hydroxyketo groups and C2-C3 double bonds are related to the free radical scavenging activity of catechins (Rice-Evans, Miller, & Paganga, 1997). (−)-EGCG has the highest scavenging activity of free radicals, followed by (−)-ECG, (−)-EC and (−)-EGC (Hu, Yang, & Liu, 2001). The addition of GTP and GTE significant increased the antioxidant capability of bread (Bajerska et al., 2010; Ning et al., 2017), cookies (Ahmad et al., 2015; Mildner-Szkudlarz et al., 2009; Zhang, Chen, & Wang, 2014) and cake (Lu et al., 2010; Mau et al., 2015). Acting as a natural antioxidant, tea catechins can protect human body from oxidative stress, which is one of the main contributors to chronic inflammatory reaction, cardiovascular diseases and many more.

2.5.2 Carbohydrate digestion

Tea catechins demonstrate an inhibitory activity against α-amylase and α-glucosidase, which are the two main enzymes for starch digestion and associated with postprandial hyperglycemia (Yilmazer-Musa, Griffith, Michels, Schneider, & Frei, 2012). Tea polyphenols interact with digestive enzymes through hydrogen bonding and hydrophobic interactions (Amoako & Awika, 2016). Goh et al. (2015) explored the possibility of using GTE to develop bread with a low glycaemic index (GI). Based on an in vitro digestion study, it was shown that GTE fortification at 2% of flour weight significantly reduced the rapidly digested starch (RDS) content in both baked bread and steamed bread, i.e., 8.3%–14.9% for crust and 5.4%–14.7% for crumb. Its potential in GI reduction is comparable to reported studies of reformulation (e.g., increasing amylose-lipid complexes by using coconut oil) (Lau, Zhou, & Henry, 2016) or altering processing conditions (e.g., switch from baking to steaming) (Lau, Soong, Zhou, & Henry, 2015). Hence, it is promising to develop low GI bread with a proper fortification of tea catechins.

Nevertheless, based on the limited number of in vivo studies, the glycemic lowering benefits of green tea catechins are not obvious. White bread with 0.4% GTE did not reduce glycemic response, insulin response or satiety (Coe & Ryan, 2016). Phongnarisorn et al. (2018) showed that the capillary blood glucose response (iAUC) after consuming Matcha GTP-enriched biscuits (6 g GTP/100 g flour) was slightly lower than that of the control biscuit

(89.2 mmol·min/L vs 106 mmol·min/L) without statistical significance. The response was the same among treatments within the first 30 min, suggesting that green tea catechins do not affect rapidly available glucose. A significant difference was only observed at 120 min and 150 min, suggesting that catechins have a mild effect on starch digestion. Authors speculated that the high fat content of shortbread biscuit masked the difference in starch digestion, which works by slowing gastric emptying. Further studies on the glycemic response of catechins-fortified bread and other bakery products are necessary to validate the potential of catechins in formulating low GI bakery products.

2.5.3 Fat digestion

Tea catechins inhibit gastric and pancreatic lipases, which helps to reduce the rate of fat digestion. Incubation of excessive amount of gastric and intestinal lipase with GTE (25% catechin content) showed a drastic decrease in gastric lipolysis but limited effect on pancreatic lipolysis (Juhel et al., 2000). Furthermore, (−)-GCG was shown to enlarge fat droplets through interference with emulsification and micellar solubility of lipids. The enlarged emulsion droplets could prevent the efficient emulsification of bile salt and reduce the surface area for fat digestion by lipase (Koo & Noh, 2007; Raederstorff, Schlachter, Elste, & Weber, 2003). The available evidence strongly suggests that green tea catechins have an anti-obesity effect by lowering both the absorption of dietary fat and the accumulation of body fat mass.

After consuming a standardized light meal consisting of a piece of bread (70 g) spread with 20 g butter and 10 (control), 224 (moderate dose) and 647 mg (high dose) tea catechins, subjects with mild or borderline hypertriacylglycerolemia showed a reduced incremental area (iAUC) under the plasma triacylglycerol curves by 15.1% and 28.7%, respectively (Unno et al., 2005). Phongnarisorn et al. (2018) measured the concentration of triglycerides after the consumption of biscuits fortified with matcha GTP (2–6 g/100 g flour). However, a statistically insignificant reduction in iAUC was observed for biscuit fortified with 6 g match GTP (35.6 mmol·min/L) as compared to control biscuit (44.6 mmol·min/L). This could be explained by the low catechin content in the matcha tea powder (~10% total catechins). Studies using GTE powder with higher content of catechins are required to properly assess the efficacy of tea catechins in developing anti-obesity functional bakery products.

2.5.4 Bioavailability of catechins

Studies showed that daily consumption of 400–600 mg of catechins (equivalent to 4–6 cups of green tea) and 115–300 mg of (−)-EGCG could be beneficial for healthy adults (Sanlier, Gokcen, & Altuğ, 2018). However, (−)-EGCG is unstable throughout the digestion process and has a poor small intestinal permeability and absorption (Rashidinejad et al., 2021). The bioavailability of green tea flavan-3-ols was reported to be 39% by Del Rio et al. (2010). The fraction of catechins that cannot be absorbed in the small intestine undergoes microbial fermentation in the large intestine, where they are transformed into smaller molecules that could be absorbed in large intestine (Sang, Lee, Yang, Buckley, & Yang, 2008). Nanoencapsulation is proposed to control and prolong the release of catechins and to improve their bioavailability and bioaccessibility (Dai et al., 2020). Also, co-administration with sucrose and ascorbic acid increased the bioavailability of catechins (Peters, Green, Janle, & Ferruzzi, 2010). These novel approaches could be applied to enhance the health benefits of catechin-enriched bakery products with a minimum level of fortification.

3. Anthocyanins

Anthocyanins, a subgroup of water-soluble flavonoids, are natural pigments for red, blue and purple colors of various fruits, flowers and leaves (Guo & Xia, 2018). There are more than 30 anthocyanidins that have been identified, among which six are widely distributed in nature: cyanidin, delphinidin, petunidin, peonidin, pelargonidin, and malvidin (Andersen & Markham, 2005). As shown in Fig. 3, anthocyanins have a glycosidic structure where anthocyanidins are bond to one or more sugar molecules (e.g., glucose, galactose, rhamnose, arabinose) at different hydroxylated positions.

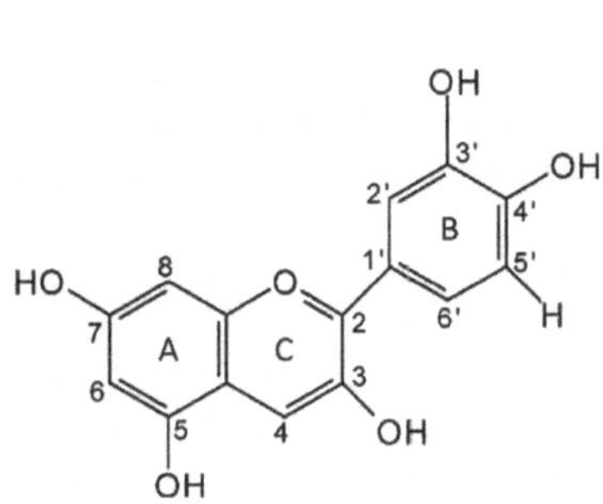

	R3'	R5'
Cyanidin	OH	H
Pelargonidin	H	H
Delphinidin	OH	OH
Peonidin	OCH_3	H
Petunidin	OCH_3	OH
Malvidin	OCH_3	OCH_3

Fig. 3 Chemical structure of anthocyanidin.

Anthocyanins could be found in blueberries, grapes, blackberries, purple cabbage, pomegranate, black rice, black carrots, purple potatoes, just to name a few. The daily consumption of anthocyanins ranged from 3 to 15 mg/person in the United States (Chun, Chung, & Song, 2007) and 8.4–23.6 mg/day in the United Kingdom (Jennings et al., 2012). This value varies depending on the different dietary choices in different regions.

Anthocyanins possess strong antioxidant, anti-inflammatory and anticarcinogenic properties, which reduce the risks of cardiovascular illnesses and various oxidative stress related diseases (He & Giusti, 2010), such as obesity (Lee et al., 2017), type 2 diabetes (Belwal, Nabavi, Nabavi, & Habtemariam, 2017), hypertension (Cassidy et al., 2016), lung cancer (Kausar et al., 2012), Alzheimer's disease and Parkinson's disease (Ali et al., 2018). These health benefits make it appealing to develop anthocyanin-rich functional bakery products which are consumed on a daily basis to increase the intake of anthocyanins.

3.1 Anthocyanins in bakery products

3.1.1 Colored wheat

Anthocyanins are naturally present in the bran (aleurone or pericarp) of pigmented wheat, making it a good candidate for developing anthocyanin-rich functional bakery products. As shown in Fig. 4, the color of the grain ranges from blue to purple to black, depending on genetic factors, growing conditions and technological processes (Ficco et al., 2014). Anthocyanins are concentrated in the pericarp layer of purple wheat, in the aleurone layer of blue wheat, and the combination of both in black wheat.

Purple tetraploid wheat was first collected in Abyssinia (present Ethiopia) in the 1870s and then introduced to Europe, where it was introgressed into hexaploid wheats (Eticha, Grausgruber, Siebenhandl-Ehn, & Berghofer, 2011). Nowadays, purple grain landraces are still cultivated in Ethiopia and can be found in the markets in the Tigray Region (Hailu Kassegn, 2018). The gene of blue aleurone was transferred from the tall wheatgrass *Agropyron elongatum* to the blue wheat (Morrison, Metzger, & Lukaszewski, 2004). The black wheat was developed in the 1990s by the Institute of Crop Genetics, Shanxi Academy of Agriculture Science after 20 years of breeding effort to cross the existing blue and purple lines (Li, Beta, Sun, & Corke, 2006; Li, Shan, Sun, Corke, & Beta, 2005; Sun et al., 1999).

Table 2 lists the major anthocyanin species identified in purple, blue and black wheat. The most common anthocyanins found in wheat were based

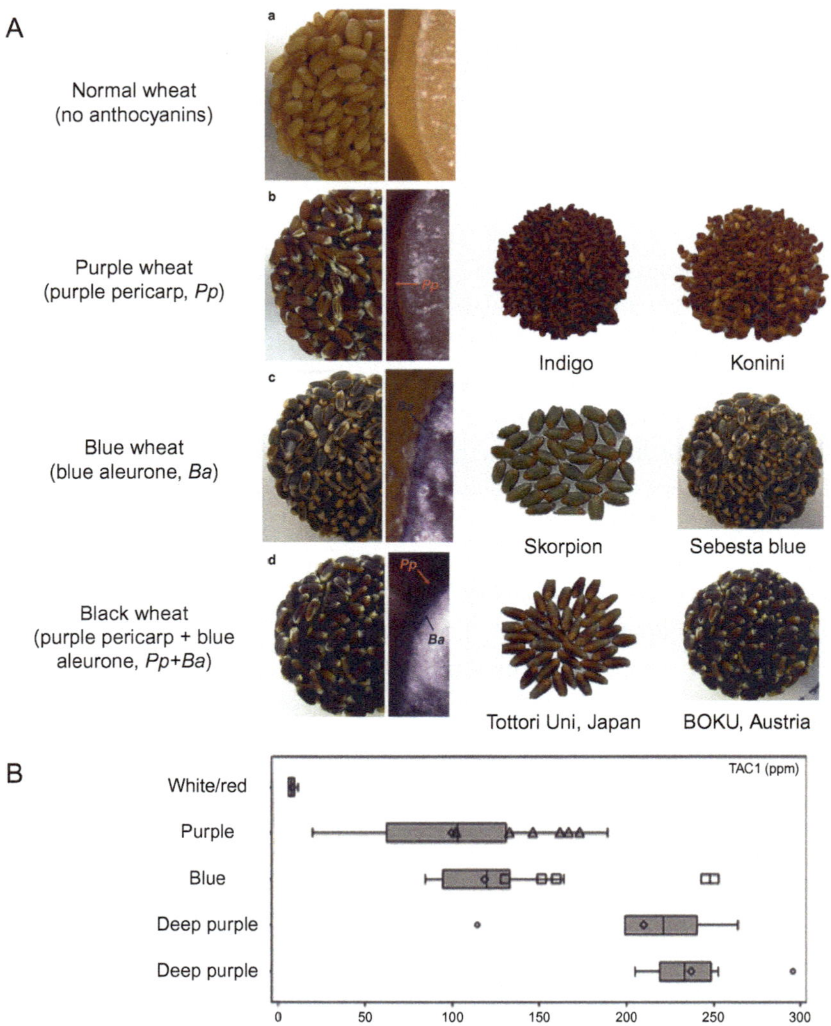

Fig. 4 (A) Wheat grain of normal wheat (no anthocyanins), purple wheat with anthocyanin in pericarp, and blue wheat with anthocyanins in aleurone (pictures extracted from Bohmdorfer, Oberlerchner, Fuchs, Rosenau, and Grausgruber (2018), Lachman, Martinek, Kotíková, Orsák, and Šulc (2017), Yu and Beta (2015) and Garg et al. (2016)); (B) Distribution of total anthocyanin content (extraction method of breeding lines (box and whisker plots) with respect to their grain color and parental genotypes (triangles: purple pericarp parents; squares: blue aleurone parents) (Syed Jaafar et al., 2013).

Table 2 Anthocyanin content and species in colored wheat.

	Genotype	Content (mg/kg)	Aglycon	Sugar	Identified major anthocyanins	References
Purple wheat	UM606a	–	Cy, Pn	Glu, Rut	Cy-Glu, Cy/Pn-Rut	Dedio, Hill, and Evans (1972)
	Konini	383 (bran)	Cy, Pn	Glu, Gal	Cy/Pn-3-Glu, Cy-3-Gal	Abdel-Aal and Hucl (2003)
	Laval (Canada) Konini (Canada)	96 (grain) 38 (grain)	Cy, Pn	Glu, Mal, Suc	Cy/Pn-3-Glu	Abdel-Aal, Young, and Rabalski (2006)
	Laval 19 (Canada)	70 (grain, db)	Cy, Pg	Glu, Rut	Cy/Pg-3-Glu, Cy-3-Rut	Abdel-Aal, Hucl, Shipp, and Rabalski (2016)
	Anthralin (Canada)	491 (grain)	Cy, Dp, Pn, Pt, Pg, Mv	Glu, Gal, Ara	Cy/Pn/Pg/Pt/Mv-3-Glu, Cy/Dp/Pn/Pg-3-ara	Hosseinian, Li, and Beta (2008)
	Abyssinskaja arrasajta (Czech)	13 (grain)	Cy, Pn, Pg	Glu, Rut	Cy/Pn/Pg-3-Glu, Cy/Pn/Pg-3-Rut	Bartl et al. (2015)
	Rosso (Austria)	102 (grain); 271 (bran)	Cy, Pn	Glu, Rut	Cy/Pn-3-Glu, Gy/Pn-3-Rut	Giordano et al. (2017)
Blue wheat	Purendo 38 (Canada)	480 (bran)	Cy, Pn	Glu	Cy/Pn-3-Glu	Abdel-Aal and Hucl (2003)
	Purendo 38 (Canada)	212 (grain)	Cy, Dp, Pt	Glu, Rut	Cy/Dp/Pt-3-Glu, Cy/Dp/Pt-3-Rut	Abdel-Aal et al. (2006)
	Spring wheat (Canada)	139 (grain) 321 (bran)	Cy, Dp, Pn, Pt, Mv	Glu, Rut	Cy/Dp-3-Glu, Cy/Dp-3-Rut	Abdel-Aal et al. (2008)

Continued

Table 2 Anthocyanin content and species in colored wheat.—cont'd

	Genotype	Content (mg/kg)	Aglycon	Sugar	Identified major anthocyanins	References
	Purendo 38 (Canada)	60 (grain, db)	Dp, Gy, Pg	Glu, Rut	Dp/Cy/Pg-3-Glu, Dp/Cy-3-Rut	Abdel-Aal et al. (2016)
	Skorpion (Czech)	9 (grain)	Cy, Dp, Pn	Glu, Rut, Rha	Cy/Dp-3-Glu, Cy/Dp/Pn-3-Rut	Bartl et al. (2015)
	Skorpion (Czech)	72 (grain) 140 (bran)	Cy, Dn, Pn	Glu, Rut	Cy/Dn-3-Glu, Cy/Dn/Pn-3-Rut	Giordano et al. (2017)
Black wheat	Hedong Wumai (China)	–	Cy, Pg, Pn	Glu, Gal, Pg	Cy/Pg/Pn-3-Glu, Cy-3-Gal	Hu, Cai, Li, and Corke (2007)
	Tottori University (Japan)	185.8 (grain)	Dp, Mv, Pg, Pn, Pt	Glu, Rut	Dp/Mv/Pg/Pn/Pt-3-Glu, Dp/Mv/Pn-3-Rut	Garg et al. (2016)
	Cross breed (Austria)	780 (grain, db)	Cy, Dp	Glu	Cy/Dp-3-Glu	Bohmdorfer et al. (2018)

Abbreviation for aglycon: Cy, Cyanidin; Pn, Peonidin; Dp, Delphinidin; Mv, Malvidin; Pg, pelargonidin; Pt, Petunidin; *Abbreviation for sugar moiety:* Glu, glucose; Rut, rutinoside; Mal, malony; Gal, galactoside; db, dry basis.

on cyanidin, but compounds containing delphinidin, peonidin and pelargonidin were also found. Glucose and rutinose are the most common sugar moieties while arabinose and galactose are also present. The anthocyanin composition is relatively simpler (i.e., fewer types of different anthocyanin glycosides) in the blue aleurone wheat as compared to the purple pericarp wheat (Abdel-Aal et al., 2006). The most abundant anthocyanin is cyanidin-3-glucoside in purple wheat (Abdel-Aal & Hucl, 2003; Hosseinian et al., 2008) and delphinidin-3-glucoside in blue wheat (Abdel-Aal et al., 2008), respectively. Cyanidin-3-glucoside is the second most predominant anthocyanin in blue wheat (~28.1% of total anthocyanins) (Abdel-Aal & Hucl, 2003). Cyanidin-3-glucoside is the predominant anthocyanin in the black wheat which also contains cyanidin-3-galactoside, pelargornidin-3-glucoside, and peonidin-3-glucoside (Hu et al., 2007). In general, black wheat has the highest anthocyanin content, followed by blue wheat and purple wheat (Saini et al., 2020). But some purple wheats contain a much higher content of anthocyanin than blue wheat (Abdel-Aal et al., 2016).

Purple and blue wheats have not been cultivated on a large scale due to lower yield in comparison with the currently available high yielding wheat varieties (Eticha et al., 2011). Over the past two decades, researchers have been working on the development of anthocyanin-rich wheat as a choice of functional raw material (Abdel-Aal & Hucl, 1999; Abdel-Aal & Hucl, 2003). Purple wheat has been grown for producing specialty breadsdx products in New Zealand, Austria and Canada (Eticha et al., 2011). Up to date, blue wheat has not been exploited commercially (Eticha et al., 2011).

Purple wheat has firstly been used as a raw material for anthocyanin extraction rather than for direct consumption (Abdel-Aal et al., 2008). Abdel-Aal, Hucl, and Rabalski (2018) reported extraction of anthocyanin from purple wheat bran into powder increased its concentration by 81- to 135-fold. The functional food products developed using purple wheat include bread (Khlestkina et al., 2017; Morgounov et al., 2020; Yu & Beta, 2015), biscuit (Pasqualone et al., 2015; Suvaluk, 2018), bun (Paznocht, Kotíková, Orsák, Lachman, & Martinek, 2019), steamed bread (Tian, Chen, & Wei, 2018), and Indian flatbread Chapatti (Kumari et al., 2020). Fig. 5 shows the pictures of the chapati developed using purple, blue and black wheat.

3.1.2 Black rice

Like wheat, pigmented rice grains can be found with light red to dark purple and black pericarp. Among them, black rice (*Oryza sativa L.*) is a special rice

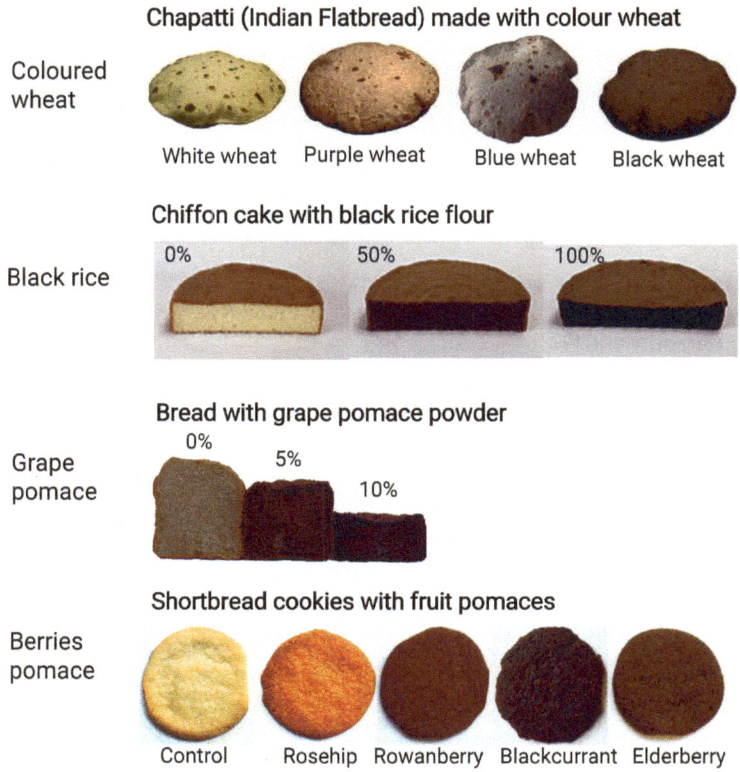

Fig. 5 Functional bakery products containing anthocyanins: chapati made from colored wheat (Kumari et al., 2020), chiffon cake made with black rice flour (Mau, Lee, Chen, & Lin, 2017), bread made with grape pomace powder (Hayta, Özuğur, Etgü, & Şeker, 2014), and shortbread cookie made with berries pomace (Tańska et al., 2016).

cultivar that has been widely consumed since ancient times in China. It contains much higher content of anthocyanins as compared to purple wheat (3276 mg/kg vs 95.8 mg/kg) (Abdel-Aal et al., 2006). There are mainly four anthocyanin species in black rice, i.e., cyanidin-3-glucoside, peonidin-3-glucoside, cyanidin-3,5-diglucoside, and cyanidin-3-rutinoside (Hou, Qin, Zhang, Cui, & Ren, 2013). The predominant anthocyanin in black rice is cyanidin-3-glucoside. Black rice flour was used to replace wheat flour (10%–100% replacement) in chiffon cake (Mau et al., 2015). White bread was fortified with anthocyanin-rich black rice extract at levels of 1%, 2% and 4% of flour weight (Sui, Zhang, & Zhou, 2016). Non-encapsulated and microencapsulated black rice extract were used in formulating biscuits (0.32% extract in the crude biscuit) and the final concentration of anthocyanin in biscuit was between 194 and 230 μg/g cookie (Papillo et al., 2018).

3.1.3 Grape pomace

Grape pomace, a by-product from the winemaking process (20%–30% of the original grape weight), is a rich source of anthocyanins. Mavidin-3-glucoside is the major anthocyanin species in grape pomace, followed by cyanidin-3-glucose, pelargonidin-3-glucoside, peonidin-3-glucoside, and delphinidin-3-glucoside (Gómez-Plaza, Miñano, & López-Roca, 2006). Grape pomace powder (GPP) was fortified at the level of 10% in bread (Hayta et al., 2014), at 4%–10% in cake (Nakov et al., 2020) and 5%–20% cookies (Maner, Sharma, & Banerjee, 2017). Grape pomace extract (GPE) is concentrated in anthocyanins (15,820 mg/kg DM) (de Sales et al., 2018) and was used to formulate biscuit which contained 140 mg malvidin-3-glc/kg (Pasqualone et al., 2014). Eminol® is a commercial GPE derived 100% from the red grape skins harvested in Spain (Karakaya et al., 2016). It contains 23.25 mg/g anthocyanins, among which malvidin-3-glucoside is the most abundant one (9.29 mg/g), followed by peonidin-3-glucoside (5.48 mg/g), cyanidin-3-glucoside (3.97 mg/g), petunidin-3-glucoside (2.74 mg/g) and delphiniodin-3-glucosie (1.23 mg/g) (Karakaya et al., 2016). Bun, breadstick and biscuit were prepared with Eminol® at the fortification levels of 1.34%, 1.6% and 1.48% (*w*/w) of the total recipes, respectively (Karakaya et al., 2016).

3.1.4 Berries

Berries are rich sources of anthocyanins too. Extensive research has been carried out to utilize anthocyanins in berries, covering from common to exotic species. Freeze-dried blue berry powder was added to bread at 12.4% (w/w) (Rodriguez-Mateos, Cifuentes-Gomez, George, & Spencer, 2014). Red raspberry (*Rubus idaeus*) juice was added to muffin at 10% (w/w) to reach 0.065 mg cyanidin-3-glucoside/g dry mass (DM) in the batter (Rosales-Soto, Powers, & Alldredge, 2012). Jakobek et al. analyzed elderberry and blackcurrant fruits juices and observed that anthocyanins constituted 66% and 56% of their total phenolics, respectively (Jakobek, Seruga, Medvidovic-Kosanovic, & Novak, 2007). The contents of anthocyanins in elderberry and black current pomaces were determined to be 4.33 and 8.34 g per 100 g of dry matter, respectively (Tańska et al., 2016). These pomaces were added into shortbread cookies at the level of 20% addition (Tańska et al., 2016). Sour cherry (*Prunus cerasus*) has a dark red color, indicating its high anthocyanin content. Sour cherry pomace extract was added to cookie to replace 10% and 15% of flour (Šaponjac et al., 2016) and at the level of

0.5 g per 100 g of batter in cake (Luca, Cilek, Hasirci, Sahin, & Sumnu, 2014). Jabuticaba, a native Brazilian berry, is a rich source of delphinidin and cyanidin glucoside (Albuquerque et al., 2020). Its extract was applied in macrons for its color and nutritional benefits (Albuquerque, Pinela, Barros, Oliveira, & Ferreira, 2020).

3.2 Stability of anthocyanins during baking

Similar to catechins, anthocyanins are highly reactive, and their stability depends on environmental factors such as pH, light, oxygen, temperature, enzymes, and metal ions (Cavalcanti, Santos, & Meireles, 2011). Anthocyanins are more stable in acidic media and when they are in high concentration. An increase in temperature, light exposure, and oxygen amplifies the degradation of anthocyanins. Food matrix can serve as a barrier to the heat effect and slow down degradation. Therefore, the retention of anthocyanin in bakery products is a combined result of thermal processing effect and food matrix effect.

The degradation of anthocyanins depends on the raw materials as well as processing conditions. Purple wheat (13.23 mg total anthocyanin content (TAC)/kg) and blue wheat (8.33 mg TAC/kg) baked at 180°C for 31 min showed a greater decrease in anthocyanin content than those baked at 240°C for 21 min (Bartl et al., 2015). Therefore, low-temperature long-time baking is less favorable than high-temperature short-time baking. Furthermore, the loss of anthocyanin was much higher in brad made from purple wheat (72.8% reduction) than those made from blue wheat (40.8% reduction), irrespective of baking temperature (Bartl et al., 2015). It was suspected that the acetylated anthocyanins identified in purple wheat but not in blue wheat (e.g., pelargonidin-3-glucoside) are more susceptible to thermal degradation.

Anthocyanins are largely unstable during food processing. Microencapsulation has been explored as a solution to improve its retention. The stability of anthocyanin-rich black rice extract was studied in baked biscuit (Papillo et al., 2018). After baking at 180°C for 25 min, the total anthocyanin content was reduced by 49% in the biscuits with unencapsulated extract and only reduced by 31%–34% in those with microencapsulated extract (Papillo et al., 2018). In particular, the amount of cyanidin-3-glucoside, as the most abundant anthocyanin species in the black rice extract, was reduced by 84%–88% after baking, regardless of whether the extract was microencapsulated or not. Cyanidin-3,5-di-glucoside was much more stable which might be attributed to its double glycosylation.

Cookie fortified with 10%–15% of sour cherry pomace extract demonstrated a 19%–22% and 34%–59% retention for whey protein and soy protein encapsulated anthocyanin, respectively (Šaponjac et al., 2016). The higher protein content of soy protein isolate might provide better interactions with anthocyanins and therefore protection.

The degradation kinetics of cyanidin-3-glucoside and cyanidin-3-rutinoside inside bread crumb and bread crust were evaluated (Sui, Yap, & Zhou, 2015). Cyanidin-3-rutinoside showed higher stability in bread crumb as compared to cyanidin-3-glucoside. Their degradation rates were lower in bread crumb as compared to the bread crust due to the lower temperature at bread crumb (65 °C vs 125 °C). Moreover, anthocyanins are much more stable in the bread matrix as compared to aqueous solutions, which could be explained by the lower level of oxygen available in the bread system. Oxygen was assimilated by yeast to produce carbon dioxide for dough development during the breadmaking process, which minimized anthocyanin oxidation.

3.3 Impact of anthocyanins on product quality and sensory properties

3.3.1 Colored wheat

Table 1 lists the details of bakery products enriched with anthocyanins. Bakery products could be made using the whole meal flour of colored wheat or with the addition of their bran. It was shown that the breadmaking quality and the organoleptic properties of bakery products made from anthocyanin-rich wheat did not concede, or in some cases were higher than the corresponding properties of products obtained from normal wheat (Khlestkina et al., 2017). Biscuits made from flour with the addition of bran of purple wheat had a darker color, more loosened shape, and slightly nutty taste (Suvaluk, 2018). The biscuit also had a lower friability and spread ratio due to the higher gluten strength in purple wheat (Pasqualone et al., 2015). The higher antioxidant activity exerted a protective action on lipid oxidation which helps to reduce the level of carboxylic acids derived from the oxidation of aldehydes and improve the volatile profile of purple wheat biscuit (Pasqualone et al., 2015). Black, purple and blue wheats were used to produced Indian flatbread chappati with acceptable sensory parameters, good taste and soft texture (Kumari et al., 2020). Bruschi, Teuber, and Dolgopolova (2015) showed that the consumer's acceptance depended on the product type. Purple wheat bread rolls were more accepted than purple wheat biscuits. Moreover, consumers exhibited a stronger willingness to pay after knowing it is an old wheat variety rather than knowing anthocyanin.

3.3.2 Black rice

Black rice extract was incorporated into white bread at 1%, 2% and 4% levels (Sui et al., 2016). Bread with black rice extract powder generally showed a smaller volume and a purple color due to the color of black rice. The bread with 4% black rice extract had a lower specific volume (2.89 cm^3/g) as compared to control bread (3.50 cm^3/g). At the same time, bread hardness increased by 38% for the bread fortified with 4% addition in comparison to that of the control bread. The denser structure of bread crumb might result from the less amount of SS bonds caused by the reducing ability of anthocyanins. Moreover, interactions between the polyphenols and the side chain amino groups of peptides happened to form additional cross-links in bread proteins. The greater mechanical strength and the reduced swelling properties of the cross-linked protein gel might result in a tight bread structure (Sui et al., 2016).

When black rice powder was used to substitute wheat flour in chiffon cake, the specific gravity of cake batter, the crumb *a* value, hardness, cohesiveness and chewiness of the cake increased with the increased level of black rice powder (Mau et al., 2015). Hedonic sensory results of control and 10–60% black rice flour substituted cakes were comparable (Mau et al., 2015). Therefore, chiffon cake formulated with partial substitution of black rice flour up to 60% was recommended as a novel product with good antioxidant activity and bioactive compounds.

3.3.3 Grape pomace

Application of grape pomace powder (GPP) significantly reduced the volume of cake and muffin due to the interference of fiber in dough structure and gas retention (Theagarajan, Malur Narayanaswamy, Dutta, Moses, & Chinnaswamy, 2019; Walker, Tseng, Cavender, Ross, & Zhao, 2014). Grape pomace also contributed to a darker color and increased hardness in bread (Hayta et al., 2014), cakes (Nakov et al., 2020), cookies (Theagarajan et al., 2019) and muffins (Walker et al., 2014). In terms of flavor profile, bread with 5%–10% of substitution of flour with grape pomace powder showed a major impact on the acidity, global flavor, astringency and the wine smell of the bread without affecting the overall bread acceptability (Tolve et al., 2021). Biscuits produced with grape marc extracts had a more intense fruity odor and sour taste than the control biscuits (Pasqualone et al., 2014). The addition of a high level of GPP (10%) showed a lower consumer acceptability of bread (Hayta et al., 2014) while addition of small quantity (4%–6%) of grape pomace to bakery products improved the nutritional quality and did not affect sensorial characteristics (e.g., taste, after taste and hardness) (Theagarajan et al., 2019).

3.3.4 Berries

Muffins prepared with raspberry juice showed a higher weight loss during baking and a smaller final volume (Rosales-Soto et al., 2012). Cookies fortified with the pomace of black current and elderberry at the level of 20% were characterized by a darker color with a higher contribution of yellowness and a harder texture compared to those without addition (Tańska et al., 2016). The cookies with pomace were sourer and had a more perceptible taste and aroma of pomace. They also had a significantly stronger antioxidant capacity based on DPPH assay than those without pomace but were ineffective in inhibiting lipid oxidation.

3.4 Health benefits of anthocyanins in bakery products

3.4.1 Antioxidant activity

Anthocyanins possess strong antioxidant activity which is retained even after thermal degradation during processing (Sui, Dong, & Zhou, 2014). Sui et al. (2015) quantified the antioxidant capacity and total phenolic content (TPC) of bread fortified with 2% black rice extract, which was shown to be 17.7 mg Trolox/100 g and 38.5 mg gallic acid/100 g for bread crumb, respectively. Biscuit prepared using purple wheat had 43% higher antioxidant capacity than those prepared using wholemeal durum wheat (Pasqualone et al., 2015). In comparison, biscuits fortified with black rice extract had a higher antioxidant activity (+78%) (Papillo et al., 2018). This is attributed to the difference in anthocyanin content in purple wheat and black rice extract (13.9 vs 213 μg/g).

3.4.2 Carbohydrate digestion

Black rice is recognized for managing diabetes in Chinese folk medicine. Anthocyanins inhibit some digestive enzymes such as α-amylase and α-glucosidase and thus suppress the increase in postprandial glucose level from digesting starch (Matsui et al., 2001). Gastrointestinal digestion could enhance α-glucosidase inhibition of anthocyanins, which might be attributed to the deglycosylation effect of alkaline conditions in the simulated intestinal environment, which decreases the steric hindrance for binding of anthocyanins with the enzyme and thus increases the inhibitory activity (Sun & Miao, 2020).

Starch digestibility of black rice fortified bread was investigated using a restricted in vitro digestion system using a dialysis tube (Sui et al., 2016). With an addition of 1%, 2% and 4% black rice extract, the rate of starch digestion in bread was reduced by 12.8%, 14.1%, and 20.5%, respectively. The incorporation of anthocyanins into bread could produce functional bread with a slower rate of starch digestion (20.5% reduction at 4% fortification), providing extra health benefits other than the strong antioxidant activity. Gowd, Jia, and Chen (2017) provided an excellent review on how anthocyanins could be used as dietary bioactive components against diabetes.

3.4.3 Bioavailability of anthocyanins

The bioavailability of anthocyanins could be determined using in vitro methods (e.g., Caco-2 cells) (Kamiloglu, Capanoglu, Grootaert, & Van Camp, 2015) and in vivo methods (e.g., isotopic labelling of cyanidin-3-glucoside) (Czank et al., 2013). Due to the large molecular weight, anthocyanin was believed to be difficult to be absorbed in the digestive tract. However, in vivo data confirmed that anthocyanins can be quickly and efficiently absorbed through the intestinal tract, transferred through the circulatory system and finally excreted through the urine (Yang, Liu, Yang, Gupta, & Jiang, 2018). The main absorption site of anthocyanin is the stomach while its absorption mechanism is still unknown (Forester, Choy, Waterhouse, & Oteiza, 2014). At the small intestine, the remaining anthocyanins are absorbed via the specific glucose transporter GLUT2 (Yoshikawa et al., 2011). The absorption of anthocyanins is influenced by aglycone structure, the type of sugar moiety, polymeric structure, and the presence of other food components (Karaś, Jakubczyk, Szymanowska, Złotek, & Zielińska, 2017). The bioaccessibility of anthocyanins in grape skin extract in biscuit and bun matrixes was reported to be 57% using in vitro digestion model (Karakaya et al., 2016). The degradation of anthocyanin observed during in vitro digestion was mainly associated with their high instability in physiological pH rather than the effect of digestive enzymes (Tagliazucchi, Verzelloni, Bertolini, & Conte, 2010).

Sui et al. (2016) assessed the release of cyanidin-3-glucoside, the most dominant anthocyanin in the black rice extract, from fortified bread. After 3 h of in vitro digestion, only a small fraction of cyanidin-3-glucoside (0.31%–2.84%) was found in the dialysates of bread fortified with 1% to 4% black rice extract. After a prolonged incubation (24 h), about 2.7%, 5.9% and 6.5% of the cyanidin-3-glucoside was detected in the dialysates of 1%, 2%

and 4% fortified bread, respectively. The low amount of released cyanidin-3-glucoside might be attributed to the binding of anthocyanins with starch and proteins in the bread matrix. Nevertheless, clinical data is required to access the bioavailability of anthocyanins from different plant sources and in different bakery products.

4. Fucoidan from *Undaria pinnatifida*

Fucoidan refers to a group of sulphated polysaccharides commonly isolated from brown seaweed species (Li, Lu, Wei, & Zhao, 2008). In recent years, fucoidan has gained increased research interest due to its numerous bio-active properties, including anti-cancer capabilities, anti-coagulant properties, antioxidant activities as well as its antiviral properties (Li et al., 2008). However, it was also reported in the literature that fucoidan isolated from different seaweed species possesses slightly different bio-activities owing to its different chemical structures (Li et al., 2008).

Undaria pinnatifida is a brown seaweed species with a high growth rate all year round and is often treated as an unwanted species in the seas of New Zealand (Fletcher & Manfredi, 1995; Stuart, 2004; Wallentinus, 2007). This makes *U. pinnatifida* an excellent source of raw material for fucoidan extraction. In addition, it has been reported that fucoidan from *U. pinnatifida* is composed of a unique backbone structure of galactose and fucose monosaccharides as shown in Fig. 6 (Koh, Lu, & Zhou, 2019). This repeated backbone structure of galactose and fucose units of the fucoidan molecule from *U. pinnatifida* has been associated with higher anti-cancer activities, thereby further enhancing the appeal of this fucoidan molecule as a functional ingredient (Vishchuk, Ermakova, & Zvyagintseva, 2011).

Fucoidan from *U. pinnatifida* has also been reported to be inhibitors of starch hydrolyzing enzymes such as α-amylase, α-glucosidase and amyloglucosidase (Koh, Lim, Lu, & Zhou, 2020). As such, fucoidan can potentially be used to reduce the glycaemic potential of carbohydrate-rich staple food products such as bread as a functional ingredient.

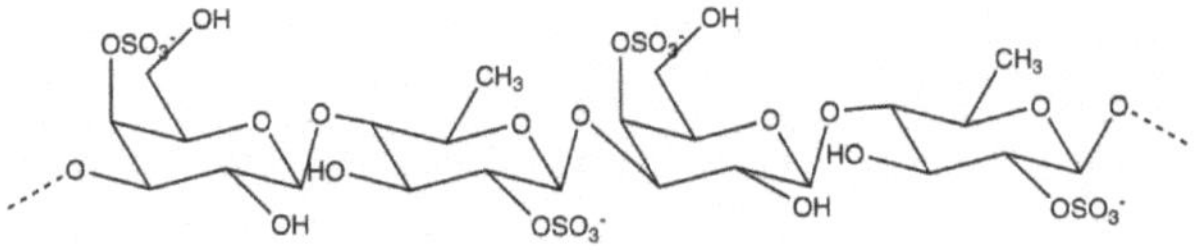

Fig. 6 Structure of fucoidan from *U. pinnatifida*.

4.1 Stability of fucoidan during baking

In order for fucoidan to be utilized as a bioactive ingredient in a food system such as baked bread, fucoidan needs to be able to withstand the high temperatures during baking for it to be retained in the final bread product to produce the desired bioactivities. Fucoidan was fortified at 0.4% and 0.8% of flour weight to produce the fucoidan-fortified bread samples. After baking, 69.07% and 71.18% of fucoidan was retained in the 0.4% and 0.8% fucoidan-fortified bread samples respectively (Koh et al., 2020). One possible reason for the loss in fucoidan, which amounted to approximately 30% during the baking process, has been proposed to be due the fermentation of fucoidan by baker's yeast during the proofing process (Eom, Lee, & Kim, 2010; Koh et al., 2020).

The other bio-activities of fucoidan including its antioxidant (DPPH free radical scavenging activity and hydroxyl radical scavenging activity) and anti-cancer properties (against MCF-7 breast cancer cell line) were also retained after baking (Koh et al., 2020). More interestingly, a higher free radical scavenging activity was reported in the fucoidan-fortified bread extract as compared to pure fucoidan samples (Koh et al., 2020). This observation implied that the baking process of fucoidan in the bread system enhanced the antioxidant activity of fucoidan. A plausible explanation for this observation is that the baking process under high heat resulted in the partial degradation of the large fucoidan polysaccharide into smaller fragments with lower molecular weight and higher sulphate to fucose ratio, thereby contributing to an increase in antioxidant activity (Wang, Zhang, et al., 2008). Therefore, the results supported the feasibility of using fucoidan as a functional ingredient in baked bread.

4.2 Impact of fucoidan fortification on baked bread

Fucoidan fortification has also resulted in improvements to bread quality. It was observed that the specific volumes of the bread samples fortified with 0.4% and 0.8% of fucoidan were significantly larger than the control bread samples (Koh et al., 2020). A significantly softer breadcrumb texture was also observed in the fucoidan-fortified bread sample as compared to the control bread sample (Koh et al., 2020). This improvement in bread volume was attributed to the favorable interactions between the sulphate groups of fucoidan and the amino group of gluten (Koh et al., 2020). It was observed that the gluten secondary structure of the fucoidan-fortified sample is composed of a significantly higher ratio of β-sheet to β-turns which in turn contributes to its larger specific volume (Gao, Koh, Tay, & Zhou, 2017; Koh et al., 2020).

4.3 Health benefits of fucoidan in baked bread

One potential health benefit of fucoidan fortification in baked bread is its ability to reduce the glycaemic potential of bread since fucoidan from *U. pinnatifida* have been reported to be inhibitors of starch digesting enzymes (Koh et al., 2020). Fucoidan fortification at higher dosages of 2.0% and 3.0% of flour weight has been reported to lower the in vitro starch digestion rate by 14.9% and 21.5%, respectively. This result highlighted the potential of fucoidan to reduce starch digestion in the bread system. In terms of the glycaemic potential, the fucoidan-fortified baked bread at 3.0 g/100 g of flour weight had a predicted glycaemic index ($pGI_{glucose}$) of 57.5 ± 4.5, which is significantly lower than GI of white wheat bread (GI = 100) and wholemeal bread (GI = 74) (Atkinson, Foster-Powell, & Brand-Miller, 2008). Therefore, the results suggested that fucoidan can be used as a novel functional ingredient to produce a "healthier" alternative of baked bread that is consumed as a staple food in many parts of the world.

5. Quercetin

Quercetin is a well-recognized antioxidant ingredient exhibiting numerous health benefits such as anti-inflammatory, cardioprotective, and hypoglycemic effects (Wang et al., 2016). Some of the food sources rich in quercetin include onions, berries, and apples. The average daily intake of flavonols and flavones has been estimated to be 20–22 mg/day in the U.S., while around 75% was contributed by the consumption of quercetin (15–18 mg/day) (Sampson, Rimm, Hollman, de Vries, & Katan, 2002). Despite having high nutritional values, the hydrophobic property of quercetin largely limits its applications in the food industry, especially for beverage products. Being semi-solid, bread could be a good food carrier for quercetin. Functional bread may therefore be produced by fortification of quercetin.

5.1 Stability of quercetin during baking

Quercetin was incorporated to the baked bread and steamed bread (Fig. 7). The retention of quercetin in both baked and steamed bread was determined

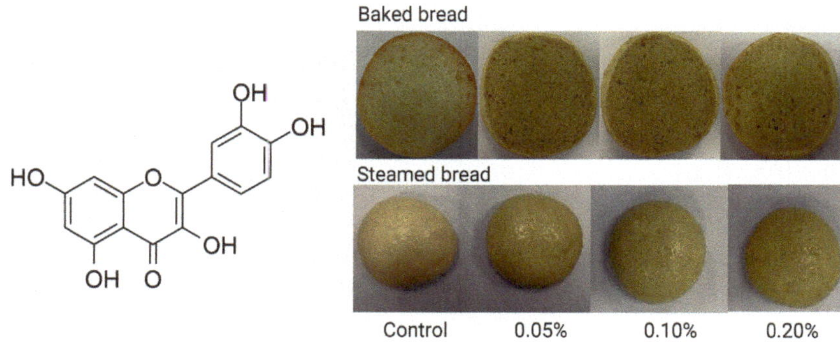

Fig. 7 (Left) chemical structure of quercetin; (Right) pictures of baked bread and steamed bread fortified with 0.05%, 0.10% and 0.20% quercetin (flour weight basis).

to reveal the stability of quercetin during breadmaking (Lin & Zhou, 2018). After baking, the average quercetin content in the bread crumb with fortification of 0.05%, 0.10%, and 0.20% was 0.35, 0.77, and 1.42 mg/g dry weight (DW), respectively, while that in the bread crust was 0.27, 0.60, and 1.20 mg/g DW, respectively. The lower amount of quercetin in the crust was attributed to the larger loss of quercetin at much higher temperature of the crust during baking. The amount of quercetin retained in the steamed bread crumb was 0.32, 0.75, and 1.55 mg/g DM with fortification of 0.05%, 0.10%, and 0.20% quercetin, respectively. These values were statistically the same as those in the bread crust (0.28, 0.66, and 1.41 mg/g bread DW, respectively) at every fortification level (Lin, Tan, Leong, & Zhou, 2018). Within the same range of fortification, at each level, the amount of quercetin retained in the baked bread and that in the steamed bread were comparable. The amount of quercetin left in bread samples ranged from 77% to 92% with respect to the amount added, showing relatively high stability of quercetin during breadmaking process.

5.2 Impact of quercetin on bread quality

The influence of quercetin fortification on the quality of both baked and steamed bread was evaluated (Lin et al., 2018; Lin & Zhou, 2018). Similar to other polyphenols with reducing capacity, quercetin incorporated in bread caused a significantly lower specific volume due to the less elastic bread structure in the presence of a smaller amount of SS bonds. Besides, less amount of CO2 produced by yeast in the dough during proofing might also result in a smaller bread volume (Lin et al., 2018). The impact of quercetin on the baked bread volume was initiated from 0.05% fortification, with the reduction by 3.2% (Lin & Zhou, 2018), while the threshold level for the steamed bread was much higher, at 1.2%, with the reduction by 6.6% (Lin et al., 2018). As the amount of quercetin increased, smaller bread volume (up to 12%) and harder bread texture (up to 32.8%) were observed in both baked (0.05%–0.2%) and steamed bread (1.2%–3.6%) by instrumental analysis. The distinction between the two types of bread is consistent with the previous studies investigating the quality of baked and steamed bread added with green tea extract. An enhanced yellow-greenish color of the bread samples with quercetin was observed. This was largely ascribed to the original color of quercetin powder and partly to unidentified colorant produced by reactions between quercetin and bread ingredients.

5.3 Health benefits of quercetin in bakery products

The nutritional properties of bread fortified with quercetin (0.05%–0.2%) have been assessed based on its antioxidant and antiglycation properties. Antiglycation capacity involves inhibition of the generation of advanced glycation endproducts (AGEs), one of the factors implicated in diabetic complications (Peng, Ma, Chen, & Wang, 2011). The accumulation of AGEs in the body occurs internally through in vivo metabolism and externally from food sources. The regular consumption of dietary AGEs (dAGEs) produced by Maillard reactions in food (e.g., bread) may pose a threat to human health, especially to diabetic patients. Quercetin could inhibit AGEs formation by the free radical scavenging activity and its capacity to trap precursors of AGEs such as methylglyoxal (MGO) (Li, Zheng, Sang, & Lv, 2014; Zhang et al., 2014). Incorporating quercetin, a natural glycation inhibitor, into bread has high potential to develop an antiglycative functional food.

The antioxidant activity was examined by ABTS and DPPH assays while the total phenolic content was also evaluated. Incorporation of quercetin (0.05%–0.2%) significantly enhanced the antioxidant capacity and the total phenolic content of the baked and steamed bread dose-dependently. The baked bread with 0.2% quercetin possessed antioxidant activity of 285.8–290.8 mg trolox per 100 g DM for the crumb and 290.8–295.2 for the crust. Significantly higher total phenolic content was found in the baked bread crust than that in the crumb across all the quercetin levels. Such difference was ascribed to the antioxidant properties of the Maillard reaction products generated in bread crust. Under the lower temperature profile during processing (i.e., steaming), the antioxidant activity and total phenolic content of the crumb and the crust of quercetin-fortified steamed bread showed no significant differences. The antioxidant profiles of a bakery product incorporated with polyphenols are dependent on various sources, such as the intrinsic polyphenols in wheat flour, the fortified phenolics, the thermal degradation products of the phenolic compounds, and the polyphenol-starch/protein complex (Sivam, Sun-Waterhouse, Quek, & Perera, 2010).

The antiglycation effects of quercetin on both AGEs in the body and the dAGEs in bread matrix have been investigated in an in vitro protein glycation model system. Results showed that the baked and steamed bread fortified with 0.2% quercetin could inhibit 40%–52% of total AGEs formed during protein glycation. IC_{50} value, revealing the specific concentration for 50% inhibition of glycation was estimated to compare the antiglycation effect between the commercial inhibitor aminoguanidine and the quercetin-fortified bread (Lin & Zhou, 2018). Smaller IC_{50} values for the baked bread with quercetin showed that it possessed a stronger antiglycation effect. Around 55%–60% of inhibition on the formation of dAGEs in the baked and steamed bread was observed with quercetin fortification at 0.20%. These results indicated that the functional bread with quercetin fortification was a promising way to reduce the risks of glycation-associated diseases.

6. Conclusions

The major efforts in developing functional bakery products have been focusing on the utilization of functional ingredients. We have demonstrated four successful examples of functional bakery products developed using phytochemicals, including green tea catechins, anthocyanins, fucoidan, and quercetin. Fortification of functional ingredients normally exerted a minimum impact on the quality and organoleptic characteristics of bakery products when they were incorporated in low dosage. With an increasing level of fortification, there are inevitable compromises on the consumer acceptance of the developed functional bakery products. Therefore, a balance between efficacy and sensorial quality needs to be searched in product development, depending on the characteristics of both active ingredients and baked goods. In particular, there is a need to improve the sensory qualities of GTE and quercetin-fortified bread in order to deliver targeted health benefits without compromising consumer acceptance. In conclusion, product optimization to maximize beneficial effects without significantly compromising the physico-sensory properties of bakery product is of primary concern when reformulating them through the incorporation of phytochemicals. Moreover, in order to fully understand how functional bakery products can better deliver health benefits, an in-depth understanding of the fate of active ingredients throughout the digestive system is essential.

References

Abdel-Aal, E.-S. M., Abou-Arab, A. A., Gamel, T. H., Hucl, P., Young, J. C., & Rabalski, I. (2008). Fractionation of blue wheat anthocyanin compounds and their contribution to antioxidant properties. *Journal of Agricultural and Food Chemistry, 56*, 11171–11177. https://doi.org/10.1021/jf802168c.

Abdel-Aal, E., & Hucl, P. (1999). A rapid method for quantifying total anthocyanins in blue aleurone and purple pericarp wheats. *Cereal Chemistry, 76*, 350–354.

Abdel-Aal, E.-S. M., & Hucl, P. (2003). Composition and stability of anthocyanins in blue-grained wheat. *Journal of Agricultural and Food Chemistry, 51*, 2174–2180.

Abdel-Aal, E.-S. M., Hucl, P., & Rabalski, I. (2018). Compositional and antioxidant properties of anthocyanin-rich products prepared from purple wheat. *Food Chemistry, 254*, 13–19.

Abdel-Aal, E. M., Hucl, P., Shipp, J., & Rabalski, I. (2016). Compositional differences in anthocyanins from blue-and purple-grained spring wheat grown in four environments in Central Saskatchewan. *Cereal Chemistry, 93*, 32–38.

Abdel-Aal, E.-S. M., Young, J. C., & Rabalski, I. (2006). Anthocyanin composition in black, blue, pink, purple, and red cereal grains. *Journal of Agricultural and Food Chemistry, 54*, 4696–4704.

Ahmad, M., Baba, W. N., Wani, T. A., Gani, A., Gani, A., Shah, U., et al. (2015). Effect of green tea powder on thermal, rheological & functional properties of wheat flour and physical, nutraceutical & sensory analysis of cookies. *Journal of Food Science and Technology, 52*, 5799–5807. https://doi.org/10.1007/s13197-014-1701-3.

Albuquerque, B. R., Pereira, C., Calhelha, R. C., José Alves, M., Abreu, R. M. V., Barros, L., et al. (2020). Jabuticaba residues (Myrciaria jaboticaba (Vell.) Berg) are rich sources of valuable compounds with bioactive properties. *Food Chemistry, 309*, 125735. https://doi.org/10.1016/j.foodchem.2019.125735.

Albuquerque, B. R., Pinela, J., Barros, L., Oliveira, M. B. P., & Ferreira, I. C. (2020). Anthocyanin-rich extract of jabuticaba epicarp as a natural colorant: Optimization of heat-and ultrasound-assisted extractions and application in a bakery product. *Food Chemistry, 316*, 126364.

Ali, T., Kim, T., Rehman, S. U., Khan, M. S., Amin, F. U., Khan, M., et al. (2018). Natural dietary supplementation of anthocyanins via PI3K/Akt/Nrf2/HO-1 pathways mitigate oxidative stress, neurodegeneration, and memory impairment in a mouse model of Alzheimer's disease. *Molecular Neurobiology, 55*, 6076–6093. https://doi.org/10.1007/s12035-017-0798-6.

Amber, S. (2013). *Tea catechins: Their stability and roles in the biscuit making process and effects on biscuit quality*. Singapore: National University of Singapore.

Amoako, D., & Awika, J. M. (2016). Polyphenol interaction with food carbohydrates and consequences on availability of dietary glucose. *Current Opinion in Food Science, 8*, 14–18.

Ananingsih, V. K., Gao, J., & Zhou, W. (2013). Impact of green tea extract and fungal alpha-amylase on dough proofing and steaming. *Food and Bioprocess Technology, 6*, 3400–3411.

Andersen, O. M., & Markham, K. R. (2005). *Flavonoids: Chemistry, biochemistry and applications*. CRC Press.

Atkinson, F. S., Foster-Powell, K., & Brand-Miller, J. C. (2008). International tables of glycemic index and glycemic load values: 2008. *Diabetes Care*, *31*, 2281. https://doi.org/10.2337/dc08-1239.

Bajerska, J., Mildner-Szkudlarz, S., Jeszka, J., & Szwengiel, A. (2010). Catechin stability, antioxidant properties and sensory profiles of rye breads fortified with green tea extracts. *Journal of Food and Nutrition Research*, *49*, 104–111.

Bartl, P., Albreht, A., Skrt, M., Tremlová, B., Ošťádalová, M., Šmejkal, K., et al. (2015). Anthocyanins in purple and blue wheat grains and in resulting bread: Quantity, composition, and thermal stability. *International Journal of Food Sciences and Nutrition*, *66*, 514–519.

Belton, P. (1999). Mini review: On the elasticity of wheat gluten. *Journal of Cereal Science*, *29*, 103–107.

Belwal, T., Nabavi, S. F., Nabavi, S. M., & Habtemariam, S. (2017). Dietary anthocyanins and insulin resistance: When food becomes a medicine. *Nutrients*, *9*. https://doi.org/10.3390/nu9101111.

Bohmdorfer, S., Oberlerchner, J. T., Fuchs, C., Rosenau, T., & Grausgruber, H. (2018). Profiling and quantification of grain anthocyanins in purple pericarp x blue aleurone wheat crosses by high-performance thin-layer chromatography and densitometry. *Plant Methods*, *14*.

Braicu, C., Ladomery, M. R., Chedea, V. S., Irimie, A., & Berindan-Neagoe, I. (2013). The relationship between the structure and biological actions of green tea catechins. *Food Chemistry*, *141*, 3282–3289. https://doi.org/10.1016/j.foodchem.2013.05.122.

Bruschi, V., Teuber, R., & Dolgopolova, I. (2015). Acceptance and willingness to pay for health-enhancing bakery products–Empirical evidence for young urban Russian consumers. *Food Quality and Preference*, *46*, 79–91.

Cassidy, A., Bertoia, M., Chiuve, S., Flint, A., Forman, J., & Rimm, E. B. (2016). Habitual intake of anthocyanins and flavanones and risk of cardiovascular disease in men. *The American Journal of Clinical Nutrition*, *104*, 587–594. https://doi.org/10.3945/ajcn.116.133132.

Cavalcanti, R. N., Santos, D. T., & Meireles, M. A. A. (2011). Non-thermal stabilization mechanisms of anthocyanins in model and food systems—An overview. *Food Research International*, *44*, 499–509. https://doi.org/10.1016/j.foodres.2010.12.007.

Chun, O. K., Chung, S. J., & Song, W. O. (2007). Estimated dietary flavonoid intake and major food sources of U.S. adults. *The Journal of Nutrition*, *137*, 1244–1252. https://doi.org/10.1093/jn/137.5.1244.

Coe, S., & Ryan, L. (2016). White bread enriched with polyphenol extracts shows no effect on glycemic response or satiety, yet may increase postprandial insulin economy in healthy participants. *Nutrition Research*, *36*(2), 193–200.

Czank, C., Cassidy, A., Zhang, Q., Morrison, D. J., Preston, T., Kroon, P. A., et al. (2013). Human metabolism and elimination of the anthocyanin, cyanidin-3-glucoside: A 13C-tracer study. *The American of Clinical Nutrition*, *97*, 995–1003.

Dai, W., Ruan, C., Zhang, Y., Wang, J., Han, J., Shao, Z., et al. (2020). Bioavailability enhancement of EGCG by structural modification and nano-delivery: A review. *Journal of Functional Foods*, *65*, 103732.

de Sales, N. F. F., Silva da Costa, L., Carneiro, T. I. A., Minuzzo, D. A., Oliveira, F. L., Cabral, L. M. C., et al. (2018). Anthocyanin-rich grape pomace extract (Vitis vinifera L.) from wine industry affects mitochondrial bioenergetics and glucose metabolism in human hepatocarcinoma HepG2 cells. *Molecules*, *23*, 611. https://doi.org/10.3390/molecules23030611.

Dedio, W., Hill, R., & Evans, L. (1972). Anthocyanins in the pericarp and coleoptiles of purple wheat. *Canadian Journal of Plant Science*, *52*, 977–980.

Del Rio, D., Calani, L., Cordero, C., Salvatore, S., Pellegrini, N., & Brighenti, F. (2010). Bioavailability and catabolism of green tea flavan-3-ols in humans. *Nutrition*, *26*, 1110–1116. https://doi.org/10.1016/j.nut.2009.09.021.

Eom, S.-H., Lee, B.-J., & Kim, Y.-M. (2010). Effect of yeast fermentation on the antioxidant and anti-inflammatory activity of sea tangle water extract. *Korean Journal of Fisheries and Aquatic Sciences*, *43*, 117–124.

Eticha, F., Grausgruber, H., Siebenhandl-Ehn, S., & Berghofer, E. (2011). Some agronomic and chemical traits of blue aleurone and purple pericarp wheat (Triticum L.). *Journal of Agricultural Science and Technology B*, *1*, 48–58.

Ficco, D. B., Mastrangelo, A. M., Trono, D., Borrelli, G. M., De Vita, P., Fares, C., et al. (2014). The colours of durum wheat: A review. *Crop and Pasture Science*, *65*, 1–15.

Fletcher, R., & Manfredi, C. (1995). The occurrence of Undaria pinnatifida (Phaeophyceae, Laminariales) on the south coast of England. *Botanica Marina*, *38*, 355–358.

Forester, S. C., Choy, Y. Y., Waterhouse, A. L., & Oteiza, P. I. (2014). The anthocyanin metabolites gallic acid, 3-O-methylgallic acid, and 2, 4, 6-trihydroxybenzaldehyde decrease human colon cancer cell viability by regulating pro-oncogenic signals. *Molecular Carcinogenesis*, *53*, 432–439.

Friedman, M., Levin, C. E., Choi, S., Kozukue, E., & Kozukue, N. (2006). HPLC analysis of catechins, theaflavins, and alkaloids in commercial teas and green tea dietary supplements: Comparison of water and 80% ethanol/water extracts. *Journal of Food Science*, *71*, C328–C337.

Gao, J., Koh, A. H. S., Tay, S. L., & Zhou, W. (2017). Dough and bread made from high-and low-protein flours by vacuum mixing: Part 1: Gluten network formation. *Journal of Cereal Science*, *74*, 288–295.

Garg, M., Chawla, M., Chunduri, V., Kumar, R., Sharma, S., Sharma, N. K., et al. (2016). Transfer of grain colors to elite wheat cultivars and their characterization. *Journal of Cereal Science*, *71*, 138–144.

Giordano, D., Locatelli, M., Travaglia, F., Bordiga, M., Reyneri, A., Coïsson, J. D., et al. (2017). Bioactive compound and antioxidant activity distribution in roller-milled and pearled fractions of conventional and pigmented wheat varieties. *Food Chemistry*, *233*, 483–491. https://doi.org/10.1016/j.foodchem.2017.04.065.

Goh, R., Gao, J., Ananingsih, V. K., Ranawana, V., Henry, C. J., & Zhou, W. (2015). Green tea catechins reduced the glycaemic potential of bread: An in vitro digestibility study. *Food Chemistry*, *180*, 203–210.

Gómez-Mascaraque, L. G., Hernández-Rojas, M., Tarancón, P., Tenon, M., Feuillère, N., Vélez Ruiz, J. F., et al. (2017). Impact of microencapsulation within electrosprayed proteins on the formulation of green tea extract-enriched biscuits. *LWT—Food Science and Technology*, *81*, 77–86. https://doi.org/10.1016/j.lwt.2017.03.041.

Gómez-Plaza, E., Miñano, A., & López-Roca, J. M. (2006). Comparison of chromatic properties, stability and antioxidant capacity of anthocyanin-based aqueous extracts from grape pomace obtained from different vinification methods. *Food Chemistry*, *97*, 87–94. https://doi.org/10.1016/j.foodchem.2005.03.025.

Gowd, V., Jia, Z., & Chen, W. (2017). Anthocyanins as promising molecules and dietary bioactive components against diabetes—A review of recent advances. *Trends in Food Science & Technology*, *68*, 1–13. https://doi.org/10.1016/j.tifs.2017.07.015.

Gramza-Michałowska, A., Kobus-Cisowska, J., Kmiecik, D., Korczak, J., Helak, B., Dziedzic, K., et al. (2016). Antioxidative potential, nutritional value and sensory profiles of confectionery fortified with green and yellow tea leaves (Camellia sinensis). *Food Chemistry*, *211*, 448–454.

Guo, H., & Xia, M. (2018). Anthocyanins and diabetes regulation. In *Polyphenols: Mechanisms of action in human health and disease* (pp. 135–145). Elsevier.

Hailu Kassegn, H. (2018). Determination of proximate composition and bioactive compounds of the Abyssinian purple wheat. *Cogent Food & Agriculture*, *4*, 1421415.

Hayta, M., Özuğur, G., Etgü, H., & Şeker, İ. T. (2014). Effect of grape (V itis Vinifera L.) pomace on the quality, Total phenolic content and anti-radical activity of bread. *Journal of Food Processing and Preservation*, *38*, 980–986.

He, J., & Giusti, M. M. (2010). Anthocyanins: Natural colorants with health-promoting properties. *Annual Review of Food Science and Technology*, *1*, 163–187.

Hoseney, R., & Rogers, D. (1994). Mechanism of sugar functionality in cookies. In *Vol. 1. The Science of cookie and cracker production* (pp. 203–225).

Hosseinian, F. S., Li, W., & Beta, T. (2008). Measurement of anthocyanins and other phytochemicals in purple wheat. *Food Chemistry*, *109*, 916–924. https://doi.org/10.1016/j.foodchem.2007.12.083.

Hou, Z., Qin, P., Zhang, Y., Cui, S., & Ren, G. (2013). Identification of anthocyanins isolated from black rice (Oryza sativa L.) and their degradation kinetics. *Food Research International*, *50*, 691–697. https://doi.org/10.1016/j.foodres.2011.07.037.

Hu, C., Cai, Y.-Z., Li, W., & Corke, H. (2007). Anthocyanin characterization and bioactivity assessment of a dark blue grained wheat (*Triticum aestivum* L. cv. Hedong Wumai) extract. *Food Chemistry*, *104*, 955–961. https://doi.org/10.1016/j.foodchem.2006.12.064.

Hu, X., Yang, X., & Liu, K. (2001). *Structure-activity relationship of tea polyphenols*.

Jakobek, L., Seruga, M., Medvidovic-Kosanovic, M., & Novak, I. (2007). Anthocyanin content and antioxidant activity of various red fruit juices. *Deutsche Lebensmittelrundschau*, *103*, 58.

Jennings, A., Welch, A. A., Fairweather-Tait, S. J., Kay, C., Minihane, A.-M., Chowienczyk, P., et al. (2012). Higher anthocyanin intake is associated with lower arterial stiffness and central blood pressure in women. *The American Journal of Clinical Nutrition*, *96*, 781–788. https://doi.org/10.3945/ajcn.112.042036.

Juhel, C., Armand, M., Pafumi, Y., Rosier, C., Vandermander, J., & Lairon, D. (2000). Green tea extract (AR25®) inhibits lipolysis of triglycerides in gastric and duodenal medium in vitro. *The Journal of Nutritional Biochemistry*, *11*, 45–51.

Kamiloglu, S., Capanoglu, E., Grootaert, C., & Van Camp, J. (2015). Anthocyanin absorption and metabolism by human intestinal Caco-2 cells—A review. *International Journal of Molecular Sciences*, *16*, 21555–21574.

Karakaya, S., Simsek, S., Eker, A. T., Pineda-Vadillo, C., Dupont, D., Perez, B., et al. (2016). Stability and bioaccessibility of anthocyanins in bakery products enriched with anthocyanins. *Food & Function*, 7, 3488–3496. https://doi.org/10.1039/C6FO00567E.

Karaś, M., Jakubczyk, A., Szymanowska, U., Złotek, U., & Zielińska, E. (2017). Digestion and bioavailability of bioactive phytochemicals. *International Journal of Food Science & Technology*, *52*, 291–305. https://doi.org/10.1111/ijfs.13323.

Kausar, H., Jeyabalan, J., Aqil, F., Chabba, D., Sidana, J., Singh, I. P., et al. (2012). Berry anthocyanidins synergistically suppress growth and invasive potential of human non-small-cell lung cancer cells. *Cancer Letters*, *325*, 54–62. https://doi.org/10.1016/j.canlet.2012.05.029.

Khan, N., & Mukhtar, H. (2013). Tea and health: Studies in humans. *Current Pharmaceutical Design*, *19*, 6141–6147.

Khlestkina, E., Usenko, N., Gordeeva, E., Stabrovskaya, O., Sharfunova, I., & Otmakhova, Y. (2017). Evaluation of wheat products with high flavonoid content: Justification of importance of marker-assisted development and production of flavonoid-rich wheat cultivars. Vavilovskij Zbreve˜urnal Genetiki i Selekcii/Vavilov. *Journal of Genetics and Breeding*, *21*, 545–553.

Koh, H. S. A., Lim, S. E. V., Lu, J., & Zhou, W. (2020). Bioactivity enhancement of fucoidan through complexing with bread matrix and baking. *LWT*, *130*, 109646.

Koh, H. S. A., Lu, J., & Zhou, W. (2019). Structure characterization and antioxidant activity of fucoidan isolated from Undaria pinnatifida grown in New Zealand. *Carbohydrate Polymers*, *212*, 178–185.

Koo, S. I., & Noh, S. K. (2007). Green tea as inhibitor of the intestinal absorption of lipids: Potential mechanism for its lipid-lowering effect. *The Journal of Nutritional Biochemistry*, *18*, 179–183.

Kumari, A., Sharma, S., Sharma, N., Chunduri, V., Kapoor, P., Kaur, S., et al. (2020). Influence of biofortified colored wheats (purple, blue, black) on physicochemical, antioxidant and sensory characteristics of chapatti (Indian flatbread). *Molecules*, *25*. https://doi.org/10.3390/molecules25215071.

Lachman, J., Martinek, P., Kotíková, Z., Orsák, M., & Šulc, M. (2017). Genetics and chemistry of pigments in wheat grain—A review. *Journal of Cereal Science*, *74*, 145–154. https://doi.org/10.1016/j.jcs.2017.02.007.

Lau, E., Soong, Y. Y., Zhou, W., & Henry, J. (2015). Can bread processing conditions alter glycaemic response? *Food Chemistry*, *173*, 250–256. https://doi.org/10.1016/j.foodchem.2014.10.040.

Lau, E., Zhou, W., & Henry, C. J. (2016). Effect of fat type in baked bread on amylose-lipid complex formation and glycaemic response. *British Journal of Nutrition*, *115*(12), 2122–2129. https://doi:10.1017/S0007114516001458.

Lee, Y.-M., Yoon, Y., Yoon, H., Park, H.-M., Song, S., & Yeum, K.-J. (2017). Dietary anthocyanins against obesity and inflammation. *Nutrients*, *9*. https://doi.org/10.3390/nu9101089.

Li, W., Beta, T., Sun, S., & Corke, H. (2006). Protein characteristics of Chinese black-grained wheat. *Food Chemistry*, *98*, 463–472.

Li, B., Lu, F., Wei, X., & Zhao, R. (2008). Fucoidan: Structure and bioactivity. *Molecules*, *13*, 1671–1695.

Li, W., Shan, F., Sun, S., Corke, H., & Beta, T. (2005). Free radical scavenging properties and phenolic content of Chinese black-grained wheat. *Journal of Agricultural and Food Chemistry*, *53*, 8533–8536.

Li, X., Zheng, T., Sang, S., & Lv, L. (2014). Quercetin inhibits advanced glycation end product formation by trapping methylglyoxal and glyoxal. *Journal of Agricultural and Food Chemistry*, *62*, 12152–12158.

Lin, J., Tan, Y. X. G., Leong, L. P., & Zhou, W. (2018). Steamed bread enriched with quercetin as an antiglycative food product: Its quality attributes and antioxidant properties. *Food & Function*, *9*, 3398–3407.

Lin, J., & Zhou, W. (2018). Role of quercetin in the physicochemical properties, antioxidant and antiglycation activities of bread. *Journal of Functional Foods*, *40*, 299–306.

Lu, T.-M., Lee, C.-C., Mau, J.-L., & Lin, S.-D. (2010). Quality and antioxidant property of green tea sponge cake. *Food Chemistry*, *119*, 1090–1095.

Luca, A., Cilek, B., Hasirci, V., Sahin, S., & Sumnu, G. (2014). Storage and baking stability of encapsulated sour cherry phenolic compounds prepared from micro- and Nano-suspensions. *Food and Bioprocess Technology*, 7, 204–211. https://doi.org/10.1007/s11947-013-1048-1.

Maner, S., Sharma, A. K., & Banerjee, K. (2017). Wheat flour replacement by wine grape pomace powder positively affects physical, functional and sensory properties of cookies. *Proceedings of the National Academy of Sciences, India Section B: Biological Sciences*, *87*, 109–113.

Massounga Bora, A. F., Ma, S., Li, X., & Liu, L. (2018). Application of microencapsulation for the safe delivery of green tea polyphenols in food systems: Review and recent advances. *Food Research International*, *105*, 241–249. https://doi.org/10.1016/j.foodres.2017.11.047.

Matsui, T., Ueda, T., Oki, T., Sugita, K., Terahara, N., & Matsumoto, K. (2001). α-Glucosidase inhibitory action of natural acylated anthocyanins. 1. Survey of natural pigments with potent inhibitory activity. *Journal of Agricultural and Food Chemistry, 49*, 1948–1951.

Mau, J.-L., Lee, C.-C., Chen, Y.-P., & Lin, S.-D. (2017). Physicochemical, antioxidant and sensory characteristics of chiffon cake prepared with black rice as replacement for wheat flour. *LWT, 75*, 434–439. https://doi.org/10.1016/j.lwt.2016.09.019.

Mau, J.-L., Lu, T.-M., Lee, C.-C., Lin, L.-Y., Cheng, C.-H., & Lin, S.-D. (2015). Physicochemical, antioxidant and sensory characteristics of chiffon cakes fortified with various tea powders. *Journal of Food Processing and Preservation, 39*, 443–450. https://doi.org/10.1111/jfpp.12249.

Mildner-Szkudlarz, S., Zawirska-Wojtasiak, R., Obuchowski, W., & Gośliński, M. (2009). Evaluation of antioxidant activity of green tea extract and its effect on the biscuits lipid fraction oxidative stability. *Journal of Food Science, 74*, S362–S370.

Morgounov, A., Karaduman, Y., Akin, B., Aydogan, S., Baenziger, P. S., Bhatta, M., et al. (2020). Yield and quality in purple-grained wheat isogenic lines. *Agronomy, 10*. https://doi.org/10.3390/agronomy10010086.

Morrison, L., Metzger, R., & Lukaszewski, A. (2004). Origin of the blue-aleurone gene in Sebesta blue wheat genetic stocks and a protocol for its use in apomixis screening. *Crop Science, 44*, 2063–2067.

Nakov, G., Brandolini, A., Hidalgo, A., Ivanova, N., Stamatovska, V., & Dimov, I. (2020). Effect of grape pomace powder addition on chemical, nutritional and technological properties of cakes. *LWT, 134*, 109950. https://doi.org/10.1016/j.lwt.2020.109950.

Namal Senanayake, S. P. J. (2013). Green tea extract: Chemistry, antioxidant properties and food applications—A review. *Journal of Functional Foods, 5*, 1529–1541. https://doi.org/10.1016/j.jff.2013.08.011.

Ning, J., Hou, G. G., Sun, J., Wan, X., & Dubat, A. (2017). Effect of green tea powder on the quality attributes and antioxidant activity of whole-wheat flour pan bread. *LWT—Food Science and Technology, 79*, 342–348. https://doi.org/10.1016/j.lwt.2017.01.052.

Papillo, V. A., Locatelli, M., Travaglia, F., Bordiga, M., Garino, C., Arlorio, M., et al. (2018). Spray-dried polyphenolic extract from Italian black rice (Oryza sativa L., var. Artemide) as new ingredient for bakery products. *Food Chemistry, 269*, 603–609. https://doi.org/10.1016/j.foodchem.2018.07.059.

Pasqualone, A., Bianco, A. M., Paradiso, V. M., Summo, C., Gambacorta, G., & Caponio, F. (2014). Physico-chemical, sensory and volatile profiles of biscuits enriched with grape marc extract. *Food Research International, 65*, 385–393.

Pasqualone, A., Bianco, A. M., Paradiso, V. M., Summo, C., Gambacorta, G., Caponio, F., et al. (2015). Production and characterization of functional biscuits obtained from purple wheat. *Food Chemistry, 180*, 64–70. https://doi.org/10.1016/j.foodchem.2015.02.025.

Pasrija, D., Ezhilarasi, P. N., Indrani, D., & Anandharamakrishnan, C. (2015). Microencapsulation of green tea polyphenols and its effect on incorporated bread quality. *LWT—Food Science and Technology, 64*, 289–296. https://doi.org/10.1016/j.lwt.2015.05.054.

Paznocht, L., Kotíková, Z., Orsák, M., Lachman, J., & Martinek, P. (2019). Carotenoid changes of colored-grain wheat flours during bun-making. *Food Chemistry, 277*, 725–734.

Peng, X., Ma, J., Chen, F., & Wang, M. (2011). Naturally occurring inhibitors against the formation of advanced glycation end-products. *Food & Function, 2*, 289–301.

Perumalla, A. V. S., & Hettiarachchy, N. S. (2011). Green tea and grape seed extracts—Potential applications in food safety and quality. *Food Research International, 44*, 827–839. https://doi.org/10.1016/j.foodres.2011.01.022.

Peters, C. M., Green, R. J., Janle, E. M., & Ferruzzi, M. G. (2010). Formulation with ascorbic acid and sucrose modulates catechin bioavailability from green tea. *Food Research International, 43*, 95–102. https://doi.org/10.1016/j.foodres.2009.08.016.

Phongnarisorn, B., Orfila, C., Holmes, M., & Marshall, L. J. (2018). Enrichment of biscuits with matcha green tea powder: Its impact on consumer acceptability and acute metabolic response. *Food, 7*, 17.

Raederstorff, D. G., Schlachter, M. F., Elste, V., & Weber, P. (2003). Effect of EGCG on lipid absorption and plasma lipid levels in rats. *The Journal of Nutritional Biochemistry, 14*, 326–332.

Rashidinejad, A., Boostani, S., Babazadeh, A., Rehman, A., Rezaei, A., Akbari-Alavijeh, S., et al. (2021). Opportunities and challenges for the nanodelivery of green tea catechins in functional foods. *Food Research International, 142*, 110186.

Rice-Evans, C., Miller, N., & Paganga, G. (1997). Antioxidant properties of phenolic compounds. *Trends in Plant Science, 2*, 152–159.

Rodriguez-Mateos, A., Cifuentes-Gomez, T., George, T. W., & Spencer, J. P. E. (2014). Impact of cooking, proving, and baking on the (poly)phenol content of wild blueberry. *Journal of Agricultural and Food Chemistry, 62*, 3979–3986. https://doi.org/10.1021/jf403366q.

Rosales-Soto, M. U., Powers, J. R., & Alldredge, J. R. (2012). Effect of mixing time, freeze-drying and baking on phenolics, anthocyanins and antioxidant capacity of raspberry juice during processing of muffins. *Journal of the Science of Food and Agriculture, 92*, 1511–1518. https://doi.org/10.1002/jsfa.4735.

Saini, P., Kumar, N., Kumar, S., Mwaurah, P. W., Panghal, A., Attkan, A. K., et al. (2020). Bioactive compounds, nutritional benefits and food applications of colored wheat: A comprehensive review. *Critical Reviews in Food Science and Nutrition*. https://doi.org/10.1080/10408398.2020.1793727.

Sampson, L., Rimm, E., Hollman, P. C., de Vries, J. H., & Katan, M. B. (2002). Flavonol and flavone intakes in US health professionals. *Journal of the American Dietetic Association, 102*, 1414–1420.

Sanders, M. E. (1998). Overview of functional foods: Emphasis on probiotic bacteria. *International Dairy Journal, 8*, 341–347.

Sang, S., Lee, M., Yang, I., Buckley, B., & Yang, C. S. (2008). Human urinary metabolite profile of tea polyphenols analyzed by liquid chromatography/electrospray ionization tandem mass spectrometry with data-dependent acquisition. *Rapid Communications in Mass Spectrometry: An International Journal Devoted to the Rapid Dissemination of Up-to-the-Minute Research in Mass Spectrometry, 22*, 1567–1578.

Sanlier, N., Gokcen, B. B., & Altuğ, M. (2018). Tea consumption and disease correlations. *Trends in Food Science & Technology, 78*, 95–106. https://doi.org/10.1016/j.tifs.2018.05.026.

Šaponjac, V. T., Ćetković, G., Čanadanović-Brunet, J., Pajin, B., Djilas, S., Petrović, J., et al. (2016). Sour cherry pomace extract encapsulated in whey and soy proteins: Incorporation in cookies. *Food Chemistry, 207*, 27–33.

Sharma, A. (2013). Tea catechins: Their stability and roles in the biscuit making process and effects on biscuit quality. PhD Thesis. National University of Singapore, Singapore.

Sharma, A., & Zhou, W. (2011). A stability study of green tea catechins during the biscuit making process. *Food Chemistry, 126*, 568–573. https://doi.org/10.1016/j.foodchem.2010.11.044.

Sivam, A., Sun-Waterhouse, D., Perera, C., & Waterhouse, G. (2013). Application of FT-IR and Raman spectroscopy for the study of biopolymers in breads fortified with fibre and polyphenols. *Food Research International, 50*, 574–585.

Sivam, A. S., Sun-Waterhouse, D., Quek, S., & Perera, C. O. (2010). Properties of bread dough with added fiber polysaccharides and phenolic antioxidants: A review. *Journal of Food Science, 75*, R163–R174.

Stuart, M. D. (2004). *Review of research on Undaria pinnatifida in New Zealand and its potential impacts on the eastern coast of the South Island*. Citeseer.

Sui, X., Dong, X., & Zhou, W. (2014). Combined effect of pH and high temperature on the stability and antioxidant capacity of two anthocyanins in aqueous solution. *Food Chemistry*, *163*, 163–170.

Sui, X., Yap, P. Y., & Zhou, W. (2015). Anthocyanins during baking: Their degradation kinetics and impacts on color and antioxidant capacity of bread. *Food and Bioprocess Technology*, *8*, 983–994.

Sui, X., Zhang, Y., & Zhou, W. (2016). Bread fortified with anthocyanin-rich extract from black rice as nutraceutical sources: Its quality attributes and in vitro digestibility. *Food Chemistry*, *196*, 910–916. https://doi.org/10.1016/j.foodchem.2015.09.113.

Sun, L., & Miao, M. (2020). Dietary polyphenols modulate starch digestion and glycaemic level: A review. *Critical Reviews in Food Science and Nutrition*, *60*, 541–555. https://doi.org/10.1080/10408398.2018.1544883.

Sun, S., Sun, Y., Yuan, W., Yan, W., Pei, Z., Zhang, M., et al. (1999). Breeding and qualitative analysis for black grain wheat 76 of superior quality. *Acta Agronomica Sinica*, *25*(1), 50–54.

Suvaluk, A. (2018). Possibilities of enriching food products with anthocyanins by using new forms of cereals. *Foods and Raw Materials*, *6*, 128–135.

Syed Jaafar, S. N., Baron, J., Siebenhandl-Ehn, S., Rosenau, T., Böhmdorfer, S., & Grausgruber, H. (2013). Increased anthocyanin content in purple pericarp × blue aleurone wheat crosses. *Plant Breeding*, *132*, 546–552.

Tagliazucchi, D., Verzelloni, E., Bertolini, D., & Conte, A. (2010). In vitro bio-accessibility and antioxidant activity of grape polyphenols. *Food Chemistry*, *120*, 599–606.

Tańska, M., Roszkowska, B., Czaplicki, S., Borowska, E. J., Bojarska, J., & Dąbrowska, A. (2016). Effect of fruit pomace addition on shortbread cookies to improve their physical and nutritional values. *Plant Foods for Human Nutrition*, *71*, 307–313. https://doi.org/10.1007/s11130-016-0561-6.

Theagarajan, R., Malur Narayanaswamy, L., Dutta, S., Moses, J. A., & Chinnaswamy, A. (2019). Valorisation of grape pomace (cv. Muscat) for development of functional cookies. *International Journal of Food Science & Technology*, *54*, 1299–1305.

Tian, S., Chen, Z., & Wei, Y. (2018). Measurement of colour-grained wheat nutrient compounds and the application of combination technology in dough. *Journal of Cereal Science*, *83*, 63–67.

Tolve, R., Simonato, B., Rainero, G., Bianchi, F., Rizzi, C., Cervini, M., et al. (2021). Wheat bread fortification by grape pomace powder: Nutritional, technological, antioxidant, and sensory properties. *Food*, *10*. https://doi.org/10.3390/foods10010075.

Unno, T., Tago, M., Suzuki, Y., Nozawa, A., Sagesaka, Y., Kakuda, T., et al. (2005). Effect of tea catechins on postprandial plasma lipid responses in human subjects. *British Journal of Nutrition*, *93*(4), 543–547. https://doi.org/10.1079/BJN20041379.

Vishchuk, O. S., Ermakova, S. P., & Zvyagintseva, T. N. (2011). Sulfated polysaccharides from brown seaweeds Saccharina japonica and Undaria pinnatifida: Isolation, structural characteristics, and antitumor activity. *Carbohydrate Research*, *346*, 2769–2776.

Walker, R., Tseng, A., Cavender, G., Ross, A., & Zhao, Y. (2014). Physicochemical, nutritional, and sensory qualities of wine grape pomace fortified baked goods. *Journal of Food Science*, *79*, S1811–S1822. https://doi.org/10.1111/1750-3841.12554.

Wallentinus, I. (2007). Alien species alert: Undaria pinnatifida:(wakame or japanese kelp). *The Council*.

Wang, H., Provan, G. J., & Helliwell, K. (2000). Tea flavonoids: Their functions, utilisation and analysis. *Trends in Food Science & Technology*, *11*, 152–160.

Wang, W., Sun, C., Mao, L., Ma, P., Liu, F., Yang, J., et al. (2016). The biological activities, chemical stability, metabolism and delivery systems of quercetin: A review. *Trends in Food Science & Technology*, *56*, 21–38. https://doi.org/10.1016/j.tifs.2016.07.004.

Wang, J., Zhang, Q., Zhang, Z., & Li, Z. (2008). Antioxidant activity of sulfated polysaccharide fractions extracted from Laminaria japonica. *International Journal of Biological Macromolecules, 42*, 127–132.

Wang, R., & Zhou, W. (2004). Stability of tea catechins in the breadmaking process. *Journal of Agricultural and Food Chemistry, 52*, 8224–8229.

Wang, R., Zhou, W., & Isabelle, M. (2007). Comparison study of the effect of green tea extract (GTE) on the quality of bread by instrumental analysis and sensory evaluation. *Food Research International, 40*, 470–479. https://doi.org/10.1016/j.foodres.2006.07.007.

Wang, R., Zhou, W., & Jiang, X. (2008). Mathematical modeling of the stability of green tea catechin epigallocatechin gallate (EGCG) during bread baking. *Journal of Food Engineering, 87*, 505–513.

Wang, R., Zhou, W., Yu, H., & Chow, W. (2006). Effects of green tea extract on the quality of bread made from unfrozen and frozen dough processes. *Journal of the Science of Food and Agriculture, 86*, 857–864.

Yang, B., Liu, H., Yang, J., Gupta, V. K., & Jiang, Y. (2018). New insights on bioactivities and biosynthesis of flavonoid glycosides. *Trends in Food Science & Technology, 79*, 116–124.

Yilmaz, Y. (2006). Novel uses of catechins in foods. *Trends in Food Science & Technology, 17*, 64–71.

Yilmazer-Musa, M., Griffith, A. M., Michels, A. J., Schneider, E., & Frei, B. (2012). Grape seed and tea extracts and catechin 3-gallates are potent inhibitors of α-amylase and α-glucosidase activity. *Journal of Agricultural and Food Chemistry, 60*, 8924–8929.

Yoshikawa, T., Inoue, R., Matsumoto, M., Yajima, T., Ushida, K., & Iwanaga, T. (2011). Comparative expression of hexose transporters (SGLT1, GLUT1, GLUT2 and GLUT5) throughout the mouse gastrointestinal tract. *Histochemistry and Cell Biology, 135*, 183–194.

Yu, L., & Beta, T. (2015). Identification and antioxidant properties of phenolic compounds during production of bread from purple wheat grains. *Molecules, 20*. https://doi.org/10.3390/molecules200915525.

Zhang, X., Chen, F., & Wang, M. (2014). Antioxidant and antiglycation activity of selected dietary polyphenols in a cookie model. *Journal of Agricultural and Food Chemistry, 62*, 1643–1648.

CHAPTER SEVEN

Sugar, salt and fat reduction of bakery products

Nantawan Therdthai*
Department of Product Development, Faculty of Agro-Industry, Kasetsart University, Bangkok, Thailand
*Corresponding author: e-mail address: faginwt@ku.ac.th

Contents

Abstract

From the health viewpoint, consumers monitor their food uptake in terms of both quality and quantity, due to awareness of the link between food ingredients and health. Accordingly, many people tend to buy healthy food products that are low in or free of sugar, salt and fat. However, in baked products, the sugar, fat and salt are needed to create unique characteristics. The role of sugar, salt and fat in baked products is discussed in this chapter to understand their functions. The understanding is necessary to design proper techniques to reduce the amount of sugar, salt and fat. Ingredients and additives that can substitute for fat, sugar and salt in bakery characteristics are reviewed in terms of their advantages and disadvantages. In addition, alternative processes to reduce the use of fat, sugar and salt are proposed in this chapter.

Advances in Food and Nutrition Research, Volume 99
ISSN 1043-4526
https://doi.org/10.1016/bs.afnr.2021.11.004

1. Health concerns: The driving force to reduce sugar, salt and fat in baked products

By consuming junk foods with high calories and doing less exercise, people tend to have over-weight problems that can lead to obesity. In addition, non-communicable diseases (NCDs) are increasingly disturbing many people's lifestyle. Diabetes is one of the most common NCDs and is characterized by the restriction or absence of blood glucose absorption from blood to cells. Therefore, high blood glucose is often detected with symptoms being weight loss, ketoacidosis, blurred vision and infection. High blood pressure is a symptom found in cardiovascular disease, another common NCD. It is characterized by a blood pressure above 150 mmHg that may result from high cholesterol and stress. One way to prevent these NCDs is to balance diet and to exercise regularly. Reduced consumption of fat, sugar and salt is recommended to avoid NCDs.

Not only does excess sugar consumption adversely affect the physical function of the human body, but it also impacts mental health that may be indicated by poor cognition and depression (Bassoli, 2006). To reduce sugar intake from foods, artificial sweeteners and sugar alcohols may be used to replace sugar and thereby reduce calories. However, they might reduce the diversity of bacteria in the gut. Consequently, gene and metabolic pathway expressions may be changed and possibly result in obesity (Sahin, Zannini, Coffey, & Arendt, 2019).

Although salt contains not a high level of calories, excess salt consumption may result in hypertension, cardiovascular disease, stroke, kidney disease, stomach cancer and bone demineralization. Approximately 65–70% of the daily salt intake for humans is via meals, usually from processed food, particularly bread (Silow, Axel, Zannini, & Arendt, 2016). Another 30–35% of the daily salt intake may be via natural diets. Generally, a person's daily salt intake should not exceed 5 g. Thus, almost all nations encourage their citizens to reduce their daily intake of salt. For example, Federal Ministry for Health in Austria started reformulation strategies in 2010 to achieve a 15% reduction of salt in baked products by 2015. The Ministry of Health in Italy aimed to reduce salt in bakery and pasta products by 10% in 2012 (Belc, Smeu, Macri, Vallauri, & Flynn, 2019). Some countries, such as Hungary, have encouraged the food industry to lower the use of salt in the food formulas by introducing new regulations such as a salt tax.

Fat is an energy source with a high level of calories. Overconsumption of fat may lead to being overweight and subsequent NCDs. An emerging trend has seen consumers paying much more attention to their health by reducing or avoiding fatty food consumption. However, in many countries, bread and bread products are staple foods consumed daily. Generally, baked products must contain fat to develop desirable characteristics such as texture, structure and flavor. Reducing fat while still maintaining the desirable characteristics in baked products is a challenge for the food industry.

From the health viewpoint, consumers monitor their food uptake in terms of both quality and quantity, due to awareness of the link between food ingredients and health. Accordingly, many people tend to buy healthy food products that are low in or free of salt, fat and sugar; however, such products may not be considered tasty. Nonetheless, some ingredients and food additives can be utilized to maintain the desirable qualities of the original recipe. However, some consumers may be concerned about unfamiliar additives displayed on the label. Thus, there is a new trend for food products to have a 'clean label' indicating the natural ingredients and additives which can still perform the required function to maintain the desirable qualities of food.

Bakery products are widely consumed globally due to their desirable texture and flavor. They require continual development to increase product quality. Bread and cake are majority of bakery. They contain a high moisture content (33–38% wb.) and a high carbohydrate content (81–87% db.). Their protein content is generally in the range of 8–13% db (Fernandes & Salas-Mellado, 2017).

In some baked products such as puff pastry, cakes and biscuits, high amounts of fat, sugar and salt are needed to create unique characteristics. To reduce the amounts of fat, sugar and salt, some food additives are needed along with special techniques and modifications of the recipe. However, the roles of fat, sugar and salt in baked products must be clearly understood to select the proper additives and techniques.

Therefore, this chapter discusses the functions of fat, sugar and salt in the development of bakery characteristics. Ingredients and additives that can substitute fat, sugar and salt in bakery products are reviewed in terms of their advantages and disadvantages. In addition, alternative processes to reduce the use of fat, sugar and salt are proposed in this chapter.

2. Role of sugar, salt and fat in baked products

2.1 Sugar

Sugar is defined as monosaccharides and disaccharides, not including polyols. The sugar components commonly used in baked products are glucose (e.g., regular corn syrup), fructose (e.g., high fructose corn syrup) and sucrose (e.g., table sugar). Their characteristics and physicochemical properties vary; however, in baked products, sugar plays important roles in structure, texture, taste and shelf-life (Cauvain & Young, 2006).

2.1.1 Role of sugar during mixing

Wheat flour and water are the major components used to form a bread dough. The flour mainly consists of starches (including amylose and amylopectin) and proteins, especially gluten. These are macromolecules that can absorb water during dough making. When sugar is added to the system, it can also be bound immediately with water molecules due to the formation of hydrogen bonds. Therefore, the proteins must compete with sugar to absorb free water and thus, the addition of sugar may delay or weaken protein network formation in the dough. To freeze dough containing sugar-bound water, the temperature should be reduced or the freezing time should be extended due to the freezing point depression.

Batter containing fat and a high sugar content has a limited initial formation of the protein network that instead is formed at the end of baking (Hesso et al., 2015). Sugar and fat are initially mixed at high speed in the creaming process. The sugar can function as an emulsification enhancer by forming linkages with fat, particularly in dough with a low water content, such as in biscuits (Clemens et al., 2016). The fat prevents the sugar from dissolving in the water phase by enveloping the sugar crystals. Subsequently, the foam structure is strong, homogenous, and stable enough to hold air in the batter, resulting in a high volume of the baked product, although the protein network is weakened by the added sugar (by around 27–45%). Consequently, high volume and a soft crumb with low resilience and chewiness is obtained in cake (Cauvain & Young, 2006).

2.1.2 Role of sugar during proving

After mixing, the dough needs some time to rest and rise in a process known as 'proving' with the subsequent rising of the product known as 'leavening'. The dough rises due to yeast fermentation often involving *Saccharomyces cerevisiae*. The optimum proving temperature around 35–45 °C and a relative humidity around 80–90% can help the progression of effective yeast fermentation (Therdthai, Zhou, & Jangchud, 2007). In addition, yeast needs sugar as a carbon source for its metabolism. Generally, 5–20% sugar is added to the leavened product (Cauvain & Young, 2007). The added sugar is normally sucrose that can be converted by yeast into glucose. Then, the glucose is converted into ethyl alcohol and carbon dioxide gas through the fermentation process. With the strong protein network of dough, the produced gas is trapped, resulting in the dough rising during proving. Therefore, sugar addition is necessary for the creation of a proper dough and bakery structure. However, too much sugar addition causes a high osmotic pressure that may inhibit yeast growth and the activity of carbon dioxide gas production (Sahin et al., 2019), resulting in a bread product with dense structural characteristics.

2.1.3 Role of sugar during baking

In the process of baking, the hydrogen bonds between the sugar and water molecules must be broken before the water can boil and evaporate. Therefore, the boiling point of the sugar-bound water is increased. Furthermore, adding sugar into the starch solution to develop the dough can increase the hydration and degree of swelling of the starch. This requires a higher amount of energy obtained by increasing the baking temperature to ensure complete starch gelatinization that in turn stabilizes the crystalline regions of the starch (Clemens et al., 2016). When 50% sugar (on a flour basis) was added to the batter, the starch gelatinization temperature increased by 20 °C (from 60 to 80 °C). This change can delay protein denaturation and starch gelatinization during baking so that the product volume can properly expand in the early stage of baking (Sahin et al., 2019). This is beneficial in non-bread products such as cake. In contrast, the addition of normal amounts of sugar in the dough (generally 30–40%) does not significantly affect protein coagulation temperature.

However, in the case of cake containing high sugar content (54% sucrose w/w), the denaturing temperature of protein was increased to 13 °C (Renzetti, van den Hoek, & van der Sman, 2020). This could delay protein denaturation and starch gelatinization processes during baking. During the late baking process, water is released from the denaturation of protein, particularly from the ovalbumin in egg white, and subsequently used for polymerization of SS bonds and hydrophobic interactions (Lambrecht, Deleu, Rombouts, & Delcour, 2018). Then, the protein network is formed through the polymerization of the four free-sulfhydryl groups. Consequently, baking qualities particularly springiness and cohesiveness were possibly varied in accordance with sugar content (Deleu et al., 2015).

In recipes with high fat and sugar but low water and starch, the high temperature during baking makes sugar and fat melt, resulting in decreased dough viscosity and the dough pieces start spreading. Eventually, the spreading is inhibited when the gluten meets its glass transition temperature (Sahin et al., 2019). Therefore, sugar has an impact on the diameter and thickness of baked goods.

The added sugar, especially reducing sugars such as glucose and fructose, can react with free amino acids through the Maillard reaction during baking. An increased temperature during baking can be one of the parameters that enhances the Maillard reaction rate. The Maillard reaction generates melanoidin as the intermediate substance and develops the brown color and the flavor of baked products. The reaction rate is dependent on many parameters such as the amounts of reducing sugar and free amino acids, temperature, water activity and pH. During the late stage of baking when there is a high temperature, particularly on the crust, the sugar melts and dehydration progresses, thereby generating a caramelization reaction at temperatures above 120 °C (Kocadağlı & Gökmen, 2019). Similar to the Maillard reaction, caramelization creates both a brown color and the flavor of the baked product.

2.1.4 Role of sugar during storage

After baking, the product is cooled before packing. During cooling, sucrose recrystallization can occur and increase the hardness and fracturability which are required characteristics of biscuits. The addition of sugar into the dough can reduce the free water during storage and thereby water activity is reduced. The low water activity can promote a longer shelf-life. Notably, sugar (particularly sucrose) effectively decreases starch retrogradation because it can stabilize the amorphous and entangled state of the gelatinized starch as well as increase the glass transition temperature (Sahin et al., 2019).

2.2 Salt

Salt, commonly known in cooking as table salt, is composed of one sodium ion and one chloride ion with the two ions linked via an ionic bond. Together, these constitute the principal ions in extracellular fluids such as blood plasma. Their function is to maintain membrane potential which is related to the transmission of nerve impulses, the contraction of muscles and cardiac function. Once sodium is taken and delivered through the digestion system, the sodium ion is absorbed into the blood system in the intestine. The estimated daily requirement for salt in humans is approximately 0.2–1.5 g (Silow, Axel, et al., 2016). Therefore, the recommended daily sodium intake is less than 2 g (WHO World Health Organization, 2012). Some countries have set a lower limit for salt intake for aging people.

In 2017, Omeroglu and Ozdal (2020) surveyed the nutritional composition of baked products sold in Turkey and found that the maximum salt content in baked products was around 0.9 g/100 g, being lower than the traditional level of 1.6–2.9 g per 100 g product (Silow, Axel, et al., 2016). This could have resulted from regulations to reduce salt consumption.

Salt is one of the food additives in baked products. It not only provides saltiness but also plays an important role in dough properties, yeast fermentation, volume expansion, color, flavor and shelf-life. Salt may not have a significant effect on baked products that do not require much gluten formation, such as cake, biscuits and pastries. Conversely, the effect of salt is likely much greater in bread, pizza, fermented products and laminated products such as puff pastry and croissants. Thus, salt reduction in such items is not simple.

2.2.1 Role of salt during mixing

During mixing, flour is hydrated to initiate the development of gluten network characterized by viscoelastic behavior. Gluten is composed of gliadin which contributes to viscosity and glutenin which contributes to elasticity. The addition of salt in the dough possibly reduces water absorption by the protein in the flour. With 1–2% salt addition, the water binding activity of dough is increased, while the dough consistency is strengthened, as measured using a farinograph. However, salt delays the hydration of gluten protein because the salt can shield the charged amino acids which are responsible for hydration. Normally, only a small portion of the charged amino acid is available in the gluten protein. However, the presence of salt reduces the available charged amino acid even more. Consequently, the electrostatic repulsion between protein molecules decreases and the β-sheet structure of the protein is reduced (Wellner et al., 2005). A fibrous structure is more likely formed in the gluten network and therefore, the development of the gluten matrix takes longer than usual so that the mixing time must be extended; the longer mixing time possibly increases loaf volume and baking quality.

In contrast, insufficient salt might weaken the gluten network resulting in incomplete dough molding and lamination. In addition, dough with a reduced salt content tends to be sticky and this can disrupt the operation of doughmaking machinery. There may be various degrees of salt reduction that still maintain an acceptable quality of bakery products. Some studies have reported a possible reduction of salt content by 25% for pizza (Diler, Le-Bail, & Chevallier, 2016) and by 30% for puff pastry (Silow, Zannini, Axel, Lynch, & Arendt, 2016). Salt also plays an important role in the extensibility of the dough. Reducing the salt from 1.2% to 0.3% resulted in no significant change in dough extensibility. However, salt reduction to 0% results in significantly decreased dough extensibility. Generally, it is possible to reduce salt to 0.3–0.6% without a significant effect on the rheological properties of dough (Lynch, Dal Bello, Sheehan, Cashman, & Arendt, 2009). In addition, salt plays a role in microbial inhibition. Thus, the mixing step has to be optimized to avoid direct contact between the salt and yeast.

2.2.2 Role of salt during proving

Baking soda, also known as baking powder, is one of the sodium-based leavening agents used to increase gas bubbles and thereby improve dough expansion during proving. Production of carbon dioxide gas from yeast fermentation during proving can be reduced by salt addition. This is due to the fact that an increase in salt may increase the osmotic pressure and electrochemical potential of Na^+ and Cl^- ions on the yeast membrane. However, the salt strengthens the dough structure so that more air bubbles can be trapped within the dough; this progresses dough expansion during proving. Reducing the amount of salt weakens the gluten network resulting in a greater release of carbon dioxide from the dough. Then, the dough volume expansion might not be that much improved. Therefore, adjusting the salt content must be done carefully with the optimized proving time to avoid 'over proving'.

2.2.3 Role of salt during baking

During baking, the dough is transformed from a liquid-like to a solid-like structure to create the desirable texture, color and flavor. The dough texture is related to the volume and structure of the dough during proving. With the proper salt content, the gluten network is strengthened and collapse of the structure is unlikely to occur. Salt reduction has both positive and negative effects on the final volume of the baked product. In the absence of salt, bread volume is slightly less than for regular bread and the bread may be tasteless and spoiled by yeasty and musty flavors (Silow, Axel, et al., 2016). These undesirable flavors can be reduced by the addition of salt. Taste tests revealed that a difference in saltiness was detected when the amount of salt in the bread was reduced from 1.3% to 1.0%. The detection of saltiness is also related to crumb texture and chewing. As the Na^+ ion is bound with negatively charged amino acids from ionic interactions, chewing bread crumb properly can lead to salt being released and detected. Therefore, crumby bread from salt reduction possibly slows down the detection of saltiness during chewing (Pflaum, Konitzer, Hofmann, & Koehler, 2013). Up to 25% salt reduction could still produce acceptable bread (Girgis et al., 2003).

In addition, color is influenced by the salt content. During baking, the plasticizing effect from salt addition can increase the mobility of the reducing sugar and amino acids which are the substrates of Maillard reaction. So, the presence of salt enhances Millard reaction, leading to an improved browning index (Silow, Axel, et al., 2016). Conversely, for bread with no salt, the yeast activity is high and this possibly uses majority of the sugar for fermentation during proving. Consequently, there is only a small amount of sugar available for the Maillard reaction during baking and therefore, the color of the salt-free bread tends to be paler than regular bread containing salt. However, the reduction in the salt content might not affect baked products that do not involve fermentation.

2.2.4 Role of salt during storage

Salt plays a role in microbial inhibition by lowering the water activity. Generally, baked products have a water activity of approximately 0.96–0.98, with products containing higher salt content tending to have a lower water activity and a longer shelf-life due to the inhibition of the growth of molds (such as *Penicillium expansum*, *Fusarium culmorum* and *Aspergillus niger*) and bacteria (such as *Bacillus* species) (Silow, Axel, et al., 2016). Where less salt is used, some antimicrobial compounds should be included to maintain the same shelf-life. However, in the case of one-day items (such as puff pastry), salt reduction is possible without any serious effect on shelf-life.

2.3 Fat

A fat molecule consists of triglycerides with glycerol forming the backbone of the fat structure that is connected with fatty acids consisting of hydrocarbon chains with a methyl group end and a carboxylic acid group end. The fatty acids use the carboxylic acid group end to link with the glycerol backbone via ester linkages. Fatty acids can be classified as either saturated or unsaturated. Saturated fatty acids are characterized by having only single C—C bonds organized into a linear structure such as lauric (C12:0), myristic (C14:0), palmitic (C16:0) and stearic (C18:0) forms. The triglycerides containing a lot of saturated fatty acid likely have a crystalline structure that is stable and has a high melting point (above 20 °C) (Talbot, 2011). Consequently, saturated fat is solid at ambient temperature which makes it good for mixing in baked products. However, blood clotting, inflammation, insulin resistance and adverse impacts on the cardiovascular systems can be resulted from an increased intake of saturated fatty acids (Omeroglu & Ozdal, 2020). In addition, saturated fatty acids increase cholesterol levels in the blood. Consequently, the European Food Safety Agency Panel on Dietetic Products, Nutrition and Allergies recommends that the saturated fatty acid intake from food be as low as possible (EFSA (European Food Safety Authority), 2017).

Triglycerides containing a lot of unsaturated fatty acids have a lower melting point and likely be present in liquid form at ambient temperature. Unsaturated fatty acids can be in either '*cis*' or '*trans*' configurations. The *cis* structure is a bending structure characterized by the carbon-carbon chains being on the same side of the double bond, such as oleic acid (C18:1), linoleic acid (C18:2), and linolenic acid (C18:3). The *trans* form is a straighter chain structure characterized by the carbon-carbon chains being on opposite sides of the double bond, such as elaidic acid (C18:1 t). The *trans* form is thermodynamically more stable than the *cis* structure and has a melting point and functions in the texture of food in similar ways to saturated fatty acids (Talbot, 2011). Therefore, trans fatty acid had been used to replace saturated fatty acids for some time until a recent discovery identified that consuming food containing high trans fatty acids can increase low-density lipoprotein (LDL) cholesterol and decrease high-density lipoprotein (HDL) cholesterol in the blood. The risk of high cholesterol levels in the blood from consuming trans unsaturated fatty acids is much higher than from saturated fatty acids (Omeroglu & Ozdal, 2020). Therefore, many countries have banned trans unsaturated fatty acids in their foods. In Europe, trans fat (not including natural trans fat in food) cannot be higher than 2 g/100 g of total fat (European Union Commission, 2019). To claim to be "trans fatty acid-free", the content of trans fatty acid must be less than 1 g/100 g of total fat in the product (Omeroglu & Ozdal, 2020).

In baked products, fat is an important ingredient in developing a desirable texture and flavor. The fat is mainly from butter, margarine, shortening and vegetable oils, with these ingredients being used on their own or as a mixture. The selection of types and ratios of fat ingredients depends on the product type and the chemical and physical properties of each fat. Hydrogenated shortenings used to be a common fat ingredient widely used in baked products. However, in the last two decades, hydrogenated shortenings composed of saturated fatty acids and trans fatty acids have been confirmed to promote health problems by increasing total cholesterol and low-density lipoprotein levels (Mert & Demirkesen, 2016). Furthermore, consuming trans fatty acids can cause diabetes, cardiovascular disease, obesity, breast cancer, prostate cancer, infertility and coronary artery disease (Abd El-Aal, Mohamed Abdel-Fattah, & El-Dawy Ahmed, 2019).

Although fat tends to cause health disorders, it is useful as a source of energy and vitamins such as vitamin A, D, E and K. A maximum 35% of the total daily calorie intake should be obtained from fat (Omeroglu & Ozdal, 2020). Low fat and reduced fat baked products with desirable eating quality can be formulated to avoid saturated fatty acid and trans fatty acid by using cis-monounsaturated fatty acid (MUFA) and cis-polyunsaturated fatty acid (PUFA) (Tarancón, Fiszman, Salvador, & Tárrega, 2013).

Generally, the amounts of saturated fatty acids in baked products such as bread, cookies, pies, tarts and waffles are above 50% of total fatty acids, particularly waffles where the amount is 72.8% of total fatty acids. Cake and puff pastry seem to have slightly less saturated fatty acids than other baked products (around 44% of total fatty acids) (Omeroglu & Ozdal, 2020). All baked products contain less than 0.2% trans fatty acids of total fatty acids. The MUFA in all baked products is in a similar range (20.6–42.2% of total fatty acid) and is less than the level of saturated fatty acids (52.1–72.8% of total fatty acid). Palmitic acid (C16:0) and oleic acid (C18:1) are the main fatty acids in baked products, in the ranges of 24.5–38.6% and 20.2–41.3%, respectively, of total fatty acids. Almost all baked products contain MUFA and PUFA in the range of 4.3–9.4 g/100 g sample and 0.7–2.8 g/100 g sample, respectively. In baked products, the ratio of MUFA to saturated fatty acid is from 0.3 to 0.9, while the ratio of PUFA to saturated fatty acid is from 0 to 0.3. Bread, cake and puff pastry tend to have higher ratios of MUFA-to-saturated fatty acid and PUFA-to-saturated fatty acid than other baked products (Omeroglu & Ozdal, 2020).

2.3.1 *Role of fat during mixing*

During dough mixing, protein and starch are hydrated. With the presence of fat, the hydrolysis of protein and starch is inhibited by hydrophobic interactions between the fat and protein or between the fat and starch. For biscuits and puff pastry, fat is required to shorten the dough as an important factor in dough rheology. Improper shortening decreases the complex modulus of dough and increases the dough density. For bread, a reduction in fat can increase dough elasticity, cohesiveness and adhesiveness. For cake, fat helps to incorporate gas into the batter during the creaming phase and stabilize gas bubbles. Consequently, the baked product has a smaller volume and its eating quality deteriorates (Moriano, Cappa, & Alamprese, 2018). Generally, fat replacers are required to maintain the desirable characteristics of the dough.

2.3.2 Role of fat during baking

Fat incorporates air in the cake batter and obstructs protein from breaking the continuity of the gluten. Thus, a cake containing sufficient fat should have soft and tender crumbs. After mixing all ingredients, the batter should have all the proper rheological properties and be ready for baking. Fat is generally in a solid form at ambient temperature so that when the dough or batter is placed in the baking oven, the fat melts and the rheological properties of the dough or batter change. After the batter is heated to 50 °C, the values for *G'*, *G"* and complex viscosity increase, indicating proper viscoelastic behavior which helps to hold gas in the system (Rodríguez-García, Sahi, & Hernando, 2014). Conversely, with fat reduction, the values for *G'*, *G"* and complex viscosity might be not high enough for gas retention. Thus, gas is transferred to the product surface and disappears into the surrounding environment. Consequently, volume expansion during baking is not properly performed and the final volume of the reduced-fat cake is smaller while its crumbs are denser and firmer than regular cake (Matsakidou, Blekas, & Paraskevopoulou, 2010).

A change in the recipe may affect the thermal parameters (*To, Tp* and *Tc*) of the batter. The fat reduction causes significant decreases in thermal parameters compared to regular cake batter (sunflower oil 46 g/100 g flour). Normally, the fat melts and covers the starch granules during baking. When the fat content is reduced, the starch is less unlikely to be surrounded by the fat so that it can be freely hydrated by the water, resulting in more rapid starch gelatinization. Therefore, the batter is transformed to the solid-like structure of the crumb earlier than usual. With 50% fat replacement (sunflower oil 22 g/100 g flour) by inulin with the addition of lipase (0.003–0.006 g/100 g flour), the thermal parameters were improved and were comparable to regular batter (Rodríguez-García et al., 2014). By changing from lipase to a vegetable, all-round, alpha-gel emulsifier—consisting of glycerol monostearate (GMS; E471) and polyglycerol esters of fatty acids (PGE; E475)—the thermal parameters increase and are much higher than those of other batters because emulsifiers enhanced the formation of lipid and amylose complexes that delay water transportation into the starch granules during baking (Rodríguez-García et al., 2014).

2.3.3 Role of fat during storage

Fat can slow down starch retrogradation and maintain crumb softness during storage. By adding some improvers such as lipase, crumb hardness was maintained for 14 days due to modification of the interaction between the fat and gluten by lipase. With reduced fat content, the crumb structure and texture quickly change. The addition of improvers did not improve the softness of the cake crumb and accelerated rapid hardening within 24 h. Furthermore, the addition of lipase when baking a cake using reduced fat caused higher moisture retention and water activity than for a cake baked using the normal full-fat content. However, mold spoilage was observed on day 14 of storage. Thus, the shelf-life of the cake with reduced fat content was shorter than for the regular cake (Rodríguez-García et al., 2014).

3. Techniques to reduce sugar in baked products

Sugar is one of the ingredients that play important roles in the characteristics and quality of the dough and batter of baked goods, particularly cake and biscuits. Generally, cake and biscuits contain 30–40% sugar (van der Sman & Renzetti, 2019). Adding sugar decreases the degree of starch gelation, the development rate of gluten network (Sahin et al., 2019) and the viscosity (Kweon, Slade, Levine, & Gannon, 2014) of the dough and batter, and decreases the water activity (Baldino, Gabriele, Lupi, de Cindio, & Cicerelli, 2014) and lightness (Struck, Gundel, Zahn, & Rohm, 2016) of the baked goods. However, the crumb softness and chewiness increase (Struck et al., 2016). Sugar can help to produce a moist and fine crumb texture with a brown crust. By reducing the sugar in the recipe, it is challenging to incorporate air cells into the batter and maintain its stability.

With a low water content, sugar recrystallization improves the development of fracturability in crispy baked products. Reducing or removing some sugar from the recipe not only affects the sweetness but also the structural and textural qualities of the baked product, due to the change in biopolymer phase transitions during baking. Thus, reducing the sugar content is not easy because of its multiple functionalities in baked products (Renzetti et al., 2020). Two major approaches are proposed to deal with the development of sugarless baking. The first approach simply uses new ingredients to replace sugar, such as artificial sweeteners, sweet bulking additives and sweet proteins. The other one uses new technology to generate sugar in the product without adding more sugar to the recipe.

3.1 Reformulation using sweetener together with bulking agent

3.1.1 Sweetener

Sweetener can be briefly classified into two types: high intensity and low-caloric. High intensity sweeteners can be either artificial or natural. Artificial sweeteners such as saccharin, aspartame, acesulfame-K and sucralose should normally have greater sweetness than sugar, i.e., approximately 30–13,000 times sweeter relative to sucrose (Mooradian, Smith, & Tokuda, 2017). Thus, the amount of artificial sweetener in the food formula is very small which decreases the solid-to-liquid ratio and may affect the dough characteristics and calorie content. Sweeteners may be used together with bulking agents. Artificial sweeteners include nonnutritive sweeteners and noncaloric artificial sweeteners. Almost all sugar substitutes contain zero calories, with the exception of aspartame (4 kcal/g).

In 1958, saccharin was approved as the first high intensity artificial sweetener by the US Food and Drug Administration. Although its sweetness is markedly higher than sucrose (approximately 300–400 times), it has a metallic or bitter taste and aftertaste. It is categorized as a nonnutritive sweetener. Aspartame is not stable at high temperature and high pH, but it can still be successfully used in baked products such as cake by encapsulation before mixing in the dough or batter (Wetzel, Weese, & Bell, 1997).

The next generation of artificial high intensity sweeteners has been in continuous development, but there is still some noticeable off-flavor, metallic taste and bitterness. To avoid such undesirable characteristics, more than one sweetener can be mixed to mask the off-flavor, metallic taste and bitterness (Pinheiro & Oliveira, 2005). However, overconsumption of nonnutritive sweeteners possibly alters the gut microbiota, including the composition and function of bacterial populations in the gut and thereby causes glucose intolerance with a resulting elevation of the blood-glucose levels beyond normal levels that may encourage obesity-related metabolic conditions (O'Sullivan, 2020). Some noncaloric artificial sweeteners such as sucralose are often added to food to replace sugar and to reduce calorie intake. Sucralose is about 600 times sweeter than sucrose. And although it is synthesized from sucrose, it contains no calories. Martínez-Cervera, Sanz, Salvador, and Fiszman (2012) reported that sucralose could successfully replace sugar up to 50% in reduced-calorie muffins.

As artificial high intensity sweeteners are not natural food additives, both nonnutritive artificial sweeteners and noncaloric artificial sweeteners must be disclosed on the label, which may not be readily accepted by increasingly health-conscious consumers who prefer an uncluttered label. This has promoted the development of novel natural sweeteners such as steviol glycosides and tagatose. Sweet glycosides are nonnutritive sweeteners that are derived from plants; thus, they consist of natural compounds. Stevia is a nonnutritive natural sweetener consisting of the sweet glycosides stevioside and rebaudioside A. It can replace an undesirable bitter aftertaste and metallic sensation with sweetness (O'Sullivan, 2020). Stevia has a relative sweetness of approximately 300. Due to its heat-stable characteristic, it is suitable for applying as sugar replacers in bakery products (Manisha, Soumya, & Indrani, 2012). For example, in muffin, sucrose could be reduced by 30% with the inclusion of rebaudioside A (0.07 g/ 100 g flour) in the mixture (17.93 g/ 100 g flour) of wheat fiber and water, in order to achieve satisfactory product characteristics. Consequently, the energy density could be reduced by 5.5% from the original muffin (379.4 kcal/100 g) (Struck et al., 2016).

Low-caloric sweeteners contain three major groups: polyols (polyhydric alcohols), tagatose and trehalose. Normally, 1 g of sucrose contains 4 kcal/g while the polyols (such as erythritol, isomaltitol, lactitol, maltitol, mannitol, sorbitol, xylitol) have caloric values relative to sugar in the range 5–60%. Erythritol tends to have very low caloric value (approximately 5% relative to sugar), whereas xylitol tends to have a very high caloric value (approximately 60% relative to sugar). The caloric values of tagatose and trehalose relative to sugar are approximately 38% and 90%, respectively. In addition, the sweetness levels of polyols, tagatose and trehalose are in the range 40–100% relative to sugar (Jolly, 2008).

Polyols tended to be successfully used in muffin and cake. Both sorbitol and maltitol can replace sugar completely. The resulting sugar-free muffins tended to have smaller height and softer texture than the original muffin and there was no significant difference in acceptance, compared to the original muffins (Martínez-Cervera, Salvador, & Sanz, 2014). Similarly, the application of xylitol to replace all the sugar in chiffon cake could maintain sensory and textural qualities of crumb including springiness, adhesiveness, cohesiveness and chewiness (Kim, Park, & Shin, 2014). Sugar-free cookie products have not been very successful. Although erythritol was used to partially reduce sugar in cookies at only 50%, the reduced sugar cookies were not really acceptable, due to the hard texture (Laguna, Primo-Martín, Salvador, & Sanz, 2013).

In some cases of 100% sugar replacement such as sugar-free cookies, there was a significant reduction in consumer acceptability of color, texture, flavor and sweetness. A combination of sweeteners might be an alternative option. For example, a mixture of fructooligosaccharides and sucralose was recommended for reduced-calorie biscuits, while a mixture of maltitol and sucralose (3:1) was recommended to completely replace sucrose in sugar-free biscuits (Aggarwal, Sabikhi, & Sathish Kumar, 2016).

3.1.2 Bulking agent

A bulking agent is used to maintain the solid proportion in a dough formula when the amount of added sugar is reduced. In addition, it maintains dough characteristics such as viscosity and air holding capacity. Bulking agents can be non-sweet such as resistant starches, polydextrose, maltodextrin, hydrocolloids and dietary fibers, while examples of sweet bulking agents include inulin, polyols or sugar alcohols. Polydextrose is a glucose polymer providing only 1 kcal/g. It is widely used as a bulking agent together with stevia and inulin to improve the texture of bakery products. To obtain sugar-free biscuits, 30% polydextrose was recommended together with sugar replacers including fructo-oligosaccharide (FOS)-sucralose and maltitol to maintain the desirable sensorial quality (Aggarwal et al., 2016).

Dietary fiber is made up of indigestible carbohydrates and aids the rapid transition of wastes or toxins through the gastrointestinal tract. The recommended daily intake is approximately 25–30 g. Dietary fiber may be either soluble or insoluble. Dietary fibers from cereals likely contain insoluble fiber such as cellulose and hemicellulose. By using rebaudioside A and fiber to replace sucrose in muffins, the process of structure modification by protein denaturation and starch seemed to be delayed. Consequently, the muffin crumb firmness and volume increased whereas starch digestibility decreased. The best conditions involved replacing sucrose by 30% of a combination of rebaudioside A and wheat fiber (Struck et al., 2016).

Dietary fibers from fruits may contain soluble fiber such as pectin and gums. To use as ingredients in foods, the ratio of soluble-to-insoluble dietary fiber should be 1:2 (Figuerola, Hurtado, Estévez, Chiffelle, & Asenjo, 2005). Dietary fiber has bulking characteristics, a high-water holding capacity, is good at gel formation and functions as a thickening agent. Several studies on baked items reported that the dietary fiber from cereal bran could replace wheat flour in the range 10–30%, while replacement with dietary fibers from fruit varies in the range 5–20% (O'Sullivan, 2020). Apple pomace tended to be more popular and was more commonly used for sugar replacement than citrus fruits (O'Sullivan, 2020). By using the dietary fiber from fruits, sweetness and a high fiber content are gained. Therefore, the inclusion of dietary fibers such as arabinoxylan oligosaccharides can be an alternative strategy to achieve sugar reduction.

Several researchers have recommended sweeteners and bulking agents in the reformulation of baked items to reduce the sugar content, as shown in Table 1. Sugar in cookies could be reduced by replacing sugar with a combination of sucralose and whey protein (Marques et al., 2016) or a mixture of stevioside and maltodextrin (Yazdi et al., 2017). For short dough biscuits, 25–50% of the sugar can be replaced, depending on the type of sweetener. Inulin could be used to reduce up to 25% of the sugar, while erythritol

Table 1 Reformulation to reduce sugar in baked products.

Product	Partial replacement	Complete replacement	Recommended sweeteners	Bulking agent	Final product quality	References
Cookies		√	Sucralose	Whey protein concentrate (WPC)	Sugar-free cookies were developed using sucralose (6.25 g/100 g) as sugar replacer. The use of a bulking agent, namely 25.9 and 54.1 g WPC/100 g, could maintain consumer acceptance rates in appearance, color, flavor and texture to greater than or equal to 80% and 70%, respectively.	Marques, Sao Jose, Silva, and Silva (2016)
	√		Stevioside	Maltodextrin	A mixture of stevioside (0.43 g/100 g flour) or maltodextrin (87.07 g/100 g flour) was used to reduce sugar and energy in the Iranian cookies by 23.06% and 7.27%, respectively, without compromising sweetness and the browning index compared to regular cookies. However, the hardness was higher than in the regular cookies and tended to further increase when the content of the stevioside-maltodextrin mixture increased.	Yazdi, Hojjatoleslamy, Keramat, Jahadi, and Amani (2017)

Continued

Table 1 Reformulation to reduce sugar in baked products.—cont'd

Product	Partial replacement	Complete replacement	Recommended sweeteners	Bulking agent	Final product quality	References
Short dough biscuits	√ up to 25%		Inulin		Inulin replacing sucrose by 25% produced softer cookies. The reduced sugar cookies were characterized as open and crumby, leading to consumer acceptance.	Laguna et al. (2013)
	√ up to 50%		Erythritol		The reduced sugar cookies were characterized as hard and not crispy, leading to lower consumer acceptance than cookies containing inulin as the sugar replacer.	Laguna et al. (2013)
Muffins	√ up to 50%		Stevia	Cocoa dietary fiber	Muffin containing a mixture of sucrose and steviol glycosides and 6% cocoa fiber produced the top rank in sensorial quality. Not only stevia but also the cocoa flavor could enhance the sweetness of reduced sugar muffin.	Karp, Wyrwisz, Kurek, and Wierzbicka (2017)
	√ up to 30%		Rebaudioside	Inulin or Polydextrose	The reduced sugar muffins could be developed by using a combination of rebaudioside A and inulin or polydextrose, resulting in decreased total energy by 6 and 5 kJ/100 kJ, respectively. However, the muffins were characterized by increased crumbliness.	Zahn, Forker, Krügel, and Rohm (2013)

	√ √	Sorbitol Maltitol	Sugar-free muffins were developed by replacing sugar 100 g/100 g flour with either sorbitol or Maltitol 100 g/100 g flour resulting in decreased height and hardness, but no difference in acceptance compared to the regular muffins.	Martínez-Cervera et al. (2014)
Chiffon cake	√	Xylitol	Lightness of crust and textural crumb qualities, including springiness, adhesiveness, cohesiveness and chewiness, were not significantly different from the regular chiffon cake. Therefore, the overall liking scores for the regular and the sugar-free chiffon cake were similar.	Kim et al. (2014)
	√	Maltitol	The replacement of sucrose by maltitol significantly increased the % overrun and foaming ability of batter, resulting in sugar-free sponge cake with similar specific volume and hardness to the regular cake.	Hao et al. (2016)

could be used to reduce up to 25% of the sugar (Laguna et al., 2013). Sugar-free muffins could be developed using polyols such as sorbitol and maltitol as the sugar replacer (Martínez-Cervera et al., 2014). The use of stevia, a high intensity sweetener, together with dietary fiber could replace up to 50% sugar (Karp et al., 2017). Polyols were the best to maintain the textural and moisture characteristics of chiffon cake. Although xylitol is expensive and does not assist with the browning reaction on the crust, it has been recommended as best for sugar-free chiffon cake (Kim et al., 2014).

3.2 Sequential sugar reduction without sugar replacement

A sequential reduction approach was initially used for salt reduction (Oliveira et al., 2016). Later, some countries such as the UK improved this approach to reduce the use of sugar by 30–40%. However, it could only reduce the sugar content to 13% in chocolate-flavored milk by using a two-step sequential reduction in one year (Oliveira et al., 2016). The sequential reduction without replacing high-intensity sweeteners was possible in a product where the sugar was only used as the sweetener (Di Monaco, Miele, Cabisidan, & Cavella, 2018). In baking, sugar has many functions, so it is not easy to reduce the sugar content without any supplementation using other ingredients.

3.3 Other techniques

To reduce the sugar content, using sugar with a small particle size may increase the perceived sweetness. For example, the amount of fine particle-sized sugar (459–972 μm) was positively associated with sweetness and a moist texture, but negatively associated with hardness . Brownie biscuits containing the finest particle-sized sugar were characterized as being sweeter, softer and moister than biscuits containing regular sugar with a wider range in particle sizes (200–972 μm). Therefore, this technique has the potential to reduce the sugar content and still maintain moisture and softness in chocolate brownies. However, further study is required to specify the amount of sugar replacement (Richardson et al., 2018). Overall, multisensory integration combines sensorial characteristics such as color, viscosity, aroma and flavor to enhance the perception of sweetness through cross-modal interactions (Stieger & van de Velde, 2013). A flavor approach was successfully used for salt reduction in cheese products, but it did not enhance the sweetness much (Syarifuddin, Septier, Salles, & Thomas-Danguin, 2016) (Table 1).

4. Techniques to reduce salt in baked products

As people tend to select healthy food, the industry is encouraged to reduce and avoid the use of salt in products. To make the claim of a 'reduced sodium' product, the sodium or sodium chloride content must be 25% less than in the regular product. A 'low sodium' or 'low sodium chloride' product means the product consists of no more than 0.12 g of sodium or 0.30 g of sodium chloride per 100 g, respectively. When the sodium or sodium chloride content is reduced to no more than 0.04 or 0.10 g per 100 g, the product can be labeled as 'very low sodium' or 'very low sodium chloride', respectively. To claim 'sodium-free' or 'sodium chloride-free', the sodium and sodium chloride contents shall not be more than 0.005 and 0.013 g per 100 g sample, respectively (European Commission, 2006). One proposal to reduce salt consumption is to market bitter-taste products to consumers with the expectation of a new preferable taste. For example, instead of salty potato chips, bitter potato chips were launched on the market (Silow, Axel, et al., 2016). Once the bitter taste becomes popular, consumption of salty products should decrease. In addition, some countries use the small step reduction strategy to adjust the perception of consumers gradually. However, this has not been successful in reducing salt and maintaining overall consumer satisfaction (Silow, Axel, et al., 2016). Therefore, some other strategies might be used such as the use of salt replacers and flavor enhancers, together with the small step reduction strategy to maintain the consumer acceptance.

4.1 Reformulation using salt replacers and taste enhancers

Salt replacers should ideally have a similar saltiness to sodium chloride which is not easily achieved because of the high specificity of sodium for the epithelial Na^+ channel. The mechanism of salty taste transduction must be clearly understood to develop salt replacers (Silow, Axel, et al., 2016). Currently, potassium chloride (KCl) is one of the salt replacers commonly used. However, it can only replace NaCl by 20–30% in bread and pizza, as over-replacement causes metallic and bitter aftertastes (Mueller, Koehler, & Scherf, 2016). In addition, the use of up to 30% salt replacers did not significantly reduce antifungal activity (Samapundo, Deschuyffeleer, Van Laere, De Leyn, & Devlieghere, 2010). Other possible replacers are calcium chloride, magnesium chloride and magnesium sulfate (Kaur, Bala, Singh, & Rehal, 2011).

4.2 Reformulation using sourdough

Sourdough is made from lactic acid fermentation in a mixture of flour and water. It is commonly used for sourdough bread which has a unique flavor and crumb texture. The acidity of the dough is reduced due to the lactic acid and acetic acid produced during the fermentation. This benefits shelf-life extension of sourdough bread. Recently, bread containing sourdough was reported to be saltier than regular bread, possibly from the combined effects of acidity and proteolysis (Zhao, Kinner, Wismer, & Ganzle, 2015). Specifically, sourdough has been made using fermentation with *Lactobacillus reuteri* strains which are glutamate-accumulating, resulting in the glutamate concentration in the bread being as high as the glutamate taste threshold for bread of 0.03% monosodium glutamate (w/w). Therefore, sourdough not only has a sour taste but also it enhances the umami taste. Salt reduction in bread by sourdough addition is possible up to 33% (from 1.5 to 1.0% flour base) (Zhao et al., 2015). For other baked products, sourdough addition can be used to reduce the amount of salt in the recipe while maintaining palatability, in particular with salt-reduced puff pastry (Silow, Axel, et al., 2016). Consequently, the use of sourdough to reformulate the recipe is a promising technique.

4.3 Design of salt distribution in product to enhance the perception of saltiness

4.3.1 Salt inhomogeneous distribution

Not only does the salt content affect the saltiness of products, but also the delivery form of the salt can enhance saltiness perception. Instead of a solid form, salt delivered in solution form by continuously alternating concentrations might increase the intensity of the saltiness. Various salt solutions may be added separately in each dough layer to create a non-homogeneous laminated dough (Silow, Axel, et al., 2016). This method is not practical to implement in actual production because of the difficulty in controlling the amount of yeast and the yeast fermentation mechanism (which is related to salt concentration) equally in each part (Silow, Axel, et al., 2016).

4.3.2 Salt agglomeration

Salt agglomeration is another technique to increase the inhomogeneous distribution of salt in foods. Salt agglomerate has been successfully used in low-salt meat products such as sausages and beef patties (Gaudette, Pietrasik, & Johnston, 2019). For baked products, the use of salt agglomerate causes a random distribution of the salt and thereby may affect the dough quality because salt plays an important role in dough rheology and fermentation mechanisms. In addition, the agglomeration reduces the numbers of Na^+ and Cl^- ions and increases intramolecular and intermolecular electrostatic forces. To produce salt agglomerate, raw materials that swell rapidly and have slow retrogradation are required to form a thick and stable gel even though the salt concentration in the agglomeration is very high (Monteiro, Nakagawa, Pimentel, & Sousa, 2021). Their pasting properties should be characterized by a high viscosity peak and low final viscosity. Therefore, waxy starch is ideally suitable for salt agglomeration. A reduction of 30% salt in bread by using the salt agglomeration technique with waxy corn starch slightly increased the dough development time and decreased dough stability. Moreover, the bread quality, including hardness, cohesiveness, adhesiveness, springiness and cohesiveness, was generally maintained (Monteiro et al., 2021). Therefore, salt agglomeration is another promising technique to reduce the salt content in bakery products.

4.3.3 Salt encapsulation

For salt encapsulation, a food model with multiple layers is developed by the use of different salt concentrations between layers. This can enhance saltiness intensity compared to a food model with homogeneous salt distribution. Encapsulation of salt was tested in bread, with the results demonstrating that the salt content could be reduced from 2% to 1% without any perceived difference in the intensity of saltiness. However, the encapsulation could block salt functioning during mixing. This effect is equivalent to salt-free mixing. Then, the rheological properties of the dough are changed, resulting in poorer bread volume, structure and texture (Noort, Bult, & Stieger, 2012).

4.3.4 Salt with different particle sizes

The particle size of salt is another factor that possibly affects saltiness in baked products. Coarser-grained salt can cause non-homogenous distribution of salt in bread crumb and increase the release rate of sodium chloride into the bread crumbs (Silow, Axel, et al., 2016). Consequently, a small amount of salt content can impart a high intensity of saltiness. This technique was confirmed for up to 25% salt reduction in bread and pizza (Mueller et al., 2016).

5. Techniques to reduce fat in baked products

Fat is another ingredient in the emulsion system playing an important role in incorporating air bubbles into the dough and batter. The emulsion system is composed of the discontinuous phase (the air bubbles) and the continuous phase (a mixture of egg, sugar, fat, flour and water). To keep the air bubbles in the continuous phase of the emulsion system, individual or combinations of egg protein, fat and emulsifier are required to generate a foam structure that contributes to a spongy texture (O'Sullivan, 2020).

Although fat provides desirable qualities in baked products, excessive fat consumption possibly increases the risk of cardiovascular diseases and strokes. The recommended daily intake of fat is 15–30% of dietary energy. Saturated fat should be less than 10% of dietary energy. In 2008, the UK Food Standards Agency recommended reducing the daily saturated fat consumption from 13.3% to 11.0% of dietary energy. In general, white bread and pizza contain 2% of the recommended daily fat intake, while biscuits and cake likely contain 3 and 4% of daily fat intake, respectively. Of the total fat intake, the percentage of saturated fat intake from white bread, cake, pizza and biscuits are 14%, 38%, 43% and 50%, respectively (Talbot, 2011).

Fat reduction might cause a chewy and dry texture of bakery products. Without any fat replacer, the quality of low-fat baked products is not acceptable to consumers. However, fat-free baking is difficult to achieve. Generally, fat reduction may be up to 50% of the total fat content in baking, depending on types of fat replacers and products (Zoulias, Oreopoulou, & Tzia, 2002). Fat-reducing baking techniques have mainly focused on: (1) reformulation using fat replacers, (2) reformulation using new forms of fat or healthy fat, and (3) reduction of fat without a fat replacer using suitable baking technology.

5.1 Reformulation using fat replacers

A fat replacer can be designed with functional variations, including as a fat substitute, fat mimetic, fat analogue and fat extender. A fat substitute is a compound synthesized to have a similar structure to fat and acts as a barrier to block fat absorption (Fellows, 2017). It is designed to simply replace fat on a weight-by-weight basis. A fat mimetic or texturizing agent is designed to imitate the sensory qualities of fat, especially its taste and mouthfeel. A fat analogue is designed to have similar characteristics of fat but with different digestibility and nutritional value. A fat extender is used to optimize the functionality of fat and minimize the use of fat in the recipe (Chavan, Khedkar, & Bhatt, 2016).

Fat replacers are mainly classified by the base substance and may be fat-based, protein-based or carbohydrate-based. A fat-based substance is made from vegetable oil through the inter-esterification of triglycerides. This reduces fat digestibility and the contribution of the fat and its energy to the body. Thus, a fat-based fat replacer has lower calories than regular fat. However, it might have a negative effect on health, particularly the gastrointestinal system, such as the absorption of fat-soluble vitamins and some antioxidant compounds. Some commercial fat-based replacers are Benefat® (Salatrim), Caprenin and Olean™ (Olestra is a fat-based sucrose polyester) with energy densities of 5, 5 and 0 kcal/g, respectively. In puff pastry, Salatrim could maintain crispiness and a high specific volume for reduced-fat puff pastry, but it tended to have an undesirable aftertaste (Pimdit, Therdthai, & Jangchud, 2008).

Protein-based fat replacers such as whey protein, egg white protein and milk protein, are normally used to improve the texture in reduced-fat baked products, because they improve the porosity and pore area of the baked product (Srikanlaya, Therdthai, Ritthiruangdej, & Zhou, 2018). Commercial protein-based fat replacers common in baking include Simplesse® and Leancreme™ (both microparticulated whey proteins), Lita® (corn zein), Dairy-Lo® (modified whey protein concentrate), K-Blazer® (a blend of protein, starch, some gums and emulsifiers) and ULTRA-FREEZE™ (egg-white, milk protein, corn syrup solids and modified starch) (Chavan et al., 2016).

Carbohydrate-based fat replacers are made from starchy materials such as cellulose, dextrin, gums and dietary fibers and are commonly used to imitate the fat function. The consequent quality of reduced fat baked products depends on the structures and properties of those fat replacers. Carbohydrate-based fat replacers are commonly used for cakes and muffins. Examples of commercial carbohydrate-based fat replacers usually used in baking are Paselli A2 (carbohydrate from potato), Stellar, Maltrin and Amalean-Instant (all carbohydrates from corn) and Oatrim (carbohydrate from oat) (Chavan et al., 2016). The different types of carhbohydrate-based fat replacers for reformulation in baking are explained below.

5.1.1 Polydextrose

Polydextrose provides only 1 kcal/g. It can be used as a low-calorie bulking ingredient due to its ability to randomly link with citric acid, glucose and sorbitol. These links generate mouthfeel and creaminess in low-fat products (Moriano et al., 2018). Examples of polydextrose fat replacers are Sta-Lite which has good solubility and Litesse which is composed of a polymer of glucose, sorbitol and citric acid in the ratio of 89:10:1 (Chavan et al., 2016). In cake, polydextrose may be used to replace fat up to 25%. To reduce the fat even more to 65%, the fat replacer should be changed to oat amylodextrin (Kocer, Hicsasmaz, Bayindirli, & Katnas, 2007).

In soft dough biscuits, fat reduction from 20 to 6 g/100 g flour increased dough hardness from 20.78 to 44.08 N and the extrusion time of the biscuit dough from 43 to 167 s. Based on farinograph analysis, the dough consistency increased from 180 to 540 Brabender units. To reduce the increased hardness and consistency of the low-fat dough, polydextrose or maltodextrin could be added to replace the fat portion (Sudha, Srivastava, Vetrimani, & Leelavathi, 2007). However, full replacement is not recommended due to the negative effect on volume. Normally some emulsifiers should be applied together (Martínez-Cervera, Salvador, & Sanz, 2015). Polydextrose could replace fat in biscuits by 30%, resulting in 15.98% lower energy (Aggarwal et al., 2016). In 50–60% reduced-fat biscuits, to maintain texture, the emulsifier (5% glycerol monostearate) was recommended to be used with 10–12% maltodextrin due to its shortening action (Sudha et al., 2007). In addition, 10.69 g/100 g polydextrose could be used together with 12.38 g/100 g shortening to make biscuits with 44% fat reduction. This resulted in reduced-fat biscuits containing only 17.54 ± 0.05 g fat/100 g sample (Moriano et al., 2018). For puff pastry, maltodextrin was recommended to maintain a comparable specific volume and sensorial quality to regular puff pastry but with 37% fat reduction (Pimdit et al., 2008).

The use of carbohydrate-based fat replacers alone might not be sufficient to maintain the desirable characteristics of the baked product. In some cases, emulsifiers and other minor ingredients such as starch, sugar and improver should be used together to obtain a suitable batter and cake end product. Lipase is one improver widely used in bread, while its use in cake has been recent. It can replace emulsifiers to reduce the surface tension between the air and water interface in the batter and to stabilize the air trapped in the batter, resulting in volume expansion not being interrupted during baking (Rodríguez-García et al., 2014).

5.1.2 Dietary fiber

Dietary fiber can be an alternative ingredient to reduce fat in baked products. However, the dietary fiber might negatively affect product quality. For example, an increase in the use of carrot pomace powder (consisting of 3.9% pectin, 12.3% hemicellulose, 51.6% cellulose and 32.1% lignin) from 0 to 15 g/100 g deteriorated the texture and specific volume of doughnuts, while retaining a high moisture content (Nouri, Nasehi, Samavati, & Mehdizadeh, 2017). In muffins, cocoa fiber was used as a fat replacer to decrease the hardness, chewiness and resilience of the reduced-fat muffin to comparable levels in full-fat muffins. Sponginess and springiness of the reduced-fat muffin were maintained with a fat reduction of up to 25%, whereas a greater fat replacement by the cocoa fiber caused increased cohesiveness and difficulty in chewing and swallowing (Martínez-Cervera, Salvador, Muguerza, Moulay, & Fiszman, 2011). Dietary fiber from coffee silverskin, a by-product from the coffee roasting process, can be used as a fat replacer in baked products without significant effects on specific volume and weight loss. However, increased coffee silverskin increased the hardness and chewiness but decreased the springiness and cohesiveness of reduced-fat cake. In addition, there was some impact on color as the cake crumbs seemed to be darker with decreased lightness (L^*) and increased redness (a^*) values. Thus, coffee silverskin could replace fat in cake by up to 30% without producing undesirable characteristics (Ateş & Elmacı, 2018).

Fiber from chia seed can be used in the form of mucilage. Chia seed is generally composed of 18–30% fibers. To produce the mucilage, the seeds are immersed in water to form a gel that can be used as a fat replacer because of its high water hydration capacity and shear thinning property. An increase in the shear rate reduced the apparent viscosity and G'' was higher than G' at low frequencies. At higher frequencies, a crossover of the G' and G'' curves occurred. The mucilage maintained water retention and softness in low-fat baked items such as cake, bread and cookies. Although the mucilage did not have a negative effect on the technical characteristics of products, it could only substitute fat up to 30% in cookies as over substitution reduced the sensorial quality below the acceptable range (Punia & Dhull, 2019).

5.1.3 Inulin

Inulin is a polydisperse carbohydrate composed mainly of β-(2–1) fructosyl-fructose links. The fructosyl-fructose glycosidic linkage is not well digested in the upper gastrointestinal tract, so as dietary fiber, it contributes to prebiotic activity (Rodríguez-García et al., 2014). Due to its health benefits and ease of use, inulin has been studied to replace shortening in biscuits by up to 40% (Laguna, Primo-Martín, Varela, Salvador, & Sanz, 2014). In cake, replacing fat with inulin caused a significant decrease in the G' and G'' values of the batter during mixing, due to the dilution effect of the additional liquid portion. During the initial phase of baking at a product temperature of 30–45 °C, both viscoelastic moduli and the complex viscosity of batter hardly changed. However, G', G'' and complex viscosity values increased later in baking at 50 °C. The high complex viscosity demonstrated more elastic behavior so that gas retention was high. The gas diffused slowly into the occluded air cells with subsequent volume expansion being maintained. Although 50% fat replacement with inulin resulted in the cake volume being less than that of a full-fat cake, the cake crumb structure was not much different from the full-fat cake in terms of total cell area. However, a further fat reduction to 70% increased the total cell area, possibly due to the modification of setting mechanisms (Rodríguez-García et al., 2014). Similarly, Agave fructans is another prebiotic that can be used as a fat replacer. An increase in agave fructans tended to produce a brittle flour dough characterized by elastic behavior; therefore, about 20% fat reduction with fructans was recommended for cookies to ensure a similar quality to the regular cookies. In addition, there were possible additional benefits from the low calories and high prebiotic function of the fructans (Santiago-García, Mellado-Mojica, Leon-Martínez, & Lopez, 2017).

5.1.4 Gum

Persian gum is an anionic hydrocolloid from almond trees, majorly composed of arabinose and galactose. It is a clear to semi-cloudy gum that is low in protein but has a high-water absorption capacity. Its emulsifying properties are similar to those of Arabic gum (Khalesi, Emadzadeh, Kadkhodaee, & Fang, 2016). In reduced-fat doughnuts, the texture may be excessively firm, so the addition of 1.2 g/100 g Persian gum helped to decrease the firmness in the reduced-fat doughnuts (Nouri et al., 2017). Hydroxypropyl methylcellulose (HPMC) is a hydrocolloid that can act as an emulsifier and increase water absorption and gas retention capacity in gluten-free baking. In rice-flour bread, the addition of HPMC at 4 g per 100 g flour improved viscoelasticity by increasing tan δ (G''/G_0), to a comparable level with regular wheat dough (Srikanlaya et al., 2018). Guar gum is a source of galactomannans and is usually used to improve the moistness in high moisture content products. However, guar gum (0.2%) was used together with maltodextrin (10%) in soft dough biscuits and did not affect the farinograph consistency of the dough but did reduce the hardness and cohesiveness of the reduced-fat biscuits (Sudha et al., 2007).

As mentioned above, fat replacers have been developed and studied for many years. Some fat replacers are commercially available. In summary, their functionality in baking is to modify the texture, maintain the moisture content in crumbs, act as a bulking agent and to ensure a comparable mouthfeel as full-fat products. Some fat replacers such as Salatrim and Caprenin have even been developed to mimic fat properties to maintain desirable characteristics in the reduced-fat bakery products that are comparable to the full-fat product. Fat replacers for each functionality are shown in Fig. 1.

5.2 Reformulation using new form of fat, healthy fat, or other additives

Reformulation of fat ingredients can be achieved by replacing unhealthy fat with new forms of healthy oil/fat such as emulsion, oleogel and hydrogel. Alternatively, some healthy oil/fat from new sources might be used. Although this technique might not greatly reduce the fat content and energy, it can reduce the level of unhealthy fat and develop healthier baked products.

5.2.1 Emulsion

An emulsion, which is suitable as a fat substitute, consists of vegetable oil, water and has cellulose ether thermogelling ability. Cellulose ether is a hydrocolloid with a high-water retention capacity. Its emulsion has a semi-solid phase with consistency. The degree of fat replacement may be up to 100%, which is much higher than conventional fat replacers. The use of cellulose emulsion as a fat replacer in muffins reduced the starch gelatinization

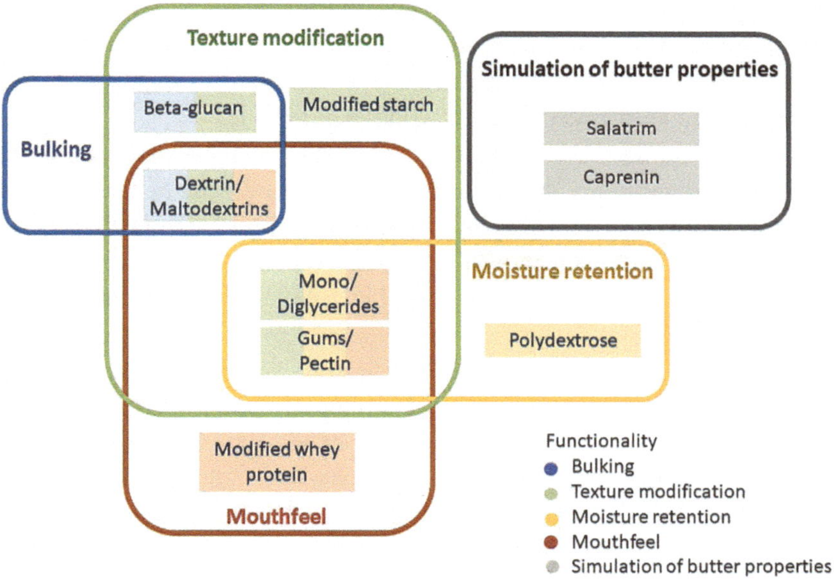

Fig. 1 Fat replacers and their functionality in reduced fat baking.

temperatures (T_{onset} and T_{peak}) and improved elasticity during baking. The decreased gelatinization temperature was possibly due to the high amount of water available for gelatinization. By reducing the fat by up to 49%, the height and crumb pore size of the reduced-fat muffins were in similar ranges as the full-fat muffins, although the overall acceptability was slightly less than for full-fat muffins due to increased hardness (Martínez-Cervera et al., 2015).

5.2.2 Oleogel

Oleogel is generated through oleogelation, which is a process that transforms liquid oil into semisolid fat by adding a small amount (less than 10%) of oleogelators to the oil such as rice bran oil and canola oil. The oleogelators are usually 12-hydroxystearic acid, ethylcellulose, plant sterols, steroids and natural waxes such as beeswax and rice bran wax (Limpimwong, Kumrungsee, Katob, Yanakab, & Thongngama, 2017). Recently, hydrocolloid-based oleogelators have been proposed with claimed better properties. The recommended hydrocolloids for this purpose include methyl cellulose, hydroxypropyl methylcellulose and chitin (Oh & Lee, 2018). Hydroxypropyl methylcellulose and ethyl cellulose have amphiphilic properties and can create oil-structuring. Thus, biopolymers, particularly polysaccharides, can be interesting oleogelators in commercial production, due to their availability, low cost and high purity (Naeli, Milani, Farmani, & Zargaraan, 2020). The mixture of oleogelators and oil is heated until it melts before being cooled. The decrease in temperature sets up a thermoreversible three-dimensional gel network (consisting of a crystalline particle network, self-assembled fibrillar network and polymer network) to trap the oil inside and becoming an oleogel (Limpimwong et al., 2017).

Oleogel can decrease the level of saturated and trans fats in foods and reduce lipid digestibility due to its network structure. The lipid metabolism of oleogel from beeswax was investigated in rats by feeding them with a combination of 50% beef tallow and 50% beeswax oleogel containing canola oil for 2 months. The body weight and LDL cholesterol of the rats reduced whereas HDL cholesterol increased. It was confirmed that fat replacement by the oleogel from the beeswax could decrease adipogenesis and improve angiogenesis in the rats (Issara, Park, Lee, Lee, & Park, 2020). People consuming oleogel tended to have low levels of triacylglycerol and free fatty acid. After 4 weeks, there was a decrease in the triacylglycerol in serum and the liver and in total cholesterol in the liver by 30%, together with an increase in excreted triacylglycerol and total cholesterol by 30% in the feces (Limpimwong et al., 2017).

Oleogel can be used as an alternative shortening in baking to replace saturated and trans fats. As mentioned above in the section on fat, shortening is an important ingredient in baking, but it generally contains a high level of saturated fatty acid which is not good for human health. An increase in the oleogel ratio in the recipe tended to decrease viscosity and the shear-thinning characteristics of the batter. Oleogel from beeswax produced a fine structure and did not affect the cake volume but tended to reduce cake crumb porosity. In addition, the use of oleogel from beeswax made the cake softer than from using oleogel from rice bran. Furthermore, the cake with oleogel from the beeswax only contained 14–17% saturated fat which was much less than in regular cake that contained approximately 58% saturated fat. Therefore, oleogel from beeswax should be considered as a new healthy fat source in baked products, especially cake (Oh, Amoah, Lim, Jeong, & Lee, 2017).

In muffins, hydroxypropyl methylcellulose oleogel containing sunflower oil was used as a shortening replacer. An increase in hydroxypropyl methylcellulose oleogel reduced the viscosity and shear-thinning behavior of the batter. After baking, the reduced-fat muffin with hydroxypropyl methylcellulose oleogel was characterized by larger air cells with an inhomogeneous distribution in the crumbs. However, up to 50% replacement by hydroxypropyl methylcellulose oleogel did not affect the specific volume, softness and chewiness of reduced-fat muffins (Oh & Lee, 2018). A mixture of ethyl cellulose and hydroxypropyl methylcellulose could be used to develop an oleogel that had a similar creamy state and plasticity to fat, due to its increased linear viscoelastic range, elastic character and strength. Therefore, an ethyl cellulose and hydroxypropyl methylcellulose-based oleogel could have potential as a replacement for fat in baked products while maintaining good texture and appearance (Naeli et al., 2020).

In soft dough cookies, solid fat is necessary for aeration during kneading. Oleogel, as a solid-like structure, is ideal as a fat replacer in this case. Oleogels formed by the crystallization of candelilla wax were used to completely replace shortening, resulting in cookies having a harder texture than in full-fat cookies. However, their hardness was still better than cookies that included oil to replace the shortening as replacement of shortening by oil would reduce air entrapment during kneading and increase the spreading rate, resulting in hard-textured cookies. Therefore, a blend of oleogel and shortening could reduce the amount of saturated fat while maintaining the desirable characteristics of the cookies (Mert & Demirkesen, 2016).

5.2.3 Hydrogel

Hydrogel is produced by dissolving a precursor (likely biopolymer) in a solvent (generally water). Common precursors for a hydrogel in food applications include gelatin, soy protein, albumin, casein, whey protein, agar, alginate, carrageenan, cellulose, curdlan, chitosan, dextran, gellan gum, gum karaya, gum tragacanth, gum arabic, gum ghatti, hyaluronan, konjac glucomannan, locust bean gum, pectin, starch, tamarind seed xyloglucan, tara gum and xanthan gum (Selvasekaran & Chidambaram, 2021). In short-dough pastries, the hydrogel from a monoglyceride-palm oil-water gel was used to replace fat with only a small negative effect on pastry quality due to the crunchier crust, higher acrylamide content and faster staling (Manzocco, Anese, Calligaris, Quarta, & Nicoli, 2012). An organogel is another kind of structural network made from vegetable oil and is classified as a complex semi-solid material. It generates a water-in-oil emulsion that can form a structure and provide consistency to water-containing dough when used to replace fat in baked products (Silva et al., 2021).

5.2.4 Seed oil

Oil from various seeds can be processed to be used as an ingredient in baking. The oil usually contains essential unsaturated n-3 and n-6 fatty acids and fat-soluble vitamins. An additional benefit from oilseeds is that they may provide some bioactive compounds such as phenolic compounds, tocopherols and phytosterols (Kotecka-Majchrzak, Sumara, Fornal, & Montowska, 2020). Healthy oil may be extracted from chia and linseed oil because of their high n-3, n-6 and low saturated fatty acid content. Chia seed is generally composed of 30–40% oil. Its oil contains polyunsaturated fatty acids, including 54–67% omega-3 fatty acids (n-3) and 12–21% omega-6 fatty acids (n-6) (Punia & Dhull, 2019). However, the n-3 PUFA-rich oil is not in solid fat form and can degrade in quality if heated (Heck et al., 2017). Furthermore, PUFAs might more easily be oxidized and so encapsulation can be used to reduce this. Rapeseed oil is another healthy oil containing unsaturated fatty acids, omega-3 and omega-6 fatty acids, fat-soluble vitamins, tocopherols, sterols, phenolic compounds, carotenoids and phospholipids. In addition, it has a low content of erucic acid that is toxic and causes damage to fatty body parenchyma and myocardial damage (Kotecka-Majchrzak et al., 2020).

In biscuits, the partially hydrogenated fats from palm oil are commonly mixed in the dough. If hydrogenated fats containing trans fatty acids are not permitted in baked products, a high unsaturated oil (such as sunflower oil) is used instead. However, the appearance of the biscuits may be oily and greasy. To resolve this issue, a low-melting unsaturated fat (such as oleine) should be blended with a high-melting saturated fat (such as stearine), rather than using only one type of fat. For example, palm oil has a wide spectrum of fats ranging from low- to high-melting unsaturated fats. Selection of some parts of the oil can be done using fractionation through enzyme-catalyzed inter-esterification. This technique is useful to create triglycerides with a desirable combination of fatty acids, structure and properties (Talbot, 2011). Sunflower oil can also be used to reduce fat in muffins by 49%. Although the elasticity of the low-fat batter was less than for the full-fat batter during mixing, it improved during baking. Consequently, the volume of the muffins was higher than the full-fat muffins due to the aerated structure (Martínez-Cervera et al., 2015).

5.2.5 Insect fat

Insect fat is a new source that has been introduced to baked products to increase functionality. For example, black soldier fly larvae are a novel fat source for use in food products. About 70% fat content in black soldier fly larvae is mainly saturated fatty acids, especially lauric acid (>40%) (Delicato, Schouteten, Dewettinck, Gellynck, & Tzompa-Sosa, 2020). Lauric acid is a fatty acid that can be more easily digested than long-chain free fatty acids. In addition, it is suitable for use as a fat in baking because it is in a solid state at a room temperature of around 20 °C, similar to dairy fat (Dayrit, 2015). Insect fat from black soldier fly larvae could replace butter in cookies and cake by 25% and up to 50% in waffles, with the variation due to differences in heating mechanisms. The cake and cookies were heated using convection whereas the waffles were heated using conduction. The higher heating rate for the waffles affected the flavor profile of samples. The insect fat had suitable functionality in the emulsion system of the batter, similar to that of butter (Delicato et al., 2020). Partial fat substitution by insect fat is possible but to achieve 100% fat replacement, deodorization is necessary to remove the off-flavor of insects. The deodorized oil from yellow mealworm could replace vegetable oil in crackers without any negative effect on appearance or the sensory attributes tested, while the crude oil decreased hardness in the crackers (Tzompa-Sosa, Dewettinck, Gellynck, & Schouteten, 2021).

5.2.6 Waxy durum wheat flour

Waxy wheat flour is composed of almost 100% amylopectin and is suitable for use as shortening in baking. The addition of 10–30% waxy durum wheat flour retarded bread going stale. Possibly, the flour slowed down water transfer from gluten to starch, leaving more water to create plasticizer in the dough system. By adding 25% waxy durum wheat flour into the dough, the bread texture was softer without any adverse effects on other bread quality parameters. While the addition of 20% waxy durum wheat flour into the dough might not present any significant difference in softness on the first day after baking, compared with regular bread, the bread containing 20% waxy durum wheat flour was significantly softer than the regular bread after 5 days of storage (Mouliney, Lavery, Sharma, & Jenner, 2011).

5.3 Reduction of fat without a fat replacer using suitable baking technology

The reduction of fat causes some negative effects on product quality. Many food additives have been studied to improve product quality to match that of the regular recipe. However, the new trends in cooking involve avoiding additives as much as possible. This makes having a "clean" label as is required for reduced-fat products challenging. To maintain the porous structure in reduced-fat baked products, rapid heating has been proposed to generate a high vapor pressure and oven-spring to ensure product height does not decrease as the fat content is reduced. Impingement involves supplying a high air flow rate directly to produce an increased heating rate. Microwaves generate rapid heat into a product by the absorption of microwave power and the conversion of the power into heat through dipolar rotation and ion movement. Therefore, the combined technology between impingement and microwaves was used to bake reduced-fat muffins. Generally, air jet impingement was used to heat the surface to create a proper crust while the microwaves were used to heat the inner layer to develop a proper crumb structure through protein denaturation and starch gelatinization. In the muffins, water replaced palm oil by up to 40%. Baking using this combined technology involving impingement and microwaves produced reduced-fat muffins with the same height as full-fat muffins. However, the reduced-fat muffin crumb seemed to be wetter and more elastic than the regular muffins. In addition, this baking technology did not increase consumer perception of an oil taste in the reduced-fat muffin without addition of a fat replacer (Carrascal et al., 2019).

Almost all techniques have been successful for the partial replacement of fat in baking. Completely fat-free baking is unlikely to be accepted by consumers, mainly due to the resultant dense structure, dry crumbs and a lack of mouthfeel. Therefore, new trends involve the use of healthy oil to replace unhealthy fat. Although the total energy might not be as low, the baked product is healthier by avoiding the consumption of saturated fat and trans-fat. A summary of the reformulation techniques to create healthier baked products is provided in Table 2.

Table 2 Reformulation to reduce unhealthy fat in baked products.

Product	Partial replacement	Recommended technique	Final product quality	References
Cookies	up to 20%	Reformulation using *Agave fructans* as fat replacer	The use of *Agave fructans* (4.8%) as a fat replacer increased the water absorption capacity of dough, resulting in a higher moisture content and hardness, a wider diameter and the same height of cookies compared to the full-fat cookies. However, the sensorial quality was still similar to the full-fat cookies, although butter was reduced from 24% to 19.2%.	Santiago-García et al. (2017)
	up to 25%	Reformulation using insect fat (black soldier fly larvae) as a healthy fat	The use of the insect fat (22.31 g/ 100 g flour) to replace butter by 25% in cookies resulted in no significant difference in the liking score between the full-fat and the reduced fat cookies. Although people reported pleasantness and satisfaction with the reduced fat cookies, they still preferred the full-fat cookies. In addition, an increased percentage of insect fat produced an off-flavor in the cookies.	Delicato et al. (2020)
	up to 30%	Reformulation using chia seed fiber as fat replacer	Chia seed fiber reduced the spread factor, resulting in a smaller diameter for the cookies biscuits. It could be used for up to 30% fat reduction in cookies, while still maintaining desirable texture and color scores greater than 8 from 9, respectively, and acceptability greater than 80%.	Punia and Dhull (2019)
Biscuits	up to 15%	Reformulation using inulin as fat replacer	The reduced fat biscuits developed using inulin at 15 g/100 g of fat as a fat replacer were characterized by a lower moisture content and water activity and being harder and crispier compared to the full-fat biscuits. Increased replacement would reduce the overall acceptability.	Laguna et al. (2014)

Continued

Table 2 Reformulation to reduce unhealthy fat in baked products.—cont'd

Product	Partial replacement	Recommended technique	Final product quality	References
Soft dough biscuits	up to 30%	Reformulation using polydextrose/ maltodextrin as fat replacer	Polydextrose was used for partial replacement of fat in soft dough biscuits, resulting in 30% fat reduction and 15.98% energy reduction. The sensory scores in color, texture, sweetness, flavor and overall were maintained at greater than 8 out of 9 points.	Aggarwal et al. (2016)
	up to 44%	Reformulation using polydextrose as fat replacer, together with shortening	The reduced-fat biscuits contained 17.54 g fat/100 g sample and were characterized by a fracture strength of 108 kPa, fracture strain of 8%, lightness (L*-value) of 62.5 and a high fiber content of 19.7 g/100 g.	Moriano et al. (2018)
	up to 50–60%	Reformulation using maltodextrin as fat replacer, together with emulsifier (glycerol monostearate)	As the fat content decreased, the dough hardness increased from 20.78 N to 44.08 N and thereby having a negative effect on texture quality. This effect could be relieved by the addition of maltodextrin together with glycerol mono stearate. Therefore, the 50–60% reduced fat biscuits were characterized by a lower spread ratio and a higher breaking strength. Sensory scores in texture, flavor and taste were less than for the full-fat biscuits.	Sudha et al. (2007)
Waffles	up to 50%	Reformulation using insect fat (black soldier fly larvae) as a healthy fat	The use of the insect fat (30.8 g/ 100 g flour) to replace butter by 50% in waffles resulted in no significant difference in the liking score between the full-fat and the reduced fat waffles. In addition, the reduced fat waffles were associated with the emotions good and calm, while the full-fat waffles were associated with the emotions glad and energetic.	Delicato et al. (2020)

Muffins	up to 25%	Reformulation using cocoa dietary fiber as fat replacer	Fat replacement by cocoa dietary fiber tended to increase color development, cohesiveness, chewiness, stickiness and bitterness. Therefore, in this case, only 25% fat reduction (11.5 g cocoa fiber and 34.5 g oil/100 g flour) was recommended to maintain sponginess and springiness, similar to the regular muffins.	Martínez-Cervera et al. (2011)
	up to 49%	Reformulation using emulsion as a new form of structured oil.	The cellulose emulsion muffins with up to 49% fat reduction had similar sensory scores in color and texture to the butter muffins. However, their taste and overall liking were less than for the full-fat muffins.	Martínez-Cervera et al. (2015)
	up to 50%	Reformulation using oleogel as a new form of structured oil.	Due to its effect of reducing the viscosity and shear-thinning behavior of batter, oleogels were recommended to replace fat at up to 50%, to avoid the decreased porosity and specific volume in reduced fat muffins. Thus, the reduced fat muffin using oleogels as a fat replacer were characterized by a soft and chewy texture.	Oh and Lee (2018)
Cakes	up to 25%	Reformulation using polydextrose as fat replacer	The use of polydextrose tended to decrease the average bubble size and increase uniformity. Fat-replacement with polydextrose was recommended at 20% to maintain the porosity, bulk density and yellowness of crust in the same ranges as for the full-fat cake.	Kocer et al. (2007)

Continued

Table 2 Reformulation to reduce unhealthy fat in baked products.—cont'd

Product	Partial replacement	Recommended technique	Final product quality	References
	up to 25%	Reformulation using insect fat (black soldier fly larvae) as a healthy fat	The use of insect fact tended to reduce the intensity for pleasantness and increase the intensity for disappointment. However, 25% fat replacement could provide a lower intensity for dissatisfied than the 50% fat replacement. Thus, insect fat (25.2 g/100 g flour) was recommended to replace butter at 25% in cakes, to have no significant effect on the liking score between the full-fat and the reduced fat cakes.	Delicato et al. (2020)
	up to 30%	Reformulation using fiber as fat replacer	An increase in the fiber content may reduce springiness, cohesiveness, lightness (L*-value) and yellowness (b*-value) of the crumb, but may increase hardness and chewiness. Thus, the fiber from coffee silverskin was recommended for use at 20–30% fat reduction in cake there were no significant changes in the specific volume, weight loss and oiliness, compared to the full-fat cakes.	Ateş and Elmacı (2018)
	up to 70%	Reformulation using inulin as fat replacer	To develop reduced fat cakes, inulin (7.5–10 g/100 g/of flour) was recommended as a fat replacer together with 0.003 g/100 g lipase or 0.5 g/100 g emulsifier, to obtain a similar volume and crumb cell structure to the full-fat cake. In addition, lipase could delay the change in hardness during storage.	Rodríguez-García et al. (2014)

6. Summary

Health concerns have forced the food industry to cease using unhealthy ingredients and additives. Sugar, salt and fat are key parameters causing non-communicable diseases. In baked products, these three ingredients play important roles in achieving the desirable characteristics and quality of products. Thus, any reduction in the levels of sugar, fat and salt in baking is challenging. Too little sugar affects the protein network during mixing, yeast fermentation during proving, protein denaturation and starch gelatinization during baking and retrogradation during storage. The common technique to partially or completely reduce sugar in baked products is to reformulate using sweeteners together with a bulking agent. Sugar-free products are likely successful in cakes and muffins. Other techniques are recommended to avoid additional additives in the recipe, such as sequential sugar reduction, hydrolysis of starch, and proper design of particle size of sugar.

Similarly, not enough salt causes difficulties in the dough development process, resulting in sticky dough that can obstruct machinery during kneading and molding, while carbon dioxide entrapment during proving is reduced, resulting in low volume expansion. A salt ratio that is too low may cause the structural collapse of the baked product. The common technique to reduce salt in bakery products is to reformulate using salt replacers and taste enhancers. When a natural additive is preferred, reformulation using sourdough should be used. In addition, the salt inhomogeneous contribution can enhance saltiness while using less salt.

Fat contains a high level of calories and in addition may contain high amounts of saturated fat and trans-fat, but fat is needed to create the softness and mouthfeel of the baked product. The fat content can be reduced by reformulation using fat replacers. However, fat-free baked products seem unacceptable to many consumers due to an excessive loss of quality, particularly texture and mouthfeel. If energy usage is not an issue, reformulation is possible using new forms of fat or 'healthy fat' to replace the unhealthy fat and still maintain the desirable functionality that fat provides. This technique tends to develop healthier baked products that have acceptable eating quality.

There are many techniques to reduce the levels of sugar, salt and fat in baking. Selection of the most suitable technique and additives to suit individual products is the key to increasing the degree of replacement while maintaining comparable quality to that of the regular product.

References

Abd El-Aal, Y. A., Mohamed Abdel-Fattah, D. M., & El-Dawy Ahmed, K. (2019). Some biochemical studies on trans fatty acid-containing diet. *Diabetes and Metabolic Syndrome: Clinical Research and Reviews*, *13*, 1753–1757.

Aggarwal, D., Sabikhi, L., & Sathish Kumar, M. H. (2016). Formulation of reduced calorie biscuits using artificial sweeteners and fat replacer with dairy multigrain approach. *NFS Journal.*, *2*, 1–7.

Ateş, G., & Elmacı, Y. (2018). Coffee silverskin as fat replacer in cake formulations and its effect on physical, chemical and sensory attributes of cakes. *LWT- Food Science and Technology*, *90*, 519–525.

Baldino, N., Gabriele, D., Lupi, F. R., de Cindio, B., & Cicerelli, L. (2014). Modeling of baking behavior of semi-sweet short dough biscuits. *Innovative Food Science & Emerging Technologies*, *25*, 40–52.

Bassoli. (2006). Chapter 14: Developing new sweeteners from natural compounds. In W. J. Spillane (Ed.), *Optimising sweet taste in foods* (pp. 327–343). Technology and Nutrition: Woodhead Publishing Series in Food Science, ISBN:9781845691646.

Belc, N., Smeu, I., Macri, A., Vallauri, D., & Flynn, K. (2019). Reformulating foods to meet current scientific knowledge about salt, sugar and fats. *Trends in Food Science & Technology*, *84*, 25–28.

Carrascal, A., Rasines, L., Ríos, Y., Rioja, P., Rodríguez, R., & Alvarez-Sabatel, S. (2019). Development of reduced-fat muffins by the application of jet-impingement microwave (JIM) technology. *Journal of Food Engineering*, *262*, 131–141.

Cauvain, S. P., Young, L. S., & Young, L. S. (2006). Key characteristics of existing bakery-product groups and typical variations within such groups. In S. P. Cauvain (Ed.), *Baked products: Science, Technology and Practice* (pp. 14–34). UK: Blackwell Publishing, ISBN:9780470995907.

Cauvain, S. P., & Young, L. S. (2007). *Technology of breadmaking*. New York: Springer US, ISBN:978-0-387-38565-5.

Chavan, R. S., Khedkar, C. D., & Bhatt, S. (2016). Fat replacer. In B. Caballero, P. M. Finglas, & F. Toldrá (Eds.), *The Encyclopedia of Food and Health* (pp. 589–595). Waltham, MA: Academic Press.

Clemens, R. A., Jones, J. M., Kern, M., Lee, S. Y., Mayhew, E. J., Slavin, J. L., et al. (2016). Functionality of sugars in foods and health. *Comprehensive Reviews in Food Science and Food Safety*, *15*(3), 433–470.

Dayrit, F. M. (2015). The properties of lauric acid and their significance in coconut oil. *Journal of the American Oil Chemists' Society*, *92*(1), 1–15.

Deleu, L. J., Wilderjans, E., Van Haesendonck, I., Courtin, C. M., Brijs, K., & Delcour, J. A. (2015). Storage induced conversion of ovalbumin into S-ovalbumin in eggs impacts the properties of pound cake and its batter. *Food Hydrocolloids*, *49*, 208–215.

Delicato, C., Schouteten, J. J., Dewettinck, K., Gellynck, X., & Tzompa-Sosa, D. A. (2020). Consumers' perception of bakery products with insect fat as partial butter replacement. *Food Quality and Preference*, *79*, 103755.

Di Monaco, R., Miele, N. A., Cabisidan, E. K., & Cavella, S. (2018). Strategies to reduce sugars in food. *Current Opinion in Food Science*, *19*, 92–97.

Diler, G., Le-Bail, A., & Chevallier, S. (2016). Salt reduction in sheeted dough: a successful technological approach. *Food Research International*, *88*, 10–15.

European Commission. (2006). Regulation (EC) No 1924/2006 of the European Parliament and of the Council of 20 December 2006 on nutrition and health claims made on foods. *Official Journal of The European Union. L*, *404*, 9–12.

EFSA (European Food Safety Authority). (2017). *Dietary reference values for nutrients: Summary report. 2017*. EFSA Supporting Publication. e15121. 92 pp. https://doi.org/10.2903/sp.efsa.2017.e15121.

European Union Commission. (2019). *Regulation (EU) 2019/649 of 24 April 2019 Amending Annex III to Regulation (EC) No 1925/2006 of the European Parliament and of the Council As Regards Trans Fat, Other Than Trans Fat Naturally Occurring in Fat of Animal Origin.*

Fellows. (2017). *Food processing technology: Principle and practice* (4th ed.). Woodhead Publishing, ISBN:9780081005231. 1152 p. eBook.

Fernandes, S., & Salas-Mellado, M.d.l. M. (2017). Addition of chia seed mucilage for reduction of fat content in bread and cakes. *Food Chemistry, 227*, 237–244.

Figuerola, F., Hurtado, M. L., Estévez, A. M., Chiffelle, I., & Asenjo, F. (2005). Fibre concentrates from apple pomace and citrus peel as potential fibre sources for food enrichment. *Food Chemistry, 91*, 395–401.

Gaudette, N. J., Pietrasik, Z., & Johnston, S. P. (2019). Application of taste contrast to enhance the saltiness of reduced sodium beef patties. *LWT- Food Science and Technology, 116*, 108585. https://doi.org/10.1016/j.lwt.2019.108585.

Girgis, S., Neal, B., Prescott, J., Prendergast, J., Dumbrell, S., Turner, C., et al. (2003). A one-quarter reduction in the salt content of bread can be made without detection. *European Journal of Clinical Nutrition, 57*, 616–620.

Hao, Y., Wang, F., Huang, W., Tang, X., Zou, Q., Li, Z., et al. (2016). Sucrose substitution by polyols in sponge cake and their effects on the foaming and thermal properties of egg protein. *Food Hydrocolloids, 57*, 153–159.

Heck, R. T., Vendruscolo, R. G., de Araújo Etchepare, M., José Cichoski, A., de Menezes, C. R., Barin, J. S., et al. (2017). Is it possible to produce a low-fat burger with a healthy n − 6/n − 3 PUFA ratio without affecting the technological and sensory properties? *Meat Science, 130*, 16–25.

Hesso, N., Marti, A., Le-Bail, P., Loisel, C., Chevallier, S., Le-Bail, A., et al. (2015). Conformational changes of polymers in model batter systems. *Food Hydrocolloids, 51*, 101–107.

Issara, U., Park, S., Lee, S., Lee, J., & Park, S. (2020). Health functionality of dietary oleogel in rats fed high-fat diet: A possibility for fat replacement in foods. *Journal of Functional Foods, 70*, 103979.

Jolly, L. (2008). Chapter 5: Alternative sweeteners. In J. Kingsman (Ed.), *Sugar trading manual* (4th ed.). Cambridge: Woodhead Publishing Limited.

Karp, S., Wyrwisz, J., Kurek, M., & Wierzbicka, A. (2017). Combined use of cocoa dietary fiber and steviol glycosides in low-calorie muffins production. *International Journal of Food Science and Technology, 52*, 944–953.

Kaur, A., Bala, R., Singh, B., & Rehal, J. (2011). Effect of replacement of sodium chloride with mineral salts on rheological characteristics of wheat flour. *American Journal of Food Technology, 6*, 674–684.

Khalesi, H., Emadzadeh, B., Kadkhodaee, R., & Fang, Y. (2016). Whey protein isolate Persian gum interaction at neutral pH. *Food Hydrocolloids, 59*, 45–49.

Kim, J. N., Park, S., & Shin, W. S. (2014). Textural and sensory characteristics of rice chiffon cake formulated with sugar alcohols instead of sucrose. *Journal of Food Quality, 37*, 281–290.

Kocadağlı, T., & Gökmen, V. (2019). Caramelization in foods: A food quality and safety perspective. *Encyclopedia of Food Chemistry.*, 18–29.

Kocer, D., Hicsasmaz, Z., Bayindirli, A., & Katnas, S. (2007). Bubble and pore formation of the high-ratio cake formulation with polydextrose as a sugar-and fat-replacer. *Journal of Food Engineering, 78*, 953–964.

Kotecka-Majchrzak, K., Sumara, A., Fornal, E., & Montowska, M. (2020). Oilseed proteins – Properties and application as a food ingredient. *Trends in Food Science & Technology, 106*, 160–170.

Kweon, M., Slade, L., Levine, H., & Gannon, D. (2014). Cookie-versus cracker-baking—what's the difference? Flour functionality requirements explored by src and alveography. *Critical Reviews in Food Science and Nutrition, 54*, 115–138.

Laguna, L., Primo-Martín, C., Salvador, A., & Sanz, T. (2013). Inulin and erythritol as sucrose replacers in short-dough cookies: sensory, fracture, and acoustic properties. *Journal of Food Science*, *78*, S777–S784.

Laguna, L., Primo-Martín, C., Varela, P., Salvador, A., & Sanz, T. (2014). HPMC and inulin as fat replacers in biscuits: Sensory and instrumental evaluation. *LWT- Food Science and Technology*, *56*, 494–501.

Lambrecht, M. A., Deleu, L. J., Rombouts, I., & Delcour, J. A. (2018). Heat-induced network formation between proteins of different sources in model systems, wheat-based noodles and pound cakes. *Food Hydrocolloids*, *79*, 352–370.

Limpimwong, W., Kumrungsee, T., Katob, N., Yanakab, N., & Thongngama, M. (2017). Rice bran wax oleogel: A potential margarine replacement and its digestibility effect in rats fed a high-fat diet. *Journal of Functional Foods*, *39*, 250–256.

Lynch, E. J., Dal Bello, F., Sheehan, E. M., Cashman, K. D., & Arendt, E. K. (2009). Fundamental studies on the reduction of salt on dough and bread characteristics. *Food Research International*, *42*, 885–891.

Manisha, G., Soumya, C., & Indrani, D. (2012). Studies on interaction between stevioside, liquid sorbitol, hydrocolloids and emulsifiers for replacement of sugar in cakes. *Food Hydrocolloids*, *29*, 363–373.

Manzocco, L., Anese, M., Calligaris, S., Quarta, B., & Nicoli, M. C. (2012). Use of Monoglyceride hydrogel for the production of low-fat short dough pastry. *Food Chemistry*, *132*, 175–180.

Marques, G. A., Sao Jose, J. F. B., Silva, D. A., & Silva, E. M. M. (2016). Whey protein as a substitute for wheat in the development of no added sugar cookies. *LWT- Food Science and Technology*, *67*, 118–126.

Martínez-Cervera, S., Salvador, A., Muguerza, B., Moulay, L., & Fiszman, S. M. (2011). Cocoa fiber and its application as a fat replacer in chocolate muffins. *LWT- Food Science and Technology*, *44*, 729–736.

Martínez-Cervera, S., Sanz, T., Salvador, A., & Fiszman, S. M. (2012). Rheological, textural and sensorial properties of low-sucrose muffins reformulated with sucralose/polydextrose. *LWT- Food Science and Technology*, *45*, 213–220.

Martínez-Cervera, S., Salvador, A., & Sanz, T. (2014). Comparison of different polyols as total sucrose replacers in muffins: thermal, rheological, texture and acceptability properties. *Food Hydrocolloids*, *35*, 1–8.

Martínez-Cervera, S., Salvador, A., & Sanz, T. (2015). Cellulose ether emulsions as fat replacers in muffins: Rheological, thermal and textural properties. *LWT- Food Science and Technology*, *63*, 1083–1090.

Matsakidou, A., Blekas, G., & Paraskevopoulou, A. (2010). Aroma and physical characteristics of cakes prepared by replacing margarine with extra virgin olive oil. *LWT- Food Science and Technology*, *43*, 949–957.

Mert, B., & Demirkesen, I. (2016). Reducing saturated fat with oleogel/shortening blends in a baked product. *Food Chemistry*, *199*, 809–816.

Monteiro, A. R. G., Nakagawa, A., Pimentel, T. C., & Sousa, I. (2021). Increasing saltiness perception and keeping quality properties of low salt bread using inhomogeneous salt distribution achieved with salt agglomerated by waxy starch. *LWT- Food Science and Technology*, *146*, 111451.

Mooradian, A. D., Smith, M., & Tokuda, M. (2017). The role of artificial and natural sweeteners in reducing the consumption of table sugar: A narrative review. *Clinical Nutrition ESPEN.*, *18*, 1–8.

Moriano, M. E., Cappa, C., & Alamprese, C. (2018). Reduced-fat soft-dough biscuits: Multivariate effects of polydextrose and resistant starch on dough rheology and biscuit quality. *Journal of Cereal Science*, *81*, 171–178.

Mouliney, M., Lavery, B., Sharma, R., & Jenner, C. (2011). Waxy durum and fat differ in their actions as improvers of bread quality. *Journal of Cereal Science*, *54*, 317–323.

Mueller, E., Koehler, P., & Scherf, K. A. (2016). Applicability of salt reduction strategies in pizza crust. *Food Chemistry*, *192*, 1116–1123.

Naeli, M. H., Milani, J. M., Farmani, J., & Zargaraan, A. (2020). Development of innovative ethyl cellulose-hydroxypropyl methylcellulose biopolymer oleogels as low saturation fat replacers: Physical, rheological and microstructural characteristics. *International Journal of Biological Macromolecules*, *156*, 792–804.

Noort, M. W. J., Bult, J. H. F., & Stieger, M. (2012). Saltiness enhancement by taste contrast in bread prepared with encapsulated salt. *Journal of Cereal Science*, *55*, 218–225.

Nouri, M., Nasehi, B., Samavati, V., & Mehdizadeh, S. A. (2017). Optimizing the effects of Persian gum and carrot pomace powder for development of low-fat donut with high fiber content. *Bioactive Carbohydrates and Dietary Fiber.*, *9*, 39–45.

Oh, I. K., Amoah, C., Lim, J., Jeong, S., & Lee, S. (2017). Assessing the effectiveness of wax-based sunflower oil oleogels in cakes as a shortening replacer. *LWT- Food Science and Technology*, *86*, 430–437.

Oh, I. M., & Lee, S. (2018). Utilization of foam structured hydroxypropyl methylcellulose for oleogels and their application as a solid fat replacer in muffins. *Food Hydrocolloids*, *77*, 796–802.

Oliveira, D., Reis, F., Deliza, R., Rosenthal, A., Giménez, A., & Ares, G. (2016). Difference thresholds for added sugar in chocolate-flavored milk: recommendations for gradual sugar reduction. *Food Research International*, *89*, 448–453.

Omeroglu, Y. P., & Ozdal, T. (2020). Fatty acid composition of sweet bakery goods and chocolate products and evaluation of overall nutritional quality in relation to the food label information. *Journal of Food Composition and Analysis*, *88*, 103438.

O'Sullivan, M. G. (2020). Chapter 2: Nutritional optimization—Reduced-sugar products and challenges. In M. G. O'Sullivan (Ed.), *Salt, fat and sugar reduction: Sensory approaches for nutritional reformulation of foods and beverages* (pp. 29–61). Woodhead Publishing Series in Food Science, Technology and Nutrition. https://doi.org/10.1016/B978-0-12-819741-7.00002-X.

Pflaum, T., Konitzer, K., Hofmann, T., & Koehler, P. (2013). Analytical and sensory studies on the release of sodium from wheat bread crumb. *Journal of Agriculture and Food Chemistry.*, *61*, 6485–6494.

Pimdit, K., Therdthai, N., & Jangchud, K. (2008). Effects of fat replacers on the physical, chemical and sensory characteristics of puff pastry. *Kasetsart Journal (Natural Science)*, *42*(4), 739–746.

Pinheiro, M., & Oliveira, M. (2005). The effect of different sweeteners in low-calorie yogurts- A review. *International Journal of Dairy Technology*, *58*, 193–199.

Punia, S., & Dhull, S. B. (2019). Chia seed (*Salvia hispanica* L.) mucilage (a heteropolysaccharide): Functional, thermal, rheological behaviour and its utilization. *International Journal of Biological Macromolecules*, *140*, 1084–1090.

Renzetti, S., van den Hoek, I. A. F., & van der Sman, R. G. M. (2020). Amino acids, polyols and soluble fibers as sugar replacers in bakery applications: Egg white proteins denaturation controlled by hydrogen bond density of solutions. *Food Hydrocolloids*, *108*, 106034.

Richardson, A. M., Tyuftin, A., Kilcawley, K. N., Gallagher, E., O'Sullivan, M. G., & Kerry, J. P. (2018). Impact of sugar particle size manipulation on the physical and sensory properties of chocolate brownies. *LWT- Food Science and Technology*, *95*, 51–57.

Rodríguez-García, J., Sahi, S. S., & Hernando, I. (2014). Functionality of lipase and emulsifiers in low-fat cakes with inulin. *LWT- Food Science and Technology*, *58*, 173–182.

Sahin, A. W., Zannini, E., Coffey, A., & Arendt, E. K. (2019). Sugar reduction in bakery products: Current strategies and sourdough technology as a potential novel approach. *Food Research International*, *126*, 108583.

Samapundo, S., Deschuyffeleer, N., Van Laere, D., De Leyn, I., & Devlieghere, F. (2010). Effect of NaCl reduction and replacement on the growth of fungi important to the spoilage of bread. *Food Microbiology, 27*, 749–756.

Santiago-García, P. A., Mellado-Mojica, E., Leon-Martínez, F. M., & Lopez, M. G. (2017). Evaluation of Agave angustifolia fructans as fat replacer in the cookies manufacture. *LWT- Food Science and Technology, 77*, 100–109.

Selvasekaran, P., & Chidambaram, R. (2021). Advances in formulation for the production of low-fat, fat-free, low-sugar, and sugar-free chocolates: An overview of the past decade. *Trends in Food Science & Technology, 113*, 315–334.

Silow, C., Axel, C., Zannini, E., & Arendt, E. K. (2016). Current status of salt reduction in bread and bakery products—A review. *Journal of Cereal Science, 72*, 135–145.

Silow, C., Zannini, E., Axel, C., Lynch, K. M., & Arendt, E. K. (2016). Effect of salt reduction on wheat-dough properties and quality characteristics of puff pastry with full and reduced fat content. *Food Research International, 89*, 330–337.

Silva, T. J., Fernandes, G. D., Bernardinelli, O. D., da Rosa Silva, E. C., Barrera-Arellano, D., & Ribeiro, A. P. B. (2021). Organogels in low-fat and high-fat margarine: A study of physical properties and shelf life. *Food Research International, 140*, 110036.

Srikanlaya, C., Therdthai, N., Ritthiruangdej, P., & Zhou, W. (2018). Effect of hydroxypropyl methylcellulose, whey protein concentrate and soy protein isolate enrichment on characteristics of gluten-free rice dough and bread. *International Journal of Food Science and Technology, 53*, 1760–1770.

Stieger, M., & van de Velde, F. (2013). Microstructure, texture and oral processing: new ways to reduce sugar and salt in foods. *Current Opinion in Colloid & Interface Science, 18*, 334–348.

Struck, S., Gundel, L., Zahn, S., & Rohm, H. (2016). Fiber enriched reduced sugar muffins made from iso-viscous batters. *LWT- Food Science and Technology, 65*, 32–38.

Sudha, M. L., Srivastava, A. K., Vetrimani, R., & Leelavathi, K. (2007). Fat replacement in soft dough biscuits: Its implications on dough rheology and biscuit quality. *Journal of Food Engineering, 80*, 922–930.

Syarifuddin, A., Septier, C., Salles, C., & Thomas-Danguin, T. (2016). Reducing salt and fat while maintaining taste: an approach on a model food system. *Food Quality and Preference, 48*, 59–69.

Talbot, G. (2011). *Reducing saturated fats in foods* (1st ed.). Woodhead Publishing Series in Food Science, Technology and Nutrition, ISBN:9780857092472.

Tarancón, P., Fiszman, S. M., Salvador, A., & Tárrega, A. (2013). Formulating biscuits with healthier fats. Consumer profiling of textural and flavor sensations during consumption. *Food Research International, 53*, 134–140.

Therdthai, N., Zhou, W., & Jangchud, K. (2007). Modeling of the effect of relative humidity and temperature on proving rate of rice-flour based dough. *LWT- Food Science and Technology, 40*, 1036–1040.

Tzompa-Sosa, D. A., Dewettinck, K., Gellynck, X., & Schouteten, J. J. (2021). Replacing vegetable oil by insect oil in food products: Effect of deodorization on the sensory evaluation. *Food Research International, 141*, 110140.

van der Sman, R. G. M., & Renzetti, S. (2019). Understanding functionality of sucrose in biscuits for reformulation purposes. *Critical Reviews in Food Science and Nutrition, 59*(14), 2225–2239.

Wellner, N., Mills, E. N. C., Brownsey, G., Wilson, R. H., Brown, N., Freeman, J., et al. (2005). Changes in protein secondary structure during gluten deformation studied by dynamic Fourier transform infrared spectroscopy. *Biomacromolecules, 6*, 255–261.

Wetzel, C. R., Weese, J. O., & Bell, L. N. (1997). Sensory evaluation of no-sugar-added cakes containing encapsulated aspartame. *Food Research International, 30*, 395–399.

WHO (World Health Organization). (2012). *Guideline: Sodium intake for adults and children* (pp. 1–56). Geneva, Switzerland: WHO Press.

Yazdi, A. G., Hojjatoleslamy, M., Keramat, J., Jahadi, M., & Amani, E. (2017). The evaluation of saccharose replacing by adding stevioside– maltodextrin mixture on the physicochemical and sensory properties of Naanberenji (an Iranian confectionary). *Food Science & Nutrition*, *5*, 845–851.

Zahn, S., Forker, A., Krügel, L., & Rohm, H. (2013). Combined use of rebaudioside A and fibers for partial sucrose replacement in muffins. *LWT- Food Science and Technology*, *50*, 695–701.

Zhao, C. J., Kinner, M., Wismer, W., & Ganzle, M. G. (2015). Effect of glutamate accumulation during sourdough fermentation with lactobacillus reuteri on the taste of bread and sodium-reduced bread. *Cereal Chemistry*, *92*, 224–230.

Zoulias, E. I., Oreopoulou, V., & Tzia, C. (2002). Textural properties of low-fat cookies containing carbohydrate- or protein-based fat replacers. *Journal of Food Engineering*, *55*, 337–342.

Kolibri 360 GmbH
Ferdinand-Jühlke-Straße 7
99095 Erfurt, Deutschland
produktsicherheit@kolibri360.de